A Photographic Atlas
For The
Biology Laboratory

Second Edition

Kent M. Van De Graaff

John L. Crawley

Morton Publishing Company
925 W. Kenyon, Unit 12
Englewood, Colorado 80110

To our teachers, colleagues,
friends, and students who share
with us a mutual love for biology.

Copyright 1994 by Morton Publishing Company

ISBN: 0-89582-272-5

10 9 8 7 6 5 4 3 2

Printed in the United States of America

Preface

Biology is an exciting, dynamic, and challenging science. It is the study of life. Students are fortunate to be living at a time when insights and discoveries in almost all aspects of biology are occurring at a very rapid pace. Much of the knowledge learned in a biology course has application in improving humanity and the quality of life. An understanding of basic biology is essential in establishing a secure foundation for more advanced courses in the life sciences or health sciences. The principles learned in a basic biology course will be of immeasurable value in dealing with the ecological and bioethical problems currently facing us.

Biology is a visually oriented science. *A Photographic Atlas for the Biology Laboratory, Second Edition,* is designed to provide you with quality photographs of organisms, similar to those you may have the opportunity to observe in a biology laboratory. It is designed to accompany any biology text or laboratory manual you may be using in the classroom. In certain courses *A Photographic Atlas for the Biology Laboratory, Second Edition,* could serve as the laboratory manual.

An objective of this atlas is to provide you with a balanced visual representation of the five kingdoms of biological organisms. Great care has been taken to construct completely labeled, informative figures that are depicted clearly and accurately. The terminology used in this atlas is that found in the more commonly used college biology texts.

Several dissections of plants and invertebrate and vertebrate animals were completed and photographed in the preparation of this atlas. These images are included for those students who have the opportunity to do similar dissections as part of their laboratory requirement. In addition, photographs of special dissections are also included in this atlas, such as the procedure for a sheep heart dissection.

Chapter 8 of this atlas is devoted to the biology of the human organism, which is emphasized in many biology textbooks and courses. In this chapter, you are provided with a complete set of photographs for each of the human body systems. Human cadavers have been carefully dissected and photographed to clearly depict each of the principal organs from each of the body systems. Selected radiographs (X-rays), CT scans, and MR images depict structures from living persons and thus provide an applied dimension to this portion of the atlas.

Acknowledgments

Many professionals have assisted in the preparation of the first and second editions of *A Photographic Atlas for the Biology Laboratory* and have shared our enthusiasm of its value for students of biology. We are especially appreciative of Drs. Samuel R. Rushforth, Richard A. Robison, Wilford M. Hess, J. Carter Rowley, and Richard A. Heckmann for their help in obtaining photographs and photomicrographs. The radiographs, CT scans, and MR images have been made possible through the generosity of Gary M. Watts, MD and the Department of Radiology at Utah Valley Regional Medical Center. Ryan L. Van De Graaff and Blaine Furniss spent many hours in assisting us with photography. Several students were very helpful in performing dissections. We gratefully acknowledge the assistance of Nathan A. Jacobson, Richard R. Rasmussen, Scott R. Gunn, Sandra E. Sephton, Michelle Kidder, Bradley C. Arnold, and Michael K. Visick. We are grateful to Thomas D. Metcalf for his professional talent and his dedicated effort to achieve the best possible development of each of the photographs used in this atlas. Christopher H. Creek rendered line art throughout the book. We appreciate his efforts. Joanne R. Saliger, of Ash Street Typecrafters, Inc., was superb to work with, and we appreciate her talent and interest in this project. We are indebted to Douglas Morton and the personnel at Morton Publishing Company for the opportunity, encouragement, and support to prepare this atlas. We appreciate the suggestions from users of the first edition. Delmar Vander Zee of Dordt College, Michael J. Shively of Utah Valley State College, and Edward Weiss of Christopher Newport University were especially helpful reviewers of the second edition of this atlas.

Contents

1 Cells and Tissues 1
Plant cells and tissues   2
Animal cells and tissues   5

2 Perpetuation of Life 11

3 Kingdom Monera 18
Archaebacteria
   (methanogens and thermoacidophiles)   19

4 Kingdom Protista 23
Chrysophyta (diatoms and golden algae)   24
Dinoflagellata (dinoflagellates)   25
Rhizopoda (amoebas)   25
Apicomplexa (sporozoa and *Plasmodium*)   25
Euglenophyta (euglena)   26
Ciliophora (ciliates and *Paramecium*)   26
Chlorophyta (green algae)   27
Phaeophyta (brown algae)   30
Rhodophyta (red algae)   33
Myxomycota (plasmodial slime molds)   34
Oomycota (water molds, white rusts,
   and downy mildews)   35

5 Kingdom Fungi 36
Zygomycota (conjugation fungi)   37
Ascomycota (yeasts, molds, morels, and truffles)   37
Basidiomycota (mushrooms, toadstools, rusts,
   and smuts)   40
Lichens   43

6 Kingdom Plantae 45
Bryophyta (liverworts, hornworts, and mosses)   46
Psilophyta (whisk ferns)   53
Lycophyta (clubmosses and quillworts)   55
Sphenophyta (horsetails)   58
Pterophyta (ferns)   59
Cycadophyta (cycads)   62
Ginkgophyta (ginkgo)   63
Coniferophyta (conifers)   64
Anthophyta (angiosperms; monocots and dicots)   68

7 Kingdom Animalia 84
Porifera (sponges)   85
Cnidaria (corals, hydra, and jellyfish)   87
Platyhelminthes (flatworms)   92
Nematoda (roundworms and nematodes)   96
Mollusca (mollusks; clams, snails, and squids)   98
Annelida (segmented worms)   102
Arthropoda (arachnids, crustaceans, and insects)   106
Echinodermata (echinoderms; sea stars, sand dollars,
   sea cucumbers, and sea urchins)   115
Chordata (amphioxus, amphibians, fishes, reptiles,
   birds, and mammals)   119

8 Human Biology 128
Skeletomusculature System   130
Controlling Systems and Sensory Organs   137
Cardiovascular System   141
Respiratory System   143
Digestive System   145
Urogenital System and Development   146

9 Vertebrate Dissections 149
Anatomy of Cartilagenous Fishes   149
Anatomy of Bony Fishes   154
Anatomy of Amphibians   155
Anatomy of Reptiles   162
Anatomy of Birds   166
Anatomy of Mammals   168

APPENDICES

A Glossary of Suffixes and Prefixes 186

B Glossary of Terms 188

INDEX 197

Cells and Tissues

1

All organisms are comprised of one or more cells. Cells are the basic structural and functional units of organisms. A cell is a minute, protoplasmic mass that contains specific organelles which function independently but in coordination one with another. Prokaryotic cells and eukaryotic cells are the two basic types.

Prokaryotic cells lack a membrane-bound nucleus, contain a single strand of *nucleic acid*, and contain organelles. They have a rigid or semi-rigid *cell wall* that provides shape to the cell outside the *cell (plasma) membrane*. Bacteria are examples of prokaryotic, single-celled organisms.

Eukaryotic cells contain a true nucleus with multiple chromosomes and have several types of specialized *organelles*. They have a differentially permeable *cell (plasma) membrane*. Organisms comprised of eukaryotic cells include protozoa, fungi, algae, plants, and invertebrate and vertebrate animals.

The *nucleus* is the large spheroid body within the eukaryotic cell that contains the genetic material of the cell. The nucleus is enclosed by a double membrane called the *nuclear membrane*, or *nuclear envelope*. The *nucleolus* is a dense, nonmembranous body composed of protein and RNA molecules. The chromatin consists of fibers of protein and DNA molecules. Prior to cellular division, the chromatin shortens and coils into rod shaped *chromosomes*. Chromosomes consist of DNA and structural proteins called *histones*.

The *cytoplasm* of the eukaryotic cell is the medium between the nuclear membrane and the cell membrane. *Organelles* are minute membrane-bound structures of the cytoplasm consisting of all of the cellular components other than the nucleus. The cellular functions carried out by organelles are referred to as *metabolism*. The functions of the principal organelles are listed on page 2. In order for cells to remain alive, metabolize and maintain *homeostasis*, certain requirements must be met that include having access to nutrients and oxygen, being able to eliminate wastes, and being maintained in a constant, protective environment.

The *cell membrane* is composed of phospholipid and protein molecules. The cell membrane gives form to a cell and controls the passage of material into and out of a cell. More specifically, the proteins in the cell membrane provide: 1) structural support; 2) a mechanism of molecule transport across the membrane; 3) enzymatic control of chemical reactions; 4) receptors for hormones and other regulatory molecules; and 5) cellular markers (antigens), which identify the blood and tissue type. The carbohydrate molecules: 1) repel negative objects due to their negative charge; 2) act as receptors for hormones and other regulatory molecules; 3) form specific cell markers which enable like cells to attach and aggregate into tissues; and 4) enter into immune reactions.

Plant cells differ from other eukaryotic cells in that they are surrounded by a primary and secondary cell wall that contains cellulose for stiffness. They also contain membrane-bound chloroplasts with photosynthetic pigments for photosynthesis and vacuoles for water storage.

Tissues are aggregations of similar cells that perform specific functions. The tissues of the body of a multicellular animal are classified into four principal types: 1) *epithelial tissues* cover body and organ surfaces, line body cavities and lumina (hollow portions of body tubes), and form various glands; 2) *connective tissues* bind, support, and protect body parts; 3) *muscle tissues* contract to produce movements; and 4) *nervous tissues* initiate and transmit nerve impulses.

The body of a typical flowering plant is composed of three tissue systems: 1) the *ground tissue system*, providing support, regeneration, respiration, photosynthesis, and storage; 2) the *vascular tissue system*, providing conduction passageways through the plant; and 3) the *dermal tissue system*, providing protection to the plant.

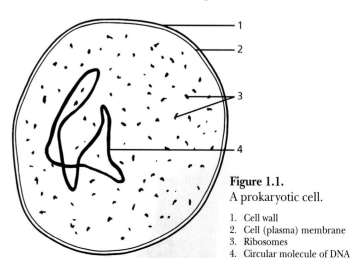

Figure 1.1.
A prokaryotic cell.

1. Cell wall
2. Cell (plasma) membrane
3. Ribosomes
4. Circular molecule of DNA

TABLE 1.1.
Structure and Function of Components of a Eukaryotic Cell

Component	Structure	Function
Cell (plasma) membrane	Composed of protein and phospholipid molecules	Provides form to cell and controls passage of materials into and out of cell
Cell wall	Cellulose fibrils	Provides structure and rigidity to plant cell
Cytoplasm	Fluid to jelly-like substance	Suspending medium of organelles
Endoplasmic reticulum	Interconnecting hollow channels	Supporting framework of cell; cell transport
Ribosomes	Granules of nucleic acid	Protein synthesis
Mitochondria	Double-layered sacs with cristae	Production of ATP
Golgi apparatus	Flattened sacs with vacuoles	Synthesizes carbohydrates and packages molecules for secretion
Lysosomes	Membrane-surrounded sacs of enzymes	Digest foreign molecules and worn cells
Centrosome	Mass of two rod-like centrioles	Organizes spindle fibers and assists mitosis
Vacuoles	Membranous sacs	Store and secrete substances within the cytoplasm
Fibrils and microtubules	Protein strands	Support cytoplasm and transport materials
Cilia and flagella	Cytoplasmic extensions from cell	Movement of particles along cell surface or move cell
Nucleus	Nuclear membrane, nucleolus, and chromatin (DNA)	Directs cell activity; forms ribosomes
Chloroplast	Inner (grana) membrane within outer membrane	Photosynthesis

PLANT CELLS AND TISSUES

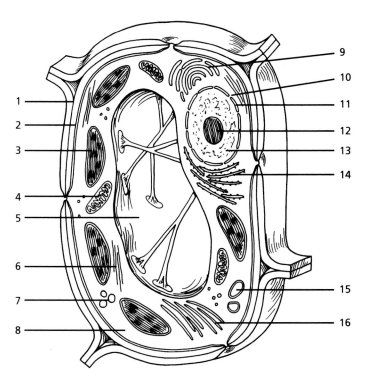

Figure 1.2. A typical plant cell.

1. Cell wall
2. Cell (plasma) membrane
3. Chloroplast
4. Mitochondrion
5. Vacuole
6. Microfilament
7. Plastid
8. Cytoplasm
9. Golgi apparatus
10. Nuclear pore
11. Nuclear membrane (envelope)
12. Nucleolus
13. Chromatin
14. Rough endoplasmic reticulum
15. Vesicle
16. Smooth endoplasmic reticulum

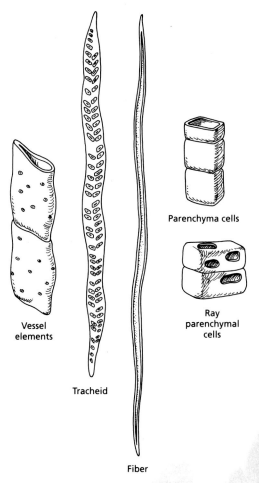

Vessel elements

Tracheid

Fiber

Parenchyma cells

Ray parenchymal cells

Figure 1.3. Examples of plant cells.

PLANT CELLS AND TISSUES

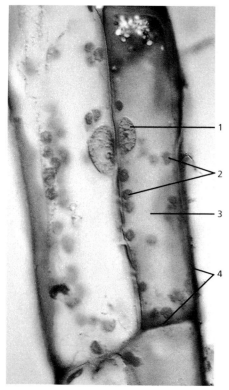

Figure 1.4. Prepared slide of *Elodea* leaf cells. (X430)

1. Nucleus   3. Vacuole
2. Chloroplasts   4. Cell wall

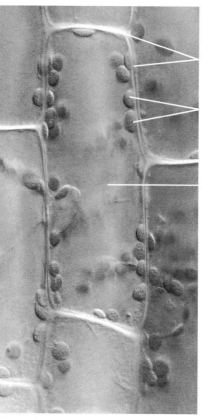

Figure 1.5. Live *Elodea* leaf cells. (X430)

1. Cell wall   2. Chloroplasts   3. Vacuole

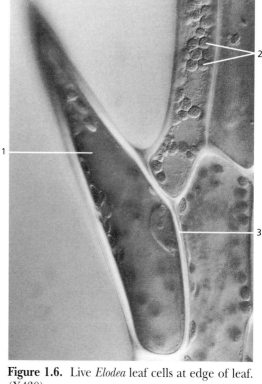

Figure 1.6. Live *Elodea* leaf cells at edge of leaf. (X430)

1. Spine-shaped cell on   2. Chloroplasts
   exposed edge of leaf   3. Cell wall

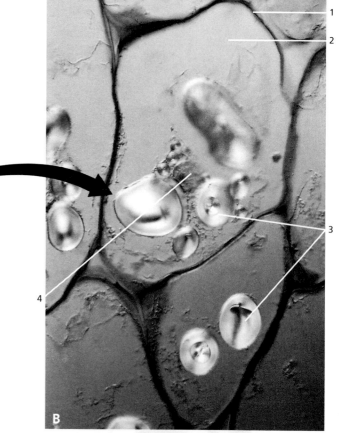

Figure 1.7. Live potato cells showing starch grains at a low magnification (a) of 430, and at a high magnification (b) of 1,000. In potato cells, food is stored as starch in organelles called pyrenoids.

1. Cell wall   3. Starch grains in pyrenoids
2. Cytoplasm   4. Nucleus

PLANT CELLS AND TISSUES

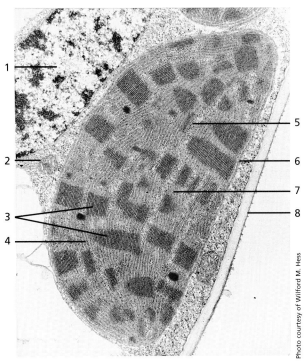

Figure 1.8. Electron micrograph of a portion of a sugar cane leaf cell.

1. Nucleus
2. Mitochondrion
3. Grana
4. Thylakoid membrane
5. Stroma
6. Chloroplast envelope
7. Chloroplast
8. Cell wall

Figure 1.9. Barley smut spore, fractured through the middle of the cell.

1. Nucleus
2. Vacuole
3. Cell wall
4. Cell membrane
5. Mitochondrion
6. Nuclear pores

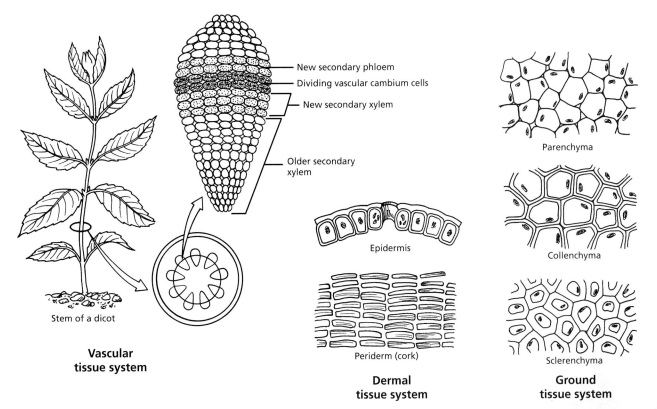

New secondary phloem
Dividing vascular cambium cells
New secondary xylem
Older secondary xylem

Stem of a dicot

Vascular tissue system

Epidermis

Periderm (cork)

Dermal tissue system

Parenchyma

Collenchyma

Sclerenchyma

Ground tissue system

Figure 1.10. Examples of plant tissues.

PLANT CELLS AND TISSUES

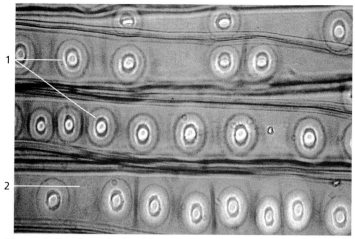

Figure 1.11. Longitudinal section through the xylem of a pine, *Pinus*, showing tracheid cells with prominent bordered pits.

1. Bordered pits   2. Tracheid cell

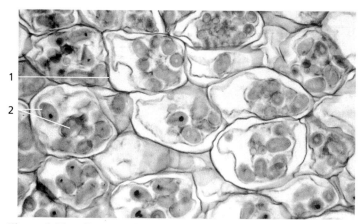

Figure 1.13. Cross section through the root of a potato, *Solanum tuberosum,* showing parenchyma cells containing numerous starch grains.

1. Cell wall   2. Starch grains

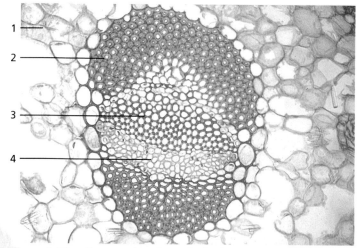

Figure 1.15. Cross section through the leaf of a yucca, *Yucca brevifolia,* showing a vascular bundle (vein). Note the prominent sclerenchyma tissue forming caps on both sides of the bundle.

1. Leaf parenchyma   2. Leaf sclerenchyma   3. Xylem   4. Phloem

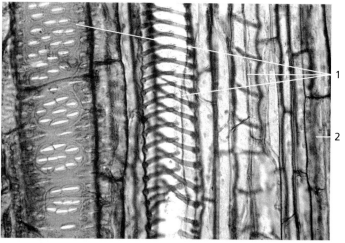

Figure 1.12. Longitudinal section through the xylem of a squash stem, *Cucurbita maxima.* The vessel elements shown here have several different patterns of wall thickenings.

1. Vessel elements   2. Parenchyma

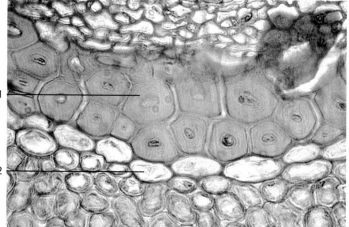

Figure 1.14. Asterosclereid in the petiole of a pondlily, *Nuphar.*

1. Parenchyma cell   2. Asterosclereid

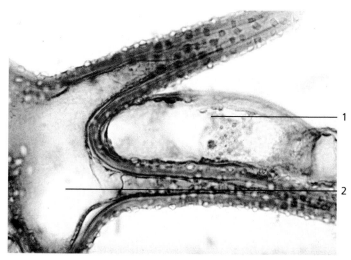

Figure 1.16. Cross section through the stem of flax, *Linum.* Note the thick-walled fibers as compared to the thin-walled parenchyma cells.

1. Fibers   2. Parenchyma cell

ANIMAL CELLS AND TISSUES ━━━━━━━━━━━

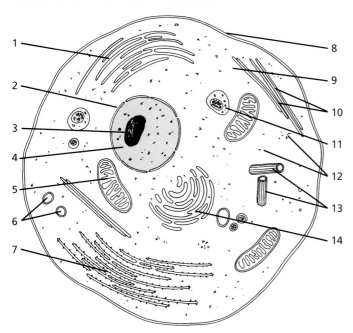

Figure 1.17. A typical animal cell.

1. Smooth endoplasmic reticulum
2. Nuclear membrane
3. Nucleolus
4. Nucleoplasm
5. Mitochondrion
6. Vesicles
7. Rough endoplasmic reticulum
8. Cell membrane
9. Cytoplasm
10. Microtubules
11. Lysosome
12. Ribosomes
13. Centrioles
14. Golgi apparatus

Photo courtesy of Scott C. Miller

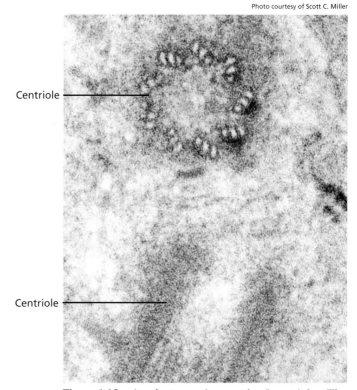

Centriole

Centriole

Figure 1.18. An electron micrograph of centrioles. The centrioles are positioned at right angles to one another.

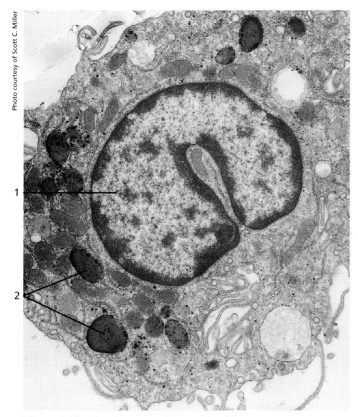

Figure 1.19. An electron micrograph of lysosomes.

1. Nucleus   2. Lysosomes

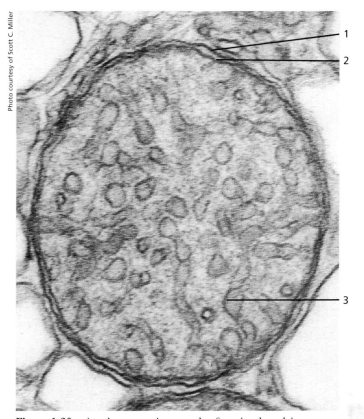

Figure 1.20. An electron micrograph of a mitochondrion.

1. Outer membrane   2. Crista   3. Inner membrane

ANIMAL CELLS AND TISSUES

Photo courtesy of Scott C. Miller

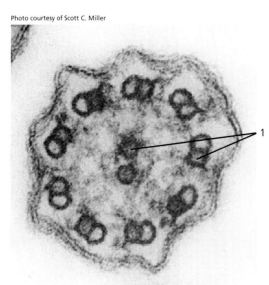

Figure 1.21. An electron micrograph of a cilium, showing the characteristic "9 + 2" arrangement of the microtubules.

1. Microtubules

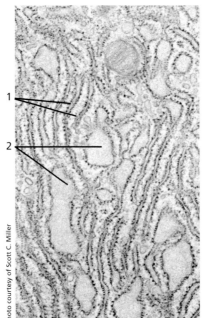

Photo courtesy of Scott C. Miller

Figure 1.22. An electron micrograph of rough endoplasmic reticulum.

1. Ribosomes   2. Cisternae

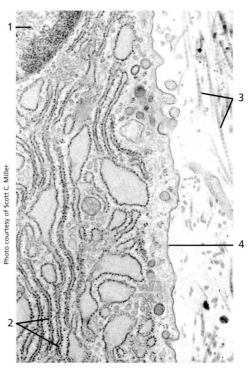

Photo courtesy of Scott C. Miller

Figure 1.23. Rough endoplasmic reticulum secreting collagenous filaments to the outside of the cell.

1. Nucleus   3. Collagenous filaments
2. Rough endoplasmic   4. Cell membrane
   reticulum

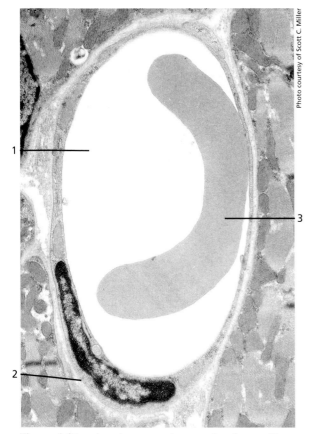

Figure 1.24. An electron micrograph of a capillary containing an erythrocyte.

1. Lumen of capillary   3. Erythrocyte
2. Endothelial cell

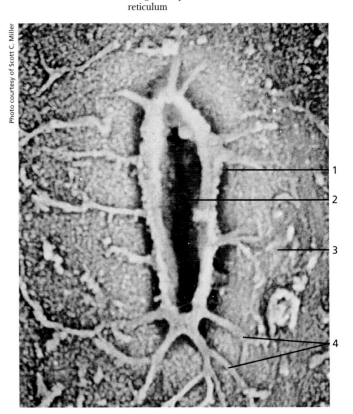

Photo courtesy of Scott C. Miller

Figure 1.25. An electron micrograph of an osteocyte in cortical bone matrix.

1. Lacuna   2. Osteocyte   3. Bone matrix   4. Canaliculi

ANIMAL CELLS AND TISSUES

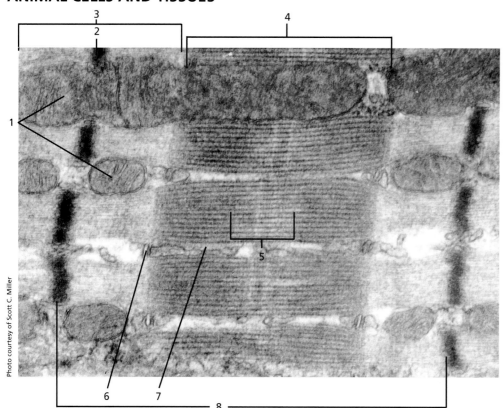

Photo courtesy of Scott C. Miller

Figure 1.26. An electron micrograph of a skeletal muscle myofibril, showing the striations.

1. Mitochondria
2. Z line
3. I band
4. A band
5. H band
6. T-tubule
7. Sarcoplasmic reticulum
8. Sarcomere

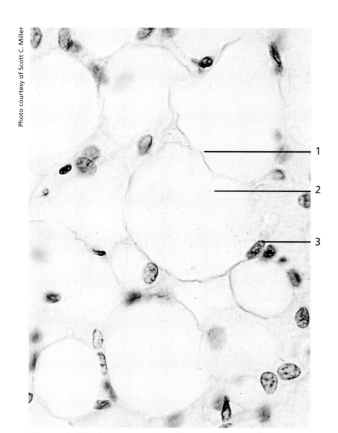

Photo courtesy of Scott C. Miller

Figure 1.27. Adipocytes in adipose tissue. (X200)

1. Cell membrane of adipocyte
2. Lipid-filled vacuole of adipocyte
3. Nucleus

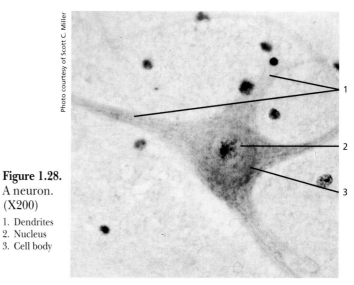

Photo courtesy of Scott C. Miller

Figure 1.28. A neuron. (X200)

1. Dendrites
2. Nucleus
3. Cell body

Figure 1.29. An electron micrograph of an erythrocyte (red blood cell).

ANIMAL CELLS AND TISSUES

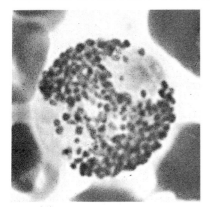

Basophil

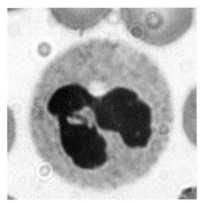

Polymorphonuclear leukocyte (PMN)

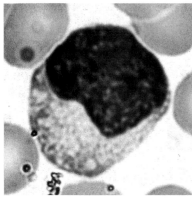

Monocyte

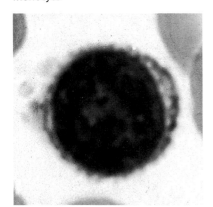

Lymphocyte

Figure 1.30. Types of leukocytes. (X200)

Epithelial Tissues

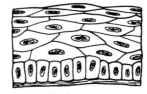

Nucleus
Cell membrane
Basement membrane

Cuboidal

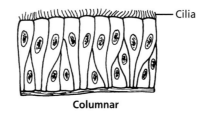

Stratified squamous

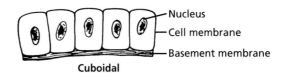

Cilia

Columnar

Nervous Tissues

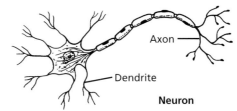

Axon

Dendrite

Neuron

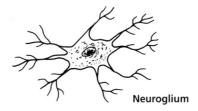

Neuroglium

Figure 1.31. Examples of animal tissues.

Connective Tissues

Dense regular

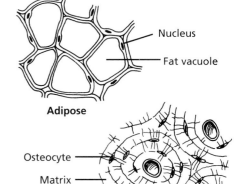

Nucleus

Fat vacuole

Adipose

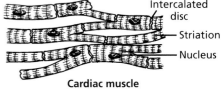

Osteocyte
Matrix

Bone

Muscle Tissues

Nucleus

Smooth muscle

Nucleus
Striation

Skeletal muscle

Intercalated disc
Striation
Nucleus

Cardiac muscle

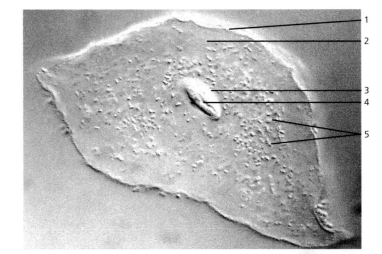

Figure 1.32.
Human epithelial cell obtained from a scraping of the inside of the cheek lining the oral cavity. (X1,000)

1. Cell membrane
2. Cytoplasm
3. Nuclear membrane
4. Nucleus
5. Organelles

1
2
3
4
5

ANIMAL CELLS AND TISSUES

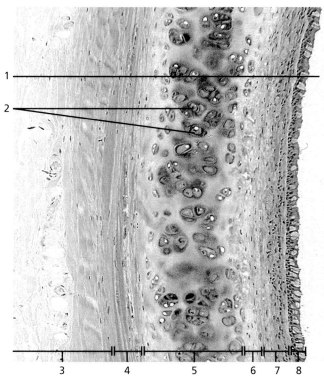

Figure 1.33. Pseudostratified ciliated columnar epithelium of the trachea. (X281)

1. Lumen of trachea
2. Chondrocytes within lacunae
3. Adventitia
4. Perichondrium
5. Hyaline cartilage
6. Perichondrium
7. Lamina propria
8. Pseudostratified ciliated columnar epithelium

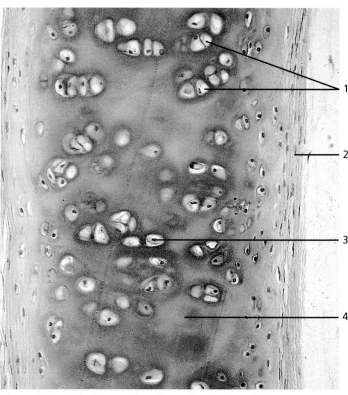

Figure 1.34. Hyaline cartilage from the trachea. (X60)

1. Chondrocytes within lacunae
2. Perichondrium
3. Nucleus of chondrocyte
4. Matrix

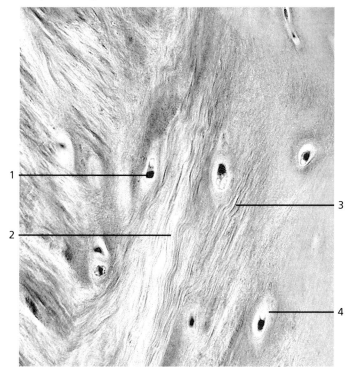

Figure 1.35. Fibrocartilage of an intervertebral disc. (X125)

1. Nucleus of chondrocyte
2. Matrix
3. Collagenous fibers
4. Chondrocyte within lacuna

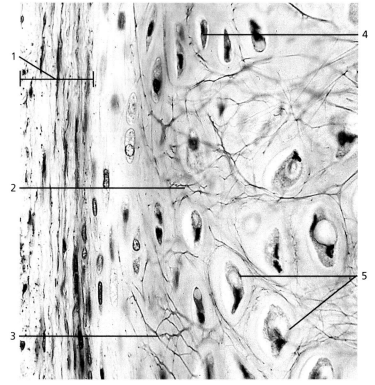

Figure 1.36. Elastic cartilage from auricle of the external ear. (X20)

1. Perichondrium
2. Matrix with elastic fibers
3. Elastic fiber
4. Nucleus of chondrocytes
5. Chondrocytes within lacunae

ANIMAL CELLS AND TISSUES

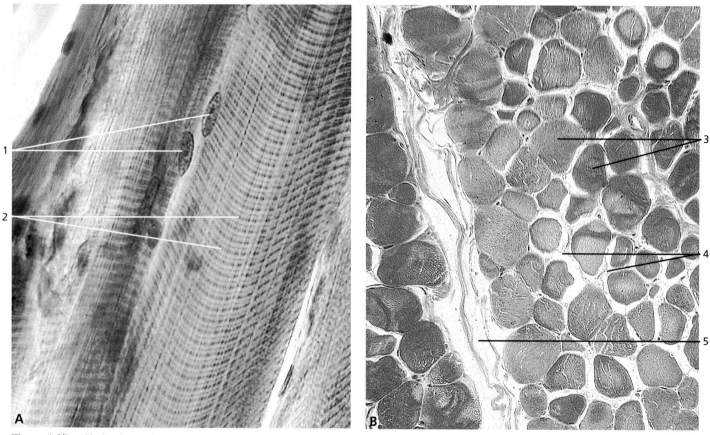

Figure 1.37.   Skeletal muscle. (A) Longitudinal section. (X200) (B) cross section. (X200)

1. Nuclei   2. Striations of skeletal muscle fibers   3. Skeletal muscle fibers   4. Endomysium   5. Perimysium

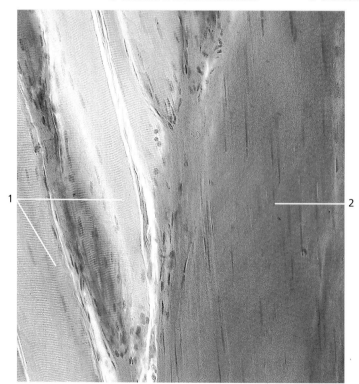

Figure 1.38.   Histology of a skeletal muscle attaching to a tendon. (X50)

1. Skeletal muscle fibers   2. Dense regular connective tissue of tendon

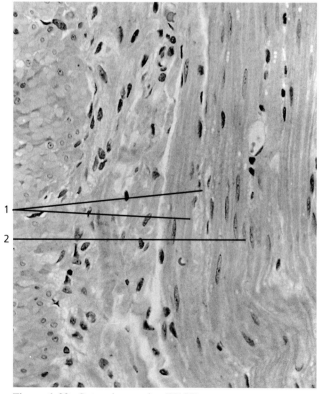

Figure 1.39. Smooth muscle. (X125)

1. Smooth myofibers   2. Nucleus of smooth myofiber

Perpetuation of Life

2

The *cell cycle* accounts for the formation of a multicellular organism, body growth, and the maintenance and repair of body tissues. The cell cycle permits each new cell to receive a complete copy of all of the genes present in the parent cell and the cytoplasmic substances and organelles to carry out hereditary instruction.

The animal cell cycle is divided into: 1) interphase, which includes G_1, S, and G_2 phases; and 2) mitosis, which includes prophase, metaphase, anaphase, and telophase. Interphase is the interval between successive cell divisions during which the cell is metabolizing and the chromosomes are directing RNA synthesis. The G_1 phase is the first growth phase, the S phase is when DNA is replicated, and the G_2 phase is the second growth phase. *Mitosis* (also *karyokinesis*) is the division of the nuclear parts of a cell to form two diploid daughter nuclei. *Cytokinesis* accompanies mitosis and is the division of the cytoplasm.

Like the animal cell cycle, the plant cell cycle consists of growth, synthesis, mitosis, and cytokinesis. *Growth* is the increase in cellular mass as the result of metabolism; *synthesis* is the production of DNA and RNA to regulate cellular activity; *mitosis* is the splitting of the nucleus and the equal separation of the chromatids; and *cytokinesis* is the division of the cytoplasm that accompanies mitosis.

Unlike animal cells, plant cells have a rigid cell wall that does not cleave during cytokinesis, but rather a new cell wall is constructed between the daughter cells. Furthermore, many land plants do not have centrioles for the attachment of spindles. The microtubules in these plants form a barrel-shaped anastral spindle at each pole. Mitosis and cytokinesis in plants follow basically the same sequence as these processes in animal cells.

Asexual reproduction is propagation without sex, that is, the production of new individuals by processes that do not involve *gametes* (sex cells). Asexual reproduction occurs in a variety of microorganisms, fungi, plants, and animals, wherein a single parent produces offspring with characteristics identical to itself. Asexual reproduction is not dependent on the presence of other individuals, and no eggs or sperm are required. In asexual reproduction, all the offspring are genetically identical (except for mutants). Examples of asexual reproduction include: *fission* — a single cell divides to form two separate cells (bacteria, protozoans, and other one-celled organisms); *sporulation* — multiple fission, many cells are formed and join together in a cystlike structure (protozoans and fungi); *budding* — buds develop organs like the parent and then detach themselves (hydras, yeast, certain plants); and *fragmentation* — organisms break into two or more parts, and each part is capable of becoming a complete organism (flatworms, echinoderms).

Sexual reproduction is propagation of new organisms through the union of genetic material from two parents. Sexual reproduction usually involves the fusion of haploid gametes (sperm and egg) during fertilization to form a zygote. The major biological difference between sexual and asexual reproduction is that sexual reproduction produces genetic variation in the offspring. The only genetic variation that can arise in asexual reproduction comes from mutations. The combining of genetic material in the sperm and egg produces offspring that are different from and contain new combinations of characteristics than either parent. This increases the ability of the species to survive environmental changes or to reproduce in new habitats.

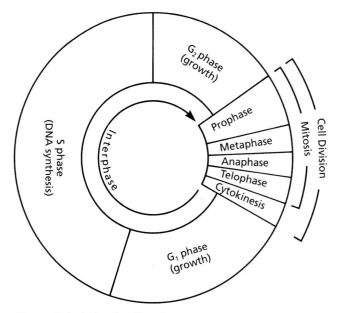

Figure 2.1. Animal cell cycle.

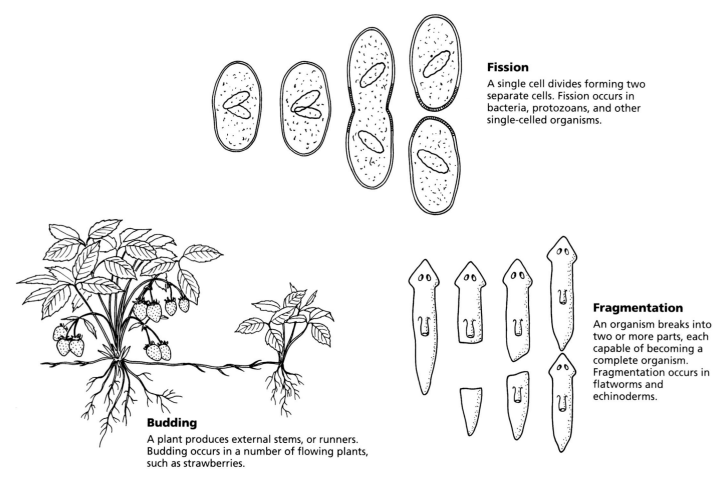

Fission

A single cell divides forming two separate cells. Fission occurs in bacteria, protozoans, and other single-celled organisms.

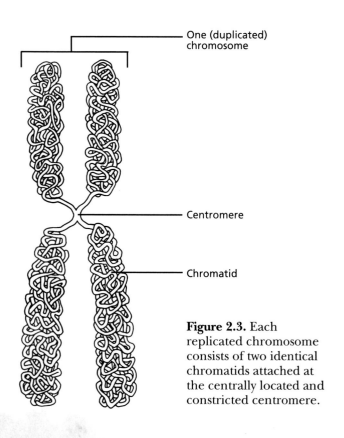

Fragmentation

An organism breaks into two or more parts, each capable of becoming a complete organism. Fragmentation occurs in flatworms and echinoderms.

Budding

A plant produces external stems, or runners. Budding occurs in a number of flowing plants, such as strawberries.

Figure 2.2. Types of asexual reproduction.

One (duplicated) chromosome

Centromere

Chromatid

Figure 2.3. Each replicated chromosome consists of two identical chromatids attached at the centrally located and constricted centromere.

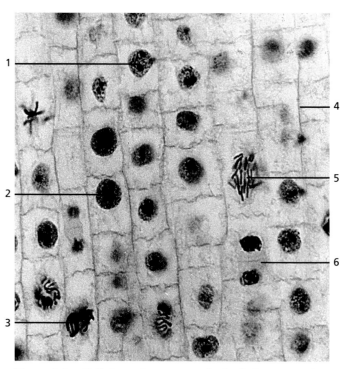

Figure 2.4.   Cells in various stages of mitosis from an onion, *Allium*, root tip. (X100)

1. Interphase	3. Metaphase	5. Anaphase
2. Prophase	4. Cell wall	6. Telophase

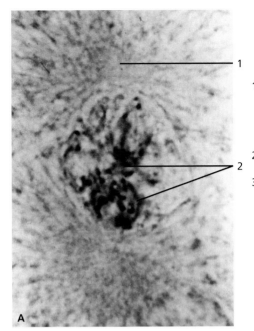

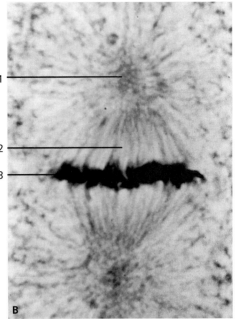

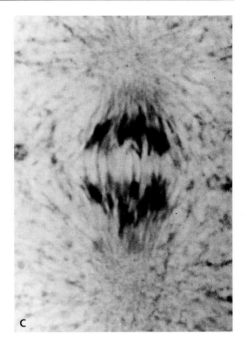

1. Centriole   2. Chromatids

1. Aster around centriole   3. Chromatids
2. Spindle fibers   at equator

Prophase
Each chromosome consists of two chromatids jointed by a centromere. Spindle fibers extend from each centriole.

Metaphase
The chromosomes are positioned at the equator. The spindle fibers from each centriole attach to the centromeres.

Anaphase
The centromeres split, and the sister chromatids separate as each is pulled to an opposite pole.

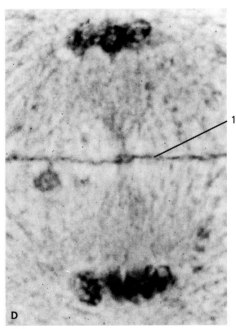

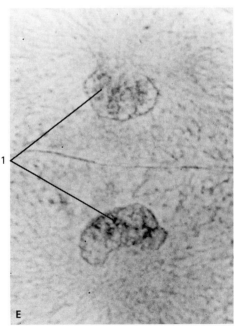

Figure 2.5. Stages of mitosis in representative animal cells. (X1000)

1. Cell membrane

1. Daughter nuclei

Telophase
The chromosomes lengthen and become less distinct. The cell membrane forms between the forming daughter cells.

Daughter cells
Cell division is complete and the newly formed cells grow and mature.

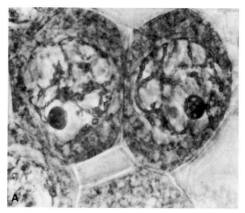

Prophase I — Each chromosome consists of two chromatids joined by a centromere. Spindle fibers extend from each centriole.

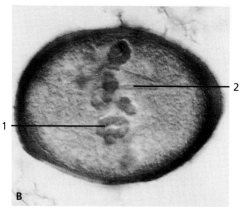

Metaphase I — The chromosomes align at the equator with their homologous partner. During this stage, called synapsis, exchange occurs between the chromosomes.

1. Chromatids at equator   2. Spindle fibers

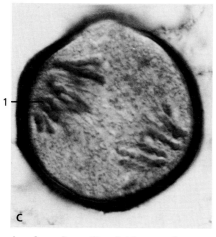

Anaphase I — No division at the centromeres occurs as the chromosomes separate, so one copy of each homologous pair of chromosomes goes to each pole.

1. Chromatids

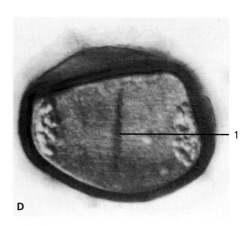

Telophase I — The chromosomes lengthen and become less distinct. The cell wall (in some plants) forms between the forming cells.

1. Cell wall

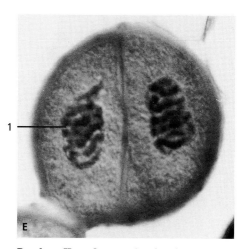

Prophase II — Once again, the chromosomes condense as in prophase I.

1. Chromatid

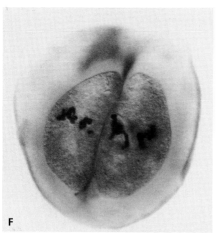

Metaphase II — The chromosomes align on the equator and the spindle fibers attach to the centromeres. This is similar to metaphase in mitosis.

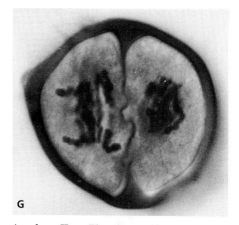

Anaphase II — The chromatids separate and each is pulled to an opposite pole.

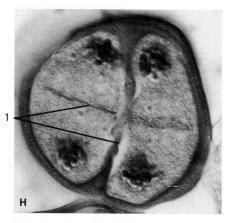

Telophase II — Cell division is complete and cell walls of four haploid cells are formed.

1. Cell walls

Figure 2.6. Stages of meiosis in lily pollen grains. (X1000)

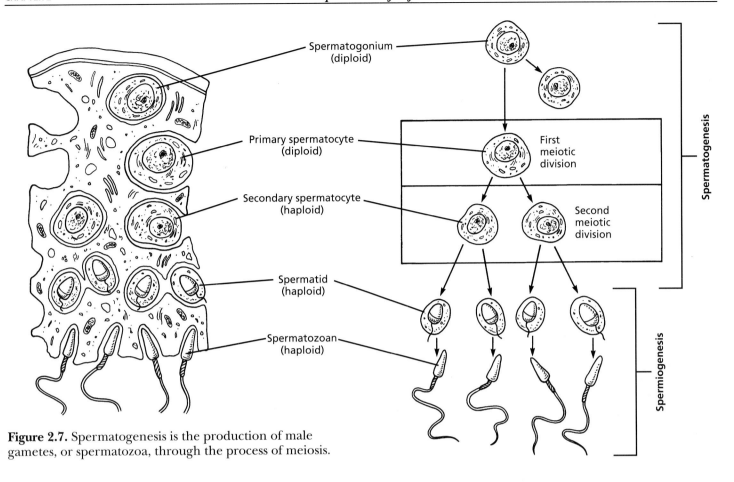

Figure 2.7. Spermatogenesis is the production of male gametes, or spermatozoa, through the process of meiosis.

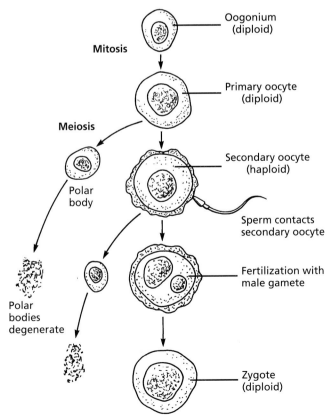

Figure 2.8. Oogenesis is the production of female gametes, or ova, through the process of meiosis.

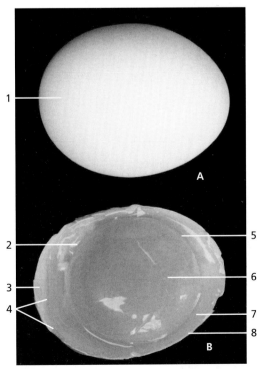

Figure 2.9. Chicken egg. (A) Intact, (B) portion of shell removed exposing internal structures.

1. Shell	4. Shell membranes	7. Albumen
2. Ovum	5. Vitelline membrane	8. Shell
3. Air space	6. Yolk	

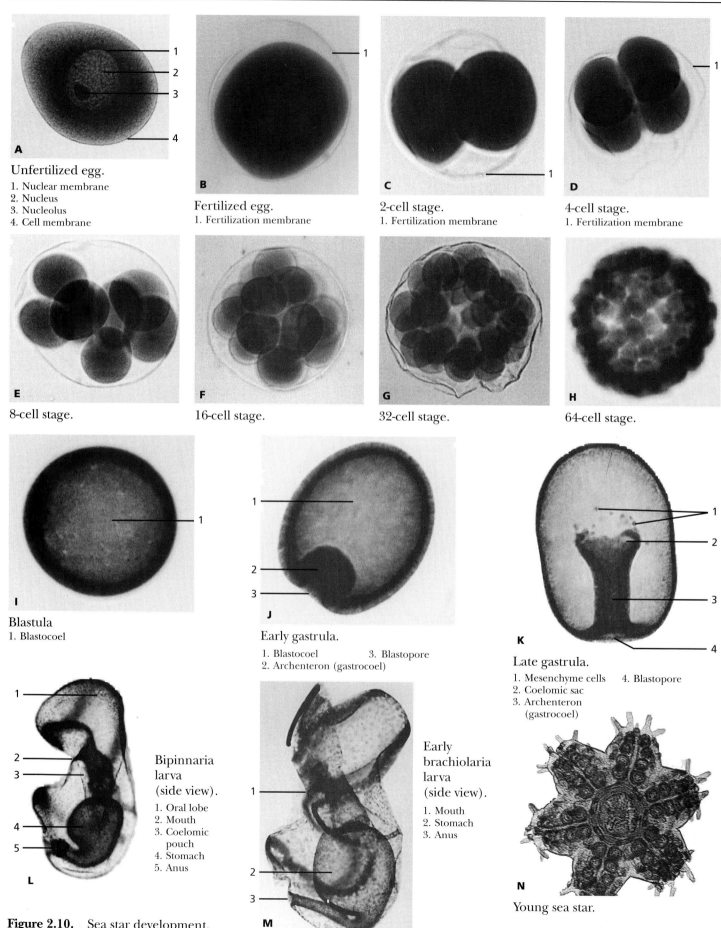

A

Unfertilized egg.
1. Nuclear membrane
2. Nucleus
3. Nucleolus
4. Cell membrane

B

Fertilized egg.
1. Fertilization membrane

C

2-cell stage.
1. Fertilization membrane

D

4-cell stage.
1. Fertilization membrane

E

8-cell stage.

F

16-cell stage.

G

32-cell stage.

H

64-cell stage.

I

Blastula
1. Blastocoel

J

Early gastrula.
1. Blastocoel   3. Blastopore
2. Archenteron (gastrocoel)

K

Late gastrula.
1. Mesenchyme cells   4. Blastopore
2. Coelomic sac
3. Archenteron
   (gastrocoel)

L

Bipinnaria
larva
(side view).
1. Oral lobe
2. Mouth
3. Coelomic
   pouch
4. Stomach
5. Anus

M

Early
brachiolaria
larva
(side view).
1. Mouth
2. Stomach
3. Anus

N

Young sea star.

Figure 2.10.   Sea star development.

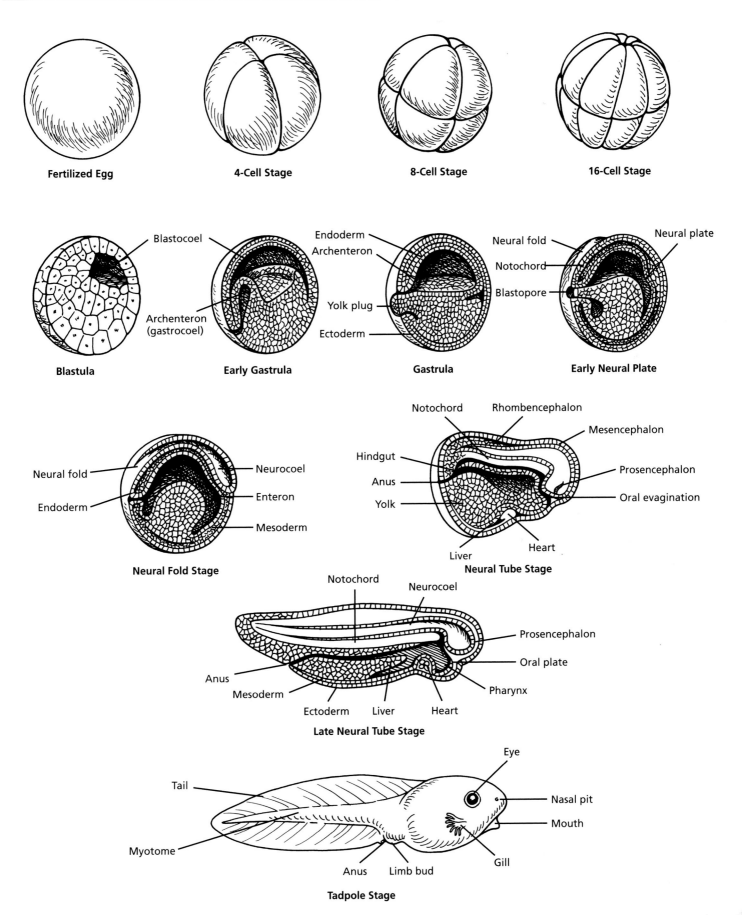

Figure 2.11. Frog development.

Kingdom Monera

The kingdom Monera contains all organisms comprised of prokaryotic cells. Prokaryotic cells were the first kinds of cells to evolve, probably about 3.5 billion years ago. The Archaebacteria and Eubacteria are the two groupings within the kingdom Monera. Archaebacteria are bacteria adapted to a limited range of extreme conditions. They include methanogens, typically found in swamps and marshes, and thermoacidophiles, found in acid hot springs and acidic soil. The cell walls of Archaebacteria lack peptidoglycan, characteristic of Eubacteria. Archaebacteria have distinctive transfer RNAs and RNA polymerases.

Methanogens are bacteria that exist in oxygen free environments and subsist on simple inorganic compounds such as CO_2, acetate, and methanol. As their name implies, *Methanobacteria* produce methane gas as a byproduct of metabolism. These organisms are typically found in organic-rich mud and sludge, particularly that which contains fecal wastes.

The thermoacidophiles are resistant to hot temperatures and high acid concentrations. The plasma membrane of these organisms contains high amounts of saturated fats, and its enzymes and other proteins are able to withstand extreme conditions without denaturation These microscopic organisms thrive in most hot springs and hot, acid soils.

Eubacteria are considered the "true" bacteria, and include the *Cyanobacteria* (formerly known as blue-green algae) and a number of other diverse types (Table 3.1). *Cyanobacteria* are photosynthetic bacteria that contain chlorophyll and release oxygen during photosynthesis. Some bacteria are obligate aerobes (require O_2) and others are facultative anaerobes (indifferent to O_2). Most bacteria are heterotrophic saprophytes, which secrete enzymes to break down surrounding molecules into absorbable compounds.

Bacteria range between 1 and 10 μm in width or diameter. The morphological appearance may be spiral (a spirillum), spherical (a coccus), or rod-shaped (a bacillus). Cocci and bacilli frequently form clusters or linear filaments, and many have cilia. Relatively few species of bacteria cause infection. Hundreds of non-pathogenic species of bacteria live on the human body and within the GI tract and are considered normal gut flora.

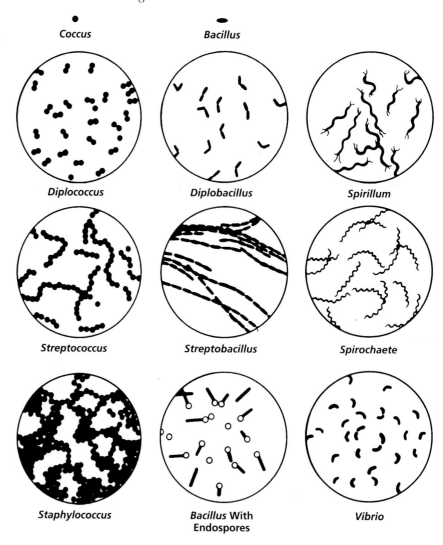

Figure 3.1. Various bacteria.

TABLE 3.1
Some Representatives of the Kingdom Monera

Categories	Representative Genera
Archaebacteria	
Methanogens	*Halobacterium, Methanobacteria*
Thermoacidophiles	*Thermoplasma, Sulfobolus*
Eubacteria	
Photosynthetic bacteria	
Cyanobacteria	*Anabaena, Oscillatoria, Spirulina, Nostoc*
Green bacteria	*Chlorobium*
Purple bacteria	*Rhodospirillum*
Gram-negative bacteria	*Proteus, Pseudomonas, Escherichia, Rhizobium, Neisseria*
Gram-positive bacteria	*Streptococcus, Staphylococcus, Bacillus, Clostridium, Lactobacillus*
Spirochaetes	*Spirochaeta, Treponema*
Actinomycetes	*Actinomyces*
Rickettsias	*Rickettsia, Chlamydia*
Mycoplasmas	*Mycoplasma*

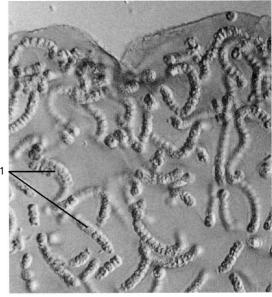

Figure 3.2. A colony of *Nostoc* filaments. Individual filaments secrete mucilage which forms a rigid matrix around the filaments. (X430)

1. Filaments

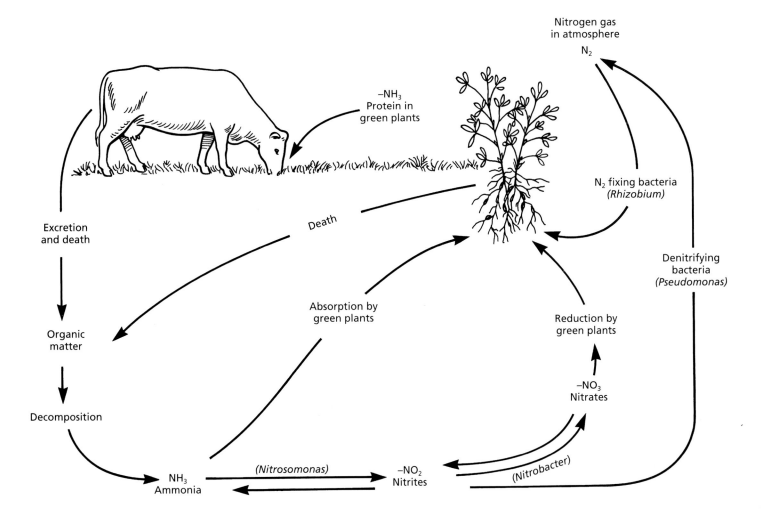

Figure 3.3. Nitrogen-fixing bacteria within the root nodules of legumes provide a usable source of nitrogen to producers.

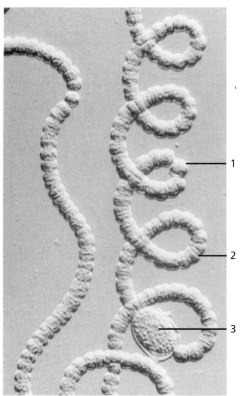

Figure 3.4. *Anabaena* filaments. This is a nitrogen-fixing cyanobacterium. Nitrogen fixation takes place within the heterocyst cells. (X430)

1. Heterocyst
2. Vegetative cell
3. Akinete (spore)

Figure 3.5. *Merismopedia,* a genus of cyanobacterium, is characterized by flattened colonies of cells. The cells are a single-layer thick and usually aligned into groups of two or four. (X430)

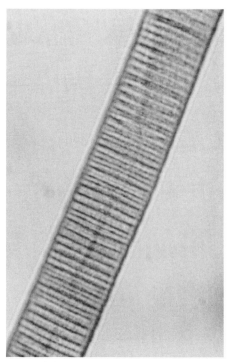

Figure 3.6. *Oscillatoria* filament. The only way this cyanobacterium can reproduce is through fragmentation of a filament. (X430)

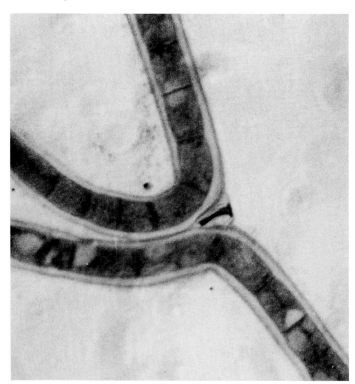

Figure 3.7. *Scytonema,* a cyanobacterium, is common on soil moistened from the spray of a waterfall or stream. Note the falsely-branched filament typical of this genus. (about X500)

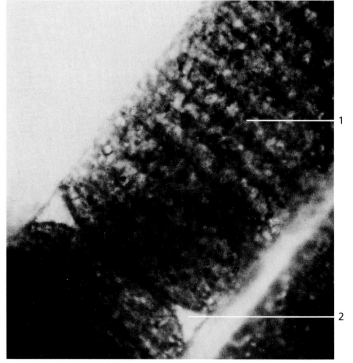

Figure 3.8. A portion of a cylindrical filament of a cyanobacterium, *Oscillatoria,* that is common in most aquatic habitats. (X750)

1. Filament segment   2. Separation disk

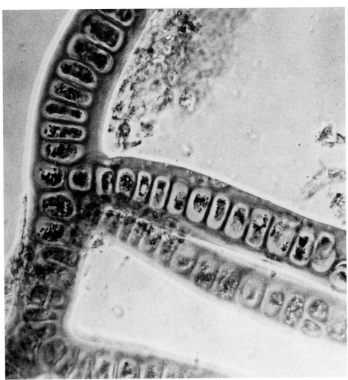

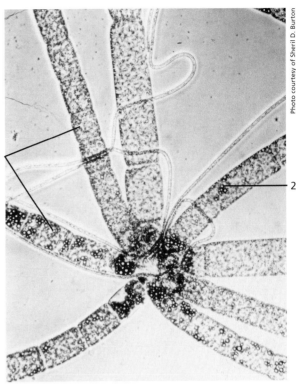

Photo courtesy of Sheril D. Burton

Figure 3.9. *Stigonema*, a cyanobacterium, is characterized by true branched filaments. (about X500)

Figure 3.10. *Thiothrix* is a genus of bacterium which forms sulfur granules in its cytoplasm. Energy for these organisms is obtained from the oxidation of H₂S. (X200)

1. Filaments   2. Sulfur granules

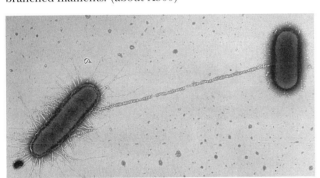

Figure 3.11. Conjugation of the bacterium *Escherichia coli*. By this process, genetic material is transferred through the conjugation tube from one cell to the other. (about X1700)

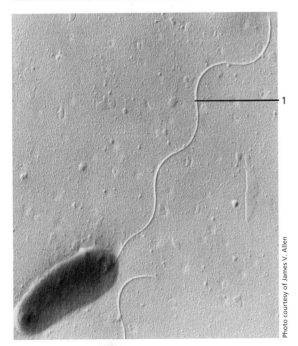

Photo courtesy of James V. Allen

Figure 3.12. A flagellated bacterium, *Pseudomonas*. (about X2500)

1. Flagellum

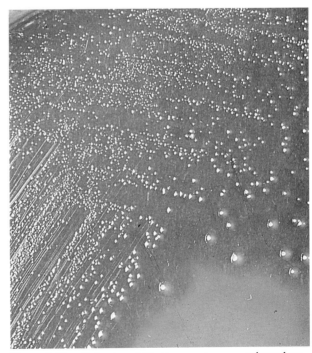

Figure 3.13. Colonies of *Streptococcus pyogenes* cultured on a nutrient agar plate. *S. pyogenes* is the organism responsible for strep throat and rheumatic fever. (X2)

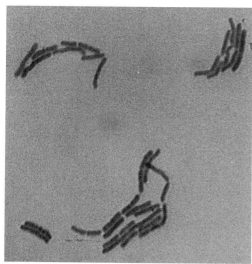

Figure 3.14. *Bacillus megaterium. Bacillus* is one of the few bacteria capable of producing endospores. This species of *Bacillus* generally remains in chains after it divides. (X1000)

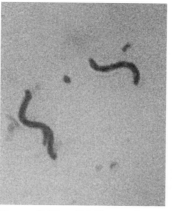

Figure 3.15. *Spirillum volutans.* Bacteria in the genus *Spirillum* are shaped like long rods twisted into rigid helices. They generally have multiple polar flagella. (X1000)

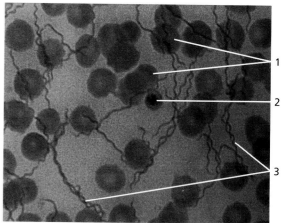

Figure 3.16. A spirochete, *Borella recurrentis.* Spirochetes are flexible rods twisted into helical shapes. This species causes relapsing fever. (X1000)

1. Red blood cells
2. White blood cells
3. Spirochete

Figure 3.17. *Streptococcus pyogenes,* the organism that causes rheumatic fever. Note the chains of coccus-shaped bacteria. (X1000)

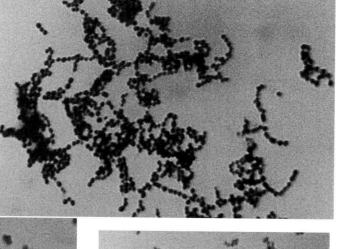

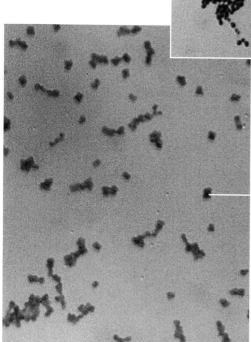

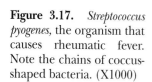

Figure 3.18. *Micrococcus luteus. Micrococcus* are gram-positive bacteria that are generally arranged as clusters or tetrads. (X1000)

1. Tetrad (4 cells)

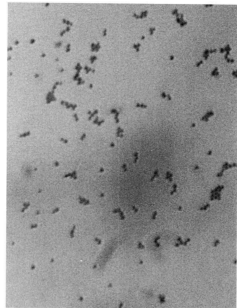

Figure 3.19. *Staphylococcus.* Cells are arranged in irregular clusters. (X1000)

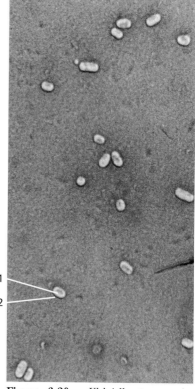

Figure 3.20 *Klebsiella pneumoniae* capsules. This bacterium is able to encapsulate itself in order to make it more resistant to host defense mechanisms.

1. Cell
2. Capsule

Kingdom Protista

Most protists are unicellular, eukaryotic organisms, but there are some species that are multicellular. As eukaryotic organisms, each of the protists has complex organization of organelles that generally include a nucleus, mitochondria, chloroplasts, endoplasmic reticulum, and Golgi apparatus. It is within this group that the processes of mitosis and meiosis arose.

Protists are abundant in water and are important constituents of plankton. Plankton are organisms that drift passively or swim slowly near the surface of ponds, lakes, and oceans. They are a major source of food for other aquatic organisms. Phytoplankton are generally considered the primary producers in aquatic ecosystems.

The algal protists include the microscopic aquatic organisms within the phylum Chrysophyta and the phylum Dinoflagellata. The chrysophytes are the yellow-green and the golden-brown algae and the diatoms. The cell wall of a diatom is composed of silica and manganese rather than cellulose. Movement is slow and gliding as cytoplasm apparently travels through the raphes of the cell wall. The dinoflagellates are single-celled algae-like organisms. In most species of dinoflagellates, the cell wall is covered by thecal plates of cellulose. They are all motile, having two flagella, one extending from a transverse groove and the other projecting from one end.

Protozoa are also protists. They are small (2 μm–1,000 μm), unicellular eukaryotic organisms that lack a cell wall. Movement of protozoa is due to flagella, cilia, or pseudopodia of various sorts. In feeding upon other organisms or organic particles, they utilize simple diffusion, pinocytosis, or phagocytosis. Although most protozoa reproduce asexually, some species may also reproduce sexually during a portion of their life cycles. Protozoa are extremely important primary consumers. Although most are harmless, some are of immense clinical concern because they are parasitic and may cause human disease, such as African sleeping sickness and malaria.

TABLE 4.1.
Some Representatives of the Kingdom Protista

Division and Representative Kinds	Characteristics
Unicellular Protists	
Phylum Chrysophyta (diatoms and golden algae)	Diatom cell walls of silica, with two halves; plastids often golden
Phylum Dinoflagellata (dinoflagellates)	Two flagella in grooves of wall; brownish plastids
Phylum Rhizopoda (amoebas)	Cytoskeleton of microtubules and microfilaments; amoeboid locomotion
Phylum Apicomplexa (sporozoa, *Plasmodium*)	Lack locomotor capabilities and contractile vacuoles; mostly parasitic
Phylum Euglenophyta (*Euglena*)	Green flagellates lacking walls
Phylum Ciliophora (ciliates, *Paramecium*)	Use cilia for motility and feeding
Algae	
Phylum Chlorophyta (green algae)	Unicellular, colonial, and multicellular forms, mostly fresh water although some marine forms; reproduce asexually and sexually; gametes often biflagellated with cup-shaped chloroplasts
Phylum Phaeophyta (brown algae, giant kelp)	Multicellular, mostly marine in the intertidal zone; most with alternation of generations
Phylum Rhodophyta (red algae)	Multicellular, mostly marine; sexual reproduction but with no flagellated cells; alternation of generations common
Protists Resembling Fungi	
Phylum Myxomycota (plasmodial slime molds)	Multinucleated continuum of cytoplasm without internal membranes; amoeboid plasmodium during feeding stage; produce asexual fruiting bodies
Phylum Acrasiomycota (cellular slime molds)	Solitary cells during feeding stage; aggregate of cells when food is scarce; produce asexual fruiting bodies
Phylum Oomycota (water molds, white rusts, downy mildews)	Decomposers or parasitic forms; walls of cellulose, dispersal by spores or flagellated zoospores

PHYLUM CHRYSOPHYTA (diatoms and golden algae)

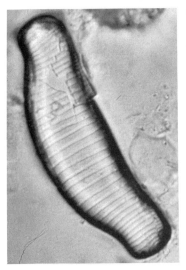

Eunotia

Navicula

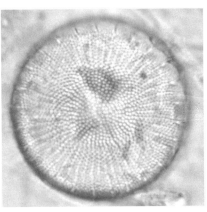

Cyclotella   *Stephanodiscus*

Figure 4.1. Examples of common freshwater diatoms. *Eunotia* species are often found in acidic waters common in forest ponds or bogs. *Navicula* is a large and widely distributed genus with species found in both marine and freshwater habitats and on wet soil. *Cyclotella* and *Stephanodiscus* are centric diatoms common in lake plankton.

Photo courtesy of Samuel R. Rushforth

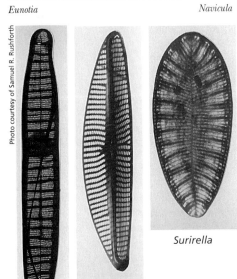

Surirella

Cymbella

Diatoma

Figure 4.2. Electron micrographs of several types of diatoms.

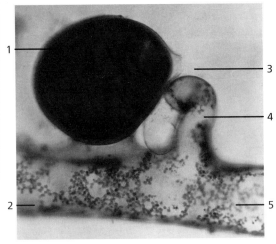

Figure 4.3. *Vaucheria*, with mature gametangia. (X430)

1. Oogonium
2. Chloroplasts
3. Fertilization pore
4. Antheridium
5. Coenocytic filament

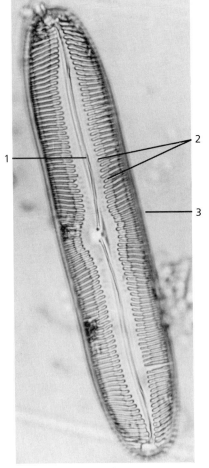

Figure 4.5. *Pinnularia*, a common diatom. (X430)

1. Raphe   2. Striae   3. Valve

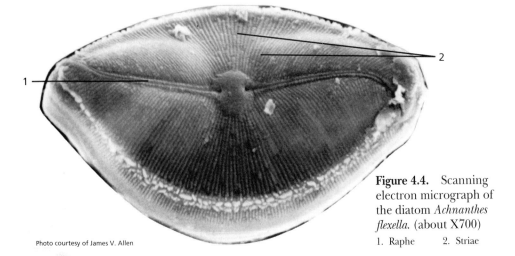

Photo courtesy of James V. Allen

Figure 4.4. Scanning electron micrograph of the diatom *Achnanthes flexella.* (about X700)

1. Raphe   2. Striae

PHYLUM DINOFLAGELLATA (dinoflagellates)

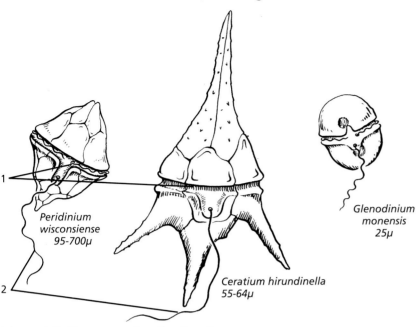

Peridinium wisconsiense 95-700μ

Ceratium hirundinella 55-64μ

Glenodinium monensis 25μ

Figure 4.6. Representative dinoflagellates.
1. Girdles   2. Flagella

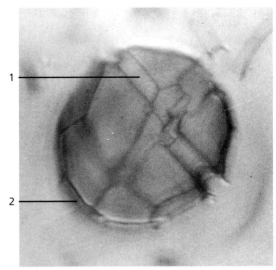

Figure 4.7. Photomicrograph of a dinoflagellate. The cell wall of many dinoflagellates is composed of overlapping plates of cellulose which are evident in this photomicrograph. (X430)

1. Transverse groove   2. Wall of cellulose plates

PHYLUM RHIZOPODA (amoebas)

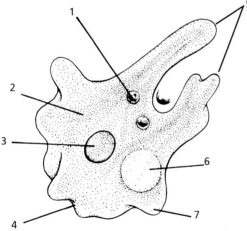

Figure 4.8. *Amoeba proteus* is a fresh water protozoan that moves by forming cytoplasmic extensions called pseudopodia.

1. Food vacuole
2. Endoplasm
3. Nucleus
4. Cell membrane
5. Pseudopodia
6. Contractile vacuole
7. Ectoplasm

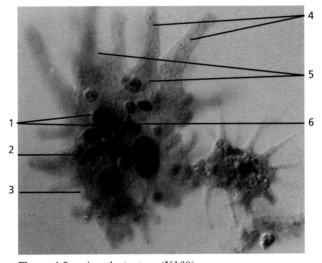

Figure 4.9. *Amoeba proteus.* (X160)

1. Food vacuoles   3. Cell membrane   5. Ectoplasm
2. Nucleus   4. Pseudopodia   6. Endoplasm

PHYLUM APICOMPLEXA (sporozoa)

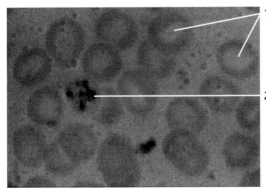

Figure 4.10. *Plasmodium vivax.* This is the protozoan that causes the disease malaria. It is transmitted by the female mosquito. (X1000)

1. Red blood cells
2. Merozites in RBC

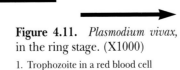

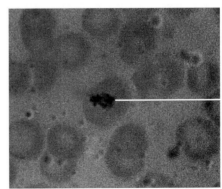

Figure 4.11. *Plasmodium vivax,* in the ring stage. (X1000)

1. Trophozoite in a red blood cell

PHYLUM EUGLENOPHYTA (*Euglena*)

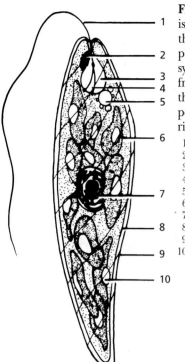

Figure 4.12. *Euglena* is a green flagellate that contains chloroplasts for photosynthesis. They are fresh-water organisms that have a flexible pellicle rather than a rigid cell wall.

1. Flagellum
2. Photoreceptor
3. Reservoir
4. Basal body
5. Contractile vacuole
6. Chloroplast
7. Nucleus
8. Pellicle
9. Cell membrane
10. Paramylon granule

Figure 4.13. Light micrograph of *Euglena*. (X430)

1. Striated pellicle
2. Chloroplast
3. Paramylon granule
4. Flagellum
5. Reservoir
6. Nucleus
7. Paramylon granule

Figure 4.14. *Euglena.* (X100)

1. Nucleus   2. Chloroplast

PHYLUM CILIOPHORA (ciliates, *Paramecium*)

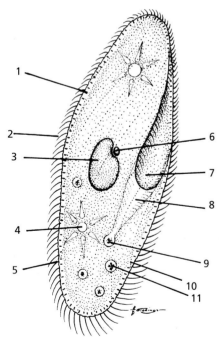

Figure 4.15. *Paramecium caudatum* is a ciliated protozoan. The poisonous trichocysts of these unicellular organisms are used for defense and capturing prey.

1. Trichocyst
2. Cilia
3. Macronucleus
4. Contractile vacuole
5. Pellicle
6. Micronucleus
7. Oral cavity
8. Gullet
9. Forming food vacuole
10. Anal pore
11. Food vacuole

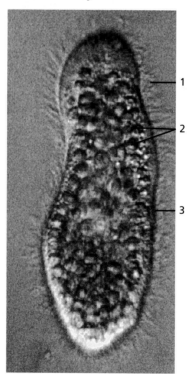

Figure 4.16. *Paramecium* is a unicellular, slipper-shaped organism. Paramecia are usually common in ponds containing decaying organic matter. (X400)

1. Cilia   3. Pellicle
2. Micronuclei

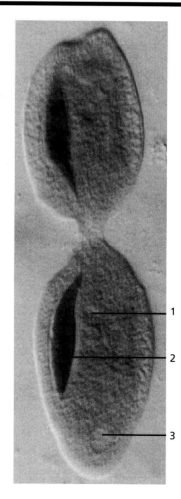

Figure 4.17. *Paramecium* in fission. (X400)

1. Micronucleus
2. Macronucleus
3. Contractile vacuole

PHYLUM CHLOROPHYTA (green algae)

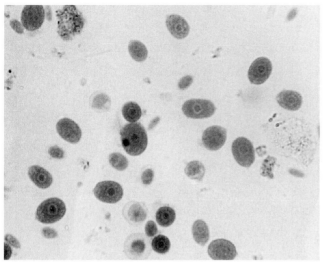

Figure 4.18. *Chlamydomonas*, a common unicellular green alga. (X800)

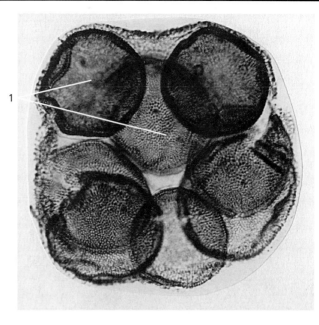

Figure 4.19. *Volvox*, a single organism with several large daughter colonies. (X100)

1. Daughter colonies

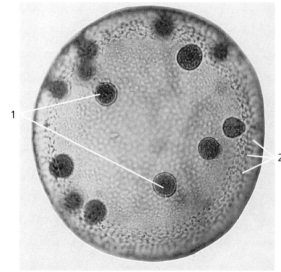

Figure 4.20. *Volvox*, a single mature specimen with several eggs and zygotes. (X100)

1. Zygotes   2. Zygotic (sexual) colonies

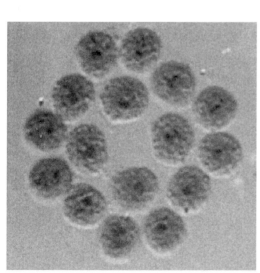

Figure 4.21. *Gonium* colony. *Gonium* is a 16-celled flat colony of *Chlamydomonas*-like cells. (X450)

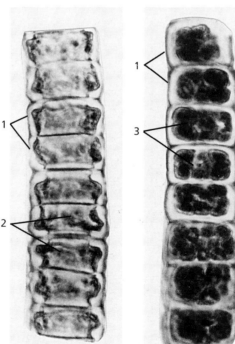

A vegetative filament.   Filament with zoospores.   Empty filament, after zoospores have been released.

Figure 4.22. The production and release of zoospores in the green alga *Ulothrix*. (all X200)

1. Filament   2. Chloroplasts   3. Zoospores   4. Empty cells

PHYLUM CHLOROPHYTA (green algae)

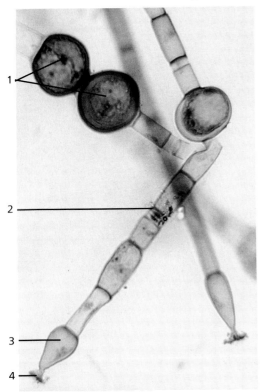

Figure 4.23. *Oedogonium,* a filamentous, unbranched green alga. (X430)

1. Oogonia
2. Antheridium
3. Basal cell
4. Holdfast

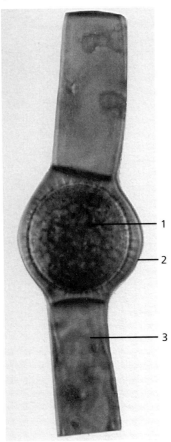

Figure 4.24. Oogonium of the green alga *Oedogonium.* (X1000)

1. Egg
2. Oogonium
3. Vegetative cell

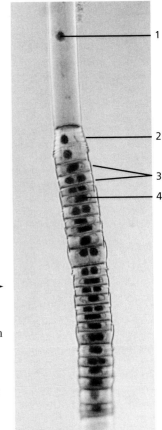

Figure 4.25. A filament of the unbranched green alga *Oedogonium.* (X430)

1. Cell nucleus
2. Annular scars from cell division
3. Antheridia
4. Sperm

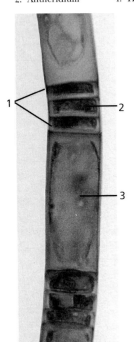

Figure 4.26. The green alga, *Oedogonium,* showing antheridia between vegetative cells. (X600)

1. Antheridia
2. Sperm
3. Vegetative cell

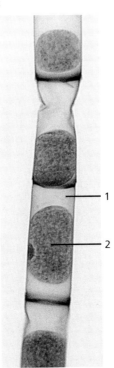

Figure 4.27. The zoosporangium of the unbranched green alga, *Oedogonium.* (X600)

1. Zoosporangium
2. Zoospore

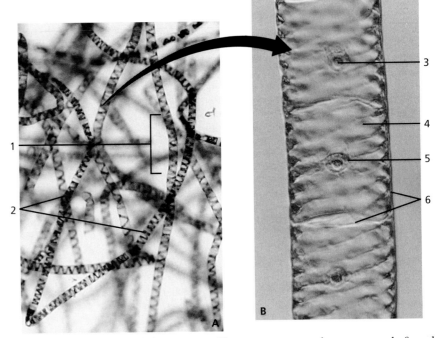

Figure 4.28. Species of *Spirogyra* are filamentous green algae commonly found in green masses on the surfaces of ponds and streams. Their chloroplasts are arranged as a spiral within the cell. (A) Several cells comprise a filament. (X65) (B) A magnified view of a single filament composed of several cells. (X430).

1. Single cell
2. Filaments
3. Nucleolus
4. Chloroplast
5. Nucleus
6. Cell wall

PHYLUM CHLOROPHYTA (green algae)

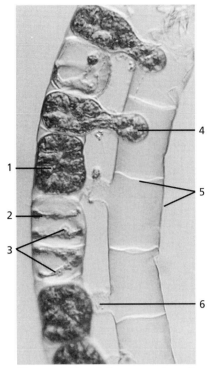

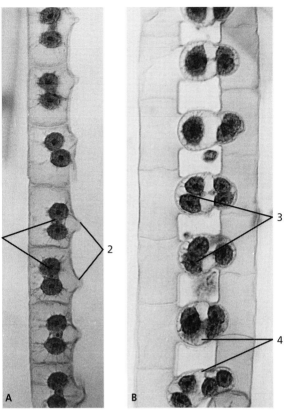

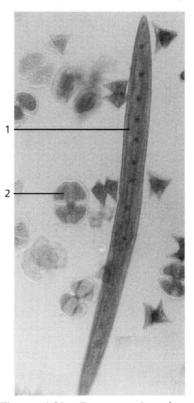

Figure 4.29. *Spirogyra* undergoing conjugation. (X200)

1. Zygote (zygospore)	4. Male gamete
2. Chloroplast	5. Cell wall
3. Pyrenoid	6. Conjugation tube

Figure 4.30. *Zygnema* undergoing conjugation.
(A) The filament is just forming conjugation tubes.
(B) Two conjugated filaments. Zygotes in this species are produced in the conjugation tubes. (X100)

1. Gametes	3. Forming zygotes
2. Forming conjugation tubes	4. Conjugation tubes

Figure 4.31. Representative desmids. Desmids are unicellular, freshwater chlorophyta. (X200)

1. *Closterium*	2. *Cosmarium*

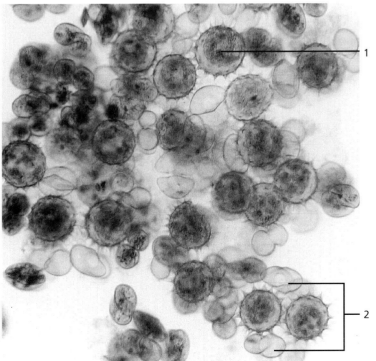

Figure 4.32. Cells of the desmid *Cosmarium* in the process of conjugation. (X450)

1. Zygote produced from conjugation
2. Parent cells that have produced gametes

Figure 4.33. Sea lettuce, *Ulva*, lives as a flat blade form in marine environments.

PHYLUM CHLOROPHYTA (green algae) ━━━━━

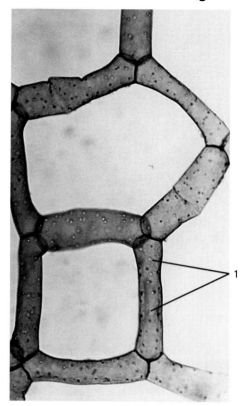

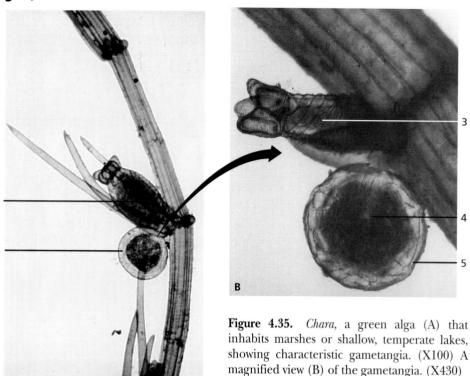

Figure 4.34. *Hydrodictyon.* The large, multi-nucleated cells form net-shaped colonies. (X430)

1. Nuclei

Figure 4.35. *Chara,* a green alga (A) that inhabits marshes or shallow, temperate lakes, showing characteristic gametangia. (X100) A magnified view (B) of the gametangia. (X430)

1. Oogonium
2. Antheridium
3. Antheridium
4. Egg
5. Oogonium

PHYLUM PHAEOPHYTA (brown algae, giant kelp) ━━━━━

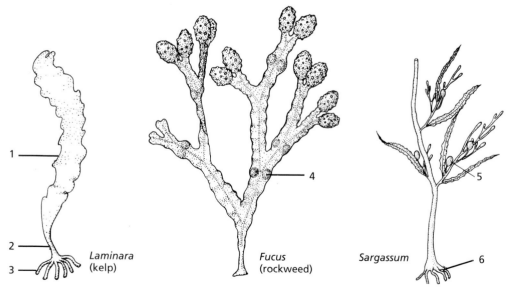

Laminara (kelp)

Fucus (rockweed)

Sargassum

Figure 4.36. Brown algae are marine organisms commonly referred to as seaweeds. They range in size from small filamentous organisms a few millimeters in length to the giant kelp between 50 and 100 meters long.

1. Blade	3. Holdfast	5. Air bladder
2. Stipe	4. Air bladder	6. Holdfast

Photo courtesy of Jack D. Brotherson

Figure 4.37. "Sea palm," *Postelsia palmaeformis,* a common brown alga found on the west coast of North America.

PHYLUM PHAEOPHYTA (brown algae, giant kelp)

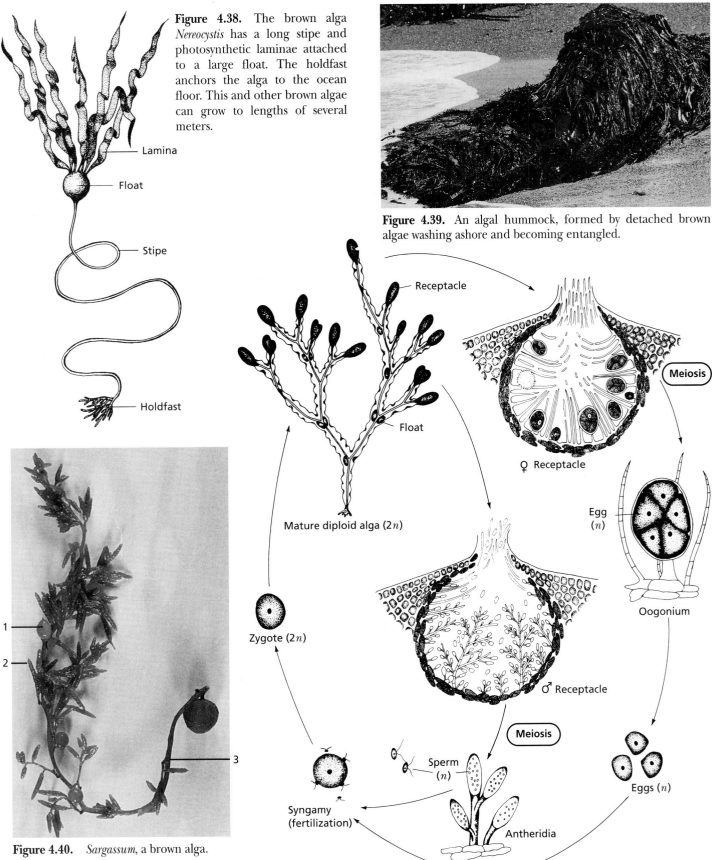

Figure 4.38. The brown alga *Nereocystis* has a long stipe and photosynthetic laminae attached to a large float. The holdfast anchors the alga to the ocean floor. This and other brown algae can grow to lengths of several meters.

Lamina

Float

Stipe

Holdfast

Figure 4.39. An algal hummock, formed by detached brown algae washing ashore and becoming entangled.

Receptacle

Float

Mature diploid alga (2*n*)

Meiosis

♀ Receptacle

Egg (*n*)

Oogonium

Zygote (2*n*)

♂ Receptacle

Meiosis

Sperm (*n*)

Eggs (*n*)

Syngamy (fertilization)

Antheridia

Figure 4.40. *Sargassum*, a brown alga.

1. Float (air-filled bladder)
2. Blade
3. Stipe

Figure 4.41. Life cycle of *Fucus*, a common brown alga.

PHYLUM PHAEOPHYTA (brown algae, giant kelp)

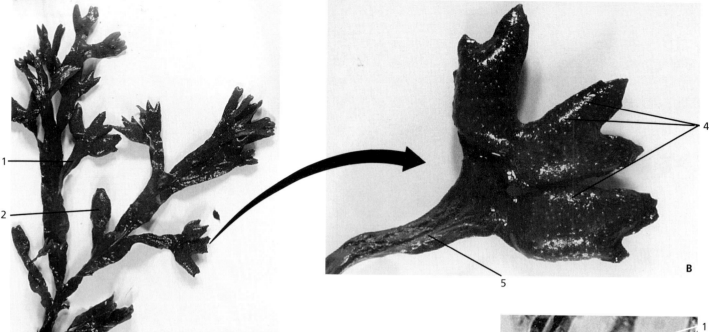

A

B

Figure 4.42. (A) *Fucus,* a brown alga, commonly called rockweed. (B) An enlargement of a blade supporting the receptacles.

1. Blade
2. Receptacle
3. Stipe
4. Conceptacles (light-colored spots) are chambers imbedded in the receptacles
5. Blade

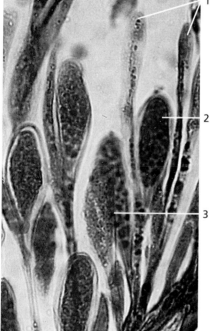

Figure 4.43. *Fucus* antheridia. (X430)

1. Paraphyses
2. Antheridium
3. Sperm within antheridium

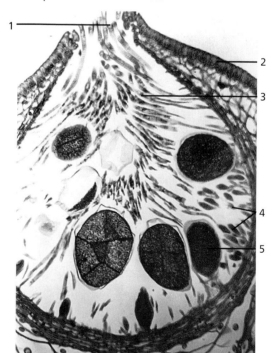

Figure 4.44. *Fucus,* conceptacle containing both antheridia and oogonia. (X200)

1. Ostiole
2. Surface of receptacle
3. Paraphyses (sterile hairs)
4. Antheridia
5. Oogonium

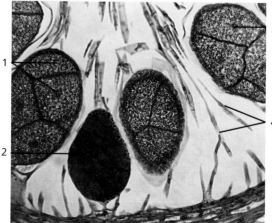

Figure 4.45. *Fucus,* closeup of female conceptacle. (X200)

1. Eggs
2. Oogonium
3. Nucleus
4. Paraphyses

RHODOPHYTA (red algae)

Porphyra

Ceramium

Polysiphonia.

Figure 4.46. Examples of common marine red algae.

Figure 4.47. The red alga, *Polysiphonia,* has alternation of three generations. (A) Female gametophyte with attached carposporophyte generation. (X100) (B) A closeup of the cysto-carps.

1. Pericarp
2. Carposporophyte producing carpospores
3. Cystocarps
4. Carpospores

Figure 4.48. *Polysiphonia,* tetrasporo-phyte. (X100)

1. Tetrasporophyte (2*n*)   2. Tetraspores (*n*)

Figure 4.49. *Polysiphonia,* male game-tophyte. Male reproductive structures, known as spermatangia, produce non-motile spermatia. (X40)

1. Spermatangia

PHYLUM MYXOMYCOTA (plasmodial slime molds)

A

B

Figure 4.50. Slime mold sporangia vary considerably in size and shape. (A) and (B) are species of *Fuligo;* (C) and (D) are species of *Lycogala.*

C

D

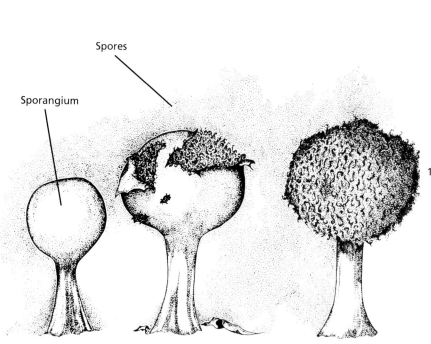

Spores

Sporangium

Figure 4.51. Diagrams of the slime mold, *Hemitrichia,* showing the release of spores from the sporangium.

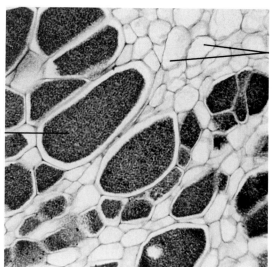

1

2

Figure 4.52. *Plasmodiophora brassicae,* a myxomycete responsible for clubroot in cabbage. Depicted is a cross section through a cabbage root showing spores of *Plasmodiophora* in the cabbage root cells.

1. Spores   2. Cabbage root cells.

PHYLUM MYXOMYCOTA (plasmodial slime molds)

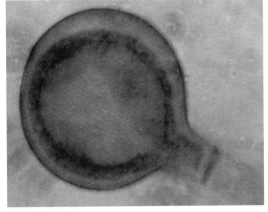

Figure 4.54. Slime mold, *Physarum*, growing on an agar culture medium. (X40)

Figure 4.53. Sporangia of the slime mold *Comatricha typhoides.*

PHYLUM OOMYCOTA (water molds, white rusts, downy mildews)

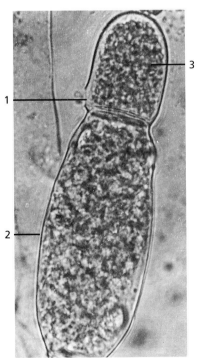

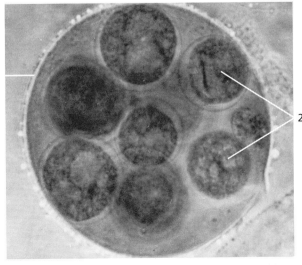

Figure 4.56. A water mold, *Saprolegnia*, showing a young oogonium before eggs have been formed. (X430)

Figure 4.55. The gametangia of the water mold *Allomyces*. Male gametes escape through exit pores. (X430)

1. Exit pore
2. Female gametangium
3. Male gametangium

Figure 4.57. A mature oogonium of the water mold *Saprolegnia*. (X430)

1. Oogonium
2. Eggs

Kingdom Fungi

The kingdom Fungi is a large diverse group of about 250,000 species. All fungi are heterotrophs since they absorb nutrients through their cell walls and cell membranes. Included in the kingdom Fungi are the typical conjugation fungi, yeasts, mushrooms, toadstools, and rusts. Most are saprobes, absorbing nutrients from dead organic material, while a few are parasitic, absorbing nutrients from living hosts.

Except for the unicellular yeasts, fungi consist of elongated filaments called *hyphae*. Hyphae begin as tubular extensions of spores that branch as they grow to form a network of hyphae called a *mycelium*. Even the body of a mushroom consists of a mass of tightly packed hyphae attached to an extensive mycelium. Most fungi are nonmotile, although their reproductive cells may be motile. Fungi reproduce by means of spores, which are produced sexually or asexually.

Fungi are important as decomposers of organic material, helping to recycle the inorganic nutrients essential for plant growth. Many species of fungi are commercially important. Some, such as mushrooms, are utilized directly as food. Others, such as yeasts, are used in making bread, cheese, beer, and wine. Fungi are also important in medicine, such as in the production of the antibiotic penicillin. Many other species of fungi are of medical and economic concern because they cause plant and animal diseases and destroy crops and stored goods.

TABLE 5.1
Some Representatives of the Kingdom Fungi

Divisions and Representative Kinds	Characteristics
Zygomycota (conjugation fungi)	Hyphae lack cross walls between nuclei
Ascomycota (yeasts, molds, morels, truffles)	Septate hyphae, reproductive structures contain eight ascospores within asci; asexual reproduction by budding
Basidiomycota (mushrooms, toadstools, rusts, smuts)	Septate hyphae; spores produced externally on cells called basidia contained in basidiocarp
Lichens (*not a division*, but rather a symbiotic association of algae and fungi)	Algal component (usually a green alga) provides food from photosynthesis; fungal component (usually an ascomycete) provides anchorage, water retention, and nutrients

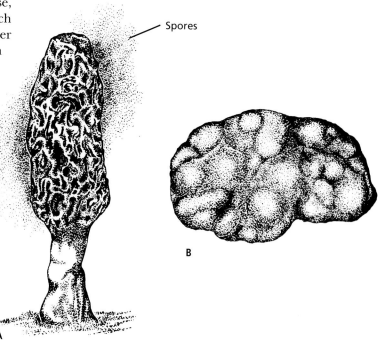

Figure 5.1 Diagrams of two commercially important ascomycetes. (A) An ascocarp of *Morchella esculenta,* the common morel prized as a gourmet food; (B) an ascocarp of the truffle *Tuber.* Truffles develop their ascocarps underground, where they are difficult to find and harvest.

DIVISION ZYGOMYCOTA (conjugation fungi)

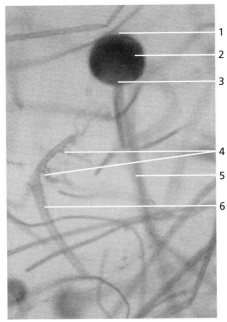

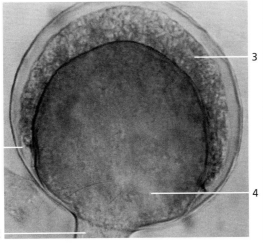

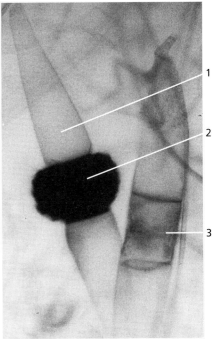

Figure 5.2. A whole mount of the bread mold *Rhizopus.* (X100)

1. Sporangium   4. Rhizoids
2. Spores   5. Sporangiophore
3. Columella   6. Stolon (hyphae)

Figure 5.3. A mature sporangium in the asexual reproductive cycle of the bread mold *Rhizopus.* (X430)

1. Sporangium
2. Sporangiophore
3. Spores
4. Columella

Figure 5.4. Conjugation and sexual fusion in the common bread mold *Rhizopus.* (X400)

1. Suspensor hypha
2. Zygosporangium
3. Fused gametangia (conjugation)

DIVISION ASCOMYCOTA (yeasts, molds, morels, truffles)

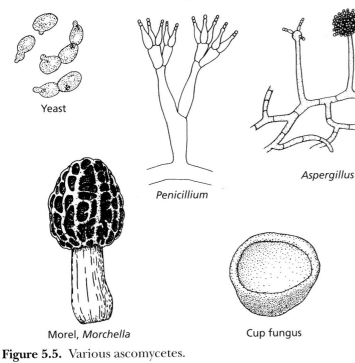

Yeast

Penicillium

Aspergillus

Morel, *Morchella*

Cup fungus

Figure 5.5. Various ascomycetes.

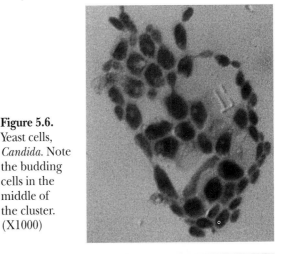

Figure 5.6. Yeast cells, *Candida.* Note the budding cells in the middle of the cluster. (X1000)

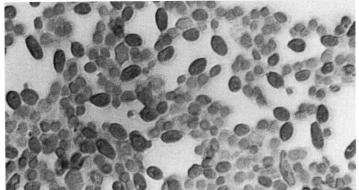

Figure 5.7. Baker's yeast, *Saccharomyces cerevisiae,* ascospores. These ascospores are characteristically spheroidal or ellipsoidal in shape. (X1000)

DIVISION ASCOMYCOTA (yeasts, molds, morels, truffles)

Morchella

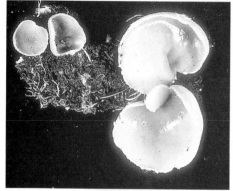

Helvella

Peziza repanda

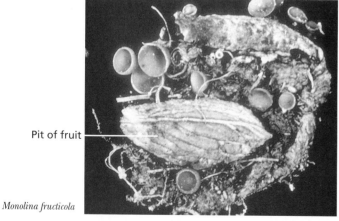

Pit of fruit

Monolina fructicola

Figure 5.8. Fruiting bodies (ascocarps) of common ascomycetes. *Morchella* is a common edible morel, *Helvella* is sometimes known as a saddle fungus since the fruiting body is thought by some to resemble a saddle, *Peziza repanda* is a common woodland cup fungus, and *Monolina fructicola* is an important plant pathogen causing brown rot of fruit.

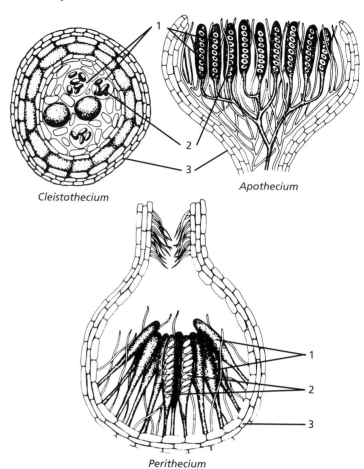

Cleistothecium

Apothecium

Perithecium

Figure 5.9. Examples of various types of ascocarps, or fruiting bodies, of ascomycetes.

1. Asci   2. Ascospores   3. Wall of ascocarp

Figure 5.10. A section through the hymenial layer of the apothecium of *Peziza*, showing asci with ascospores. (X100)

1. Hymenial layer   2. Ascus with ascospores   3. Ascocarp mycelium

DIVISION ASCOMYCOTA (yeasts, molds, morels, truffles)

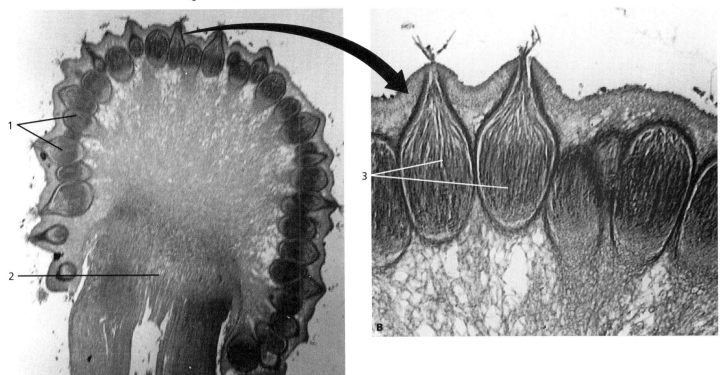

Figure 5.11. The ascomycete *Claviceps purpurea.* (A) A longitudinal section through the ascocarps. (X100) (B) An enlargement of the perithecia. (X250) This fungus causes serious plant diseases and is toxic to humans.

1. Perithecia   3. Perithecia containing asci with ascospores
2. Stroma

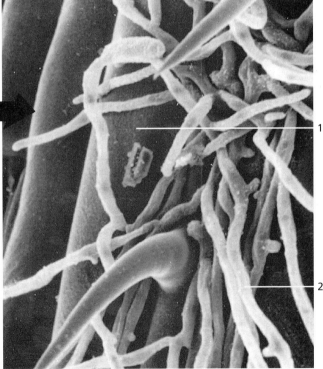

Photographs courtesy of James V. Allen

Figure 5.12. Scanning electron micrographs of the powdery mildew, *Erysiphe graminis*, on the surface of wheat. As the mycelium develops, it produces spores (conidia) that give a powdery appearance to the wheat.

1. Wheat host   2. Mycelium

DIVISION ASCOMYCOTA (yeasts, molds, morels, truffles)

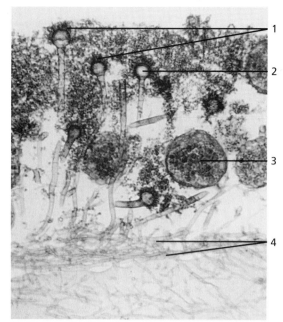

Figure 5.13 The common mold *Aspergillus.* (X200)

1. Conidia (spores)
2. Conidiophore
3. Cleistothecium
4. Hyphae

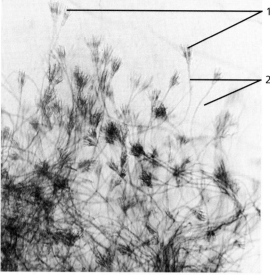

Figure 5.14. The fungus *Penicillium* causes economic damage as a mold but is also the source of important antibiotics. (X100)

1. Conidia (spores)   2. Conidiophores

Figure 5.15. Conidia of the fungus *Penicillium.* (X1000)

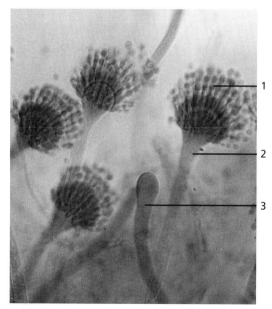

Figure 5.16. Closeup of sporangia of the mold *Aspergillus.* The conidia, or spores, of this genus are produced in a characteristic radiate pattern. (X430)

1. Conidia (spores)
2. Conidiophore
3. Developing conidiophore

DIVISION BASIDIOMYCOTA (mushrooms, toadstools, rusts, smuts)

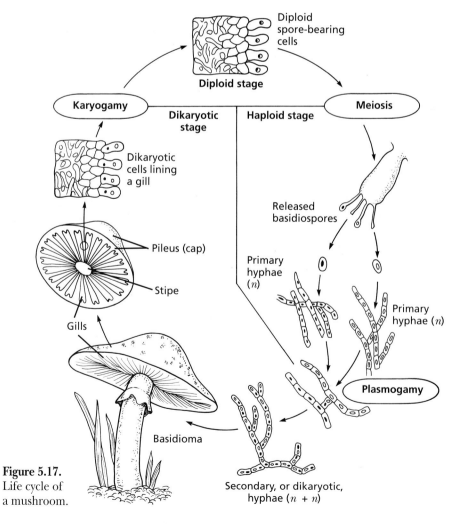

Figure 5.17. Life cycle of a mushroom.

DIVISION BASIDIOMYCOTA (mushrooms, toadstools, rusts, smuts)

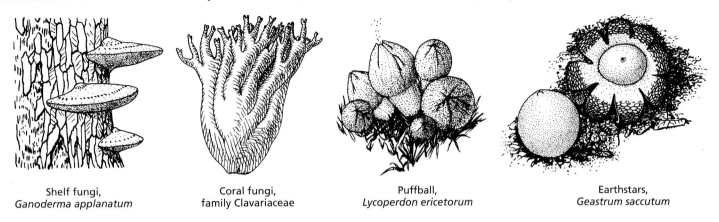

Shelf fungi,
Ganoderma applanatum

Coral fungi,
family Clavariaceae

Puffball,
Lycoperdon ericetorum

Earthstars,
Geastrum saccutum

Figure 5.18. Drawings of various basidiomycetes.

Agaricus rodmani

Cortinarius

Amanita pantherina

Boletus

Figure 5.19. Fruiting bodies (basidiocarps) of common mushrooms. *Agaricus rodmani* is a prized edible mushroom, the genus *Cortinarius* contains more mushrooms than any other in North America, *Amanita pantherina* is a deadly toxic mushroom, and *Boletus* is a mushroom with pores on the undersurface of the pileus (cap) rather than on gills.

DIVISION BASIDIOMYCOTA (mushrooms, toadstools, rusts, smuts)

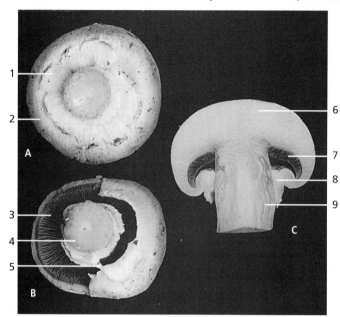

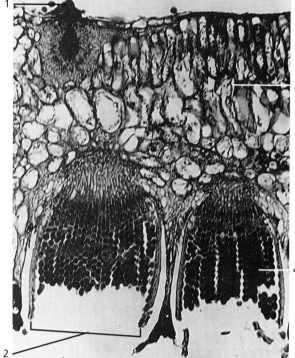

Figure 5.20. Mushroom. (A) Inferior view with the annulus intact; (B) inferior view with a portion of the annulus removed to show the gills; (C) longitudinal section.

1. Annulus	4. Stipe (stalk)	7. Gills
2. Pileus (cap)	5. Annulus	8. Annulus
3. Gills	6. Pileus (cap)	9. Stipe (stalk)

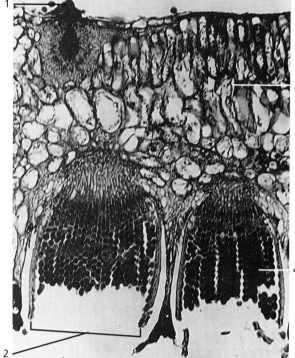

Figure 5.21. Wheat rust, *Puccinia graminis*, aecia on barberry leaf. (X100)

1. Pycnidium
2. Aecium
3. Barberry leaf
4. Aeciospores

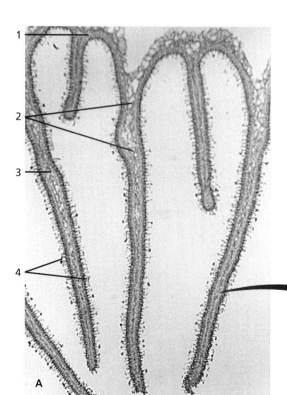

Figure 5.24. Gills of the mushroom *Coprinus*. (A) Closeup of several gills, (X40); and (B) closeup of a portion of a single gill. (X100)

1. Pileus (cap) comprised of gills	3. Gill	6. Gill (comprised of hyphae)
2. Hyphae comprising the gills	4. Basidiospores	7. Basidia
	5. Sterigma	8. Basidiopore

Figure 5.22. A photograph of a corn plant infected by the smut *Ustilago maydis*.

1. Corn stalk
2. A corn ear completely destroyed by the fungus

Figure 5.23. A photograph of a smut-infected brome grass. The grains have been destroyed by the fungus.

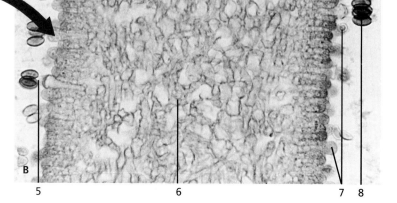

LICHENS (not a division, but rather a symbiotic association of algae and fungi)

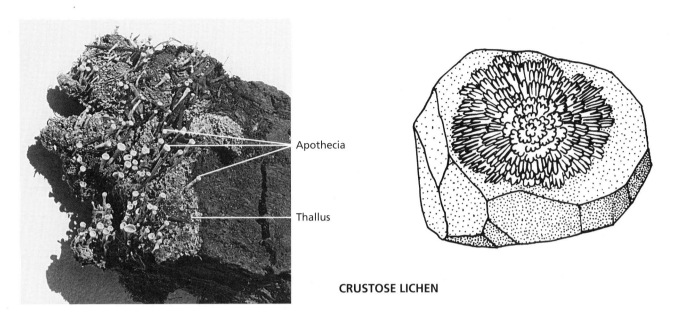

Apothecia

Thallus

CRUSTOSE LICHEN

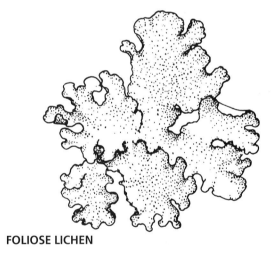

FOLIOSE LICHEN

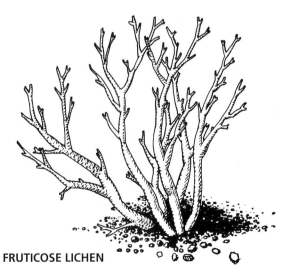

FRUTICOSE LICHEN

Figure 5.25. Although highly diverse in structure and appearance, the many kinds of lichens are grouped into three broad categories.

LICHENS (Not a division, but rather a symbiotic association of algae and fungi)

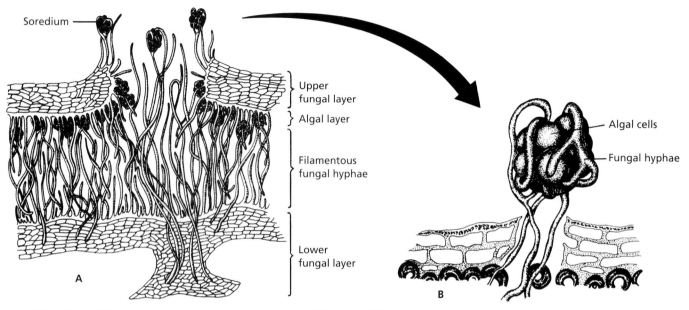

Figure 5.26.   Many lichens reproduce by producing soredia, which are small bodies containing both algal and fungal cells. (A) A diagram of a lichen thallus; (B) A diagram of a soredium.

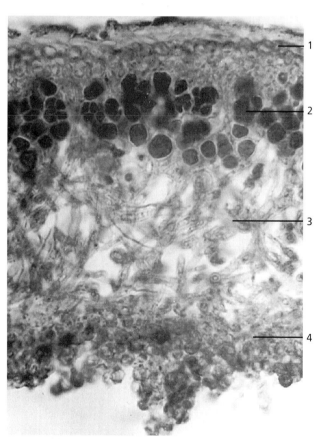

Figure 5.27.   The lichen thallus is often constructed of distinct algal and fungal layers. This cross section of a lichen thallus clearly shows these layers. (X100)

1. Upper fungal layer   3. Filamentous fungal hyphae
2. Algal layer   4. Lower fungal layer

Figure 5.28.   Closeup of a crustose lichen on the bark of a tree.
1. Lichen

Kingdom Plantae

Plants are photosynthetic multi-cellular eukaryotes. *Cellulose* in their cell walls provides protection and rigidity, while the *stomata* and *cuticle* of stems and leaves regulate gas exchange. Mitosis and meiosis are characteristic of all plants. Jacketed sex organs, called *gametangia*, protect the gametes and embryos from desiccation. Most plants have heteromorphic alternation of generations with distinctive haploid *gametophyte* and diploid *sporophyte* forms. Photosynthetic cells within plants contain *chloroplasts* with the pigments chlorophyll *a*, chlorophyll *b*, and a variety of carotenoids. Carbohydrates are produced by plants and stored in the form of starch.

Reproduction in seed plants is well adapted to a land existence. The conifers produce their seeds in protective *cones*, and the angiosperms produce their seeds in protective *fruits*. In the life cycle of a conifer, such as a pine, the mature *sporophyte* (tree) has female cones which produce *megaspores* that develop into the female gametophyte generation, and male cones which produce *microspores* that develop into the male gametophyte generation (mature pollen grains). Following fertilization, immature sporophyte generations are present in seeds located on the female cones. The female cone opens and the *seeds* (pine nuts) disperse to the ground and germinate if the conditions are right. Reproduction in angiosperms is similar to gymnosperms except that the angiosperm pollen and ovules are produced in flowers, rather than in cones and a fruit is formed.

TABLE 6.1.
Some Representatives of the Kingdom Plantae

Divisions and Representative Kinds	Characteristics
Bryophyta (liverworts, hornworts, and mosses)	Flattened and waxy leaflike cuticles that lack stomata and vascular tissue; rootlike rhizomes supporting rhizoids; homosporous
Psilotophyta (whisk ferns)	True roots and leaves are absent, but vascular tissue present; rhizome and rhizoids present
Lycophyta (clubmosses and quillworts)	Sporangia borne on sporophylls; homosporous (bisexual gametophyte); many are epiphytes
Sphenophyta (horsetails)	Epidermis embedded with silica; tips of stems bear conelike structures containing sporangia; most homosporous
Pterophyta (ferns)	Nonseed producing, vascular plants; fronds as leaves, underground rhizome as roots; homosporous
Cycadophyta (cycads)	Heterosporous, pollen and seed cones borne of different plants; large pith
Ginkgophyta (ginkgo)	Deciduous, fan-shaped leaves; seed-producing; heterosporous
Coniferophyta (conifers)	Woody plants that produce their seeds in cones; heterosporous; most have needlelike leaves that lack air spaces, and the stoma are sunken
Anthophyta (angiosperms; monocots, and dicots)	Flowering plants that produce their seeds enclosed in fruit; heterosporous; most are free-living, some are saprophytic or parasitic

DIVISION BRYOPHYTA (liverworts, hornworts, and mosses)

TABLE 6.2.
Representatives of the Division Bryophyta

Classes and Representative Kinds	Characteristics
Class Hepaticae (liverworts)	Flat or leafy gametophytes, single-celled rhizoids; simple sporophytes, most lack stomata
Class Anthocerotae (hornworts)	Flat, lobed gametophytes; more complex sporophytes with stomata
Class Musci (mosses)	Leafy gametophytes, multicellular rhizoids; sporophytes with stomata

DIVISION BRYOPHYTA ━━━━━━━━━━━━━━━━━━━━━━━━━━ **Class Hepaticae (liverworts)**

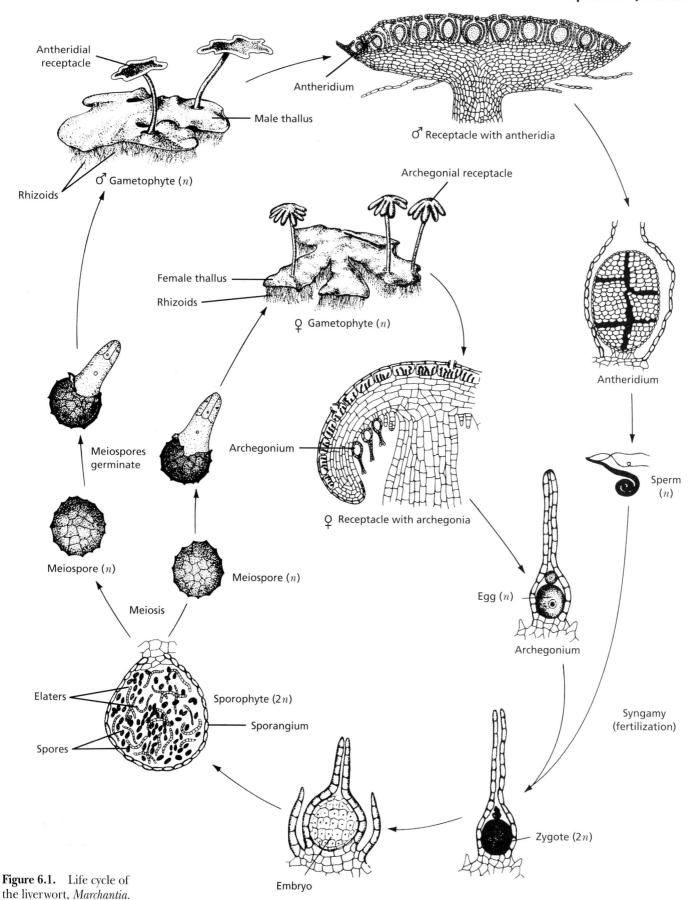

Figure 6.1.   Life cycle of the liverwort, *Marchantia*.

DIVISION BRYOPHYTA ━━━━━━━━━━━━━━━━━━━━━ Class Hepaticae (liverworts)

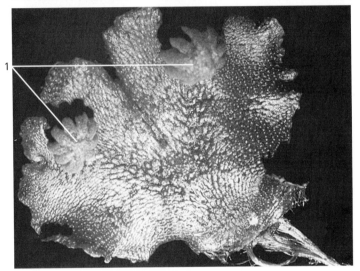

Figure 6.2. Photograph of several plants of the common liverwort *Marchantia*.

1. Antheridial receptacle
2. Gametophyte thallus (*n*)
3. Gemmae cup (*n*)

Figure 6.3. Ventral view of the common liverwort *Marchantia*, showing numerous rhizoids

1. Rhizoids

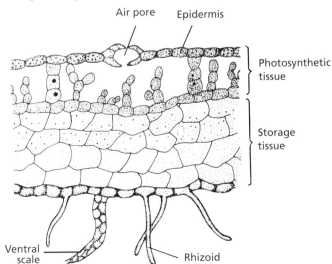

Figure 6.4. Dorsal view of a female gametophyte of the liverwort *Marchantia*.

1. Archegonial receptacle

Figure 6.5. A diagram of the thallus of *Marchantia*.

Air pore · Epidermis · Photosynthetic tissue · Storage tissue · Ventral scale · Rhizoid

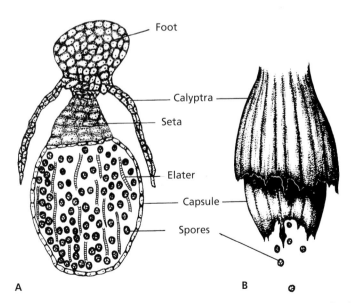

Figure 6.6. The sporophyte of *Marchantia*. (A) Longitudinal section; (B) face view showing the shedding spores.

Foot · Calyptra · Seta · Elater · Capsule · Spores

A   B

DIVISION BRYOPHYTA ━━━━━━━━━━━━━━ **Class Hepaticae (liverworts)**

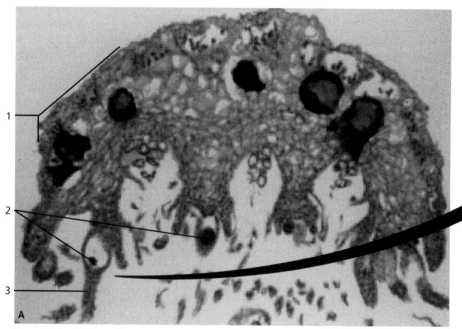

Figure 6.7. The archegonial receptacle (A) of a liverwort, *Marchantia*, in longitudinal section. (X40) The archegonium (B) showing an egg. (X240)

1. Archegonial receptacle
2. Eggs
3. Neck of archegonium
4. Stalk of archegonium
5. Egg
6. Venter of archegonium
7. Neck canal cells
8. Neck of archegonium

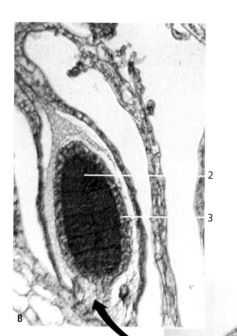

Figure 6.9. The male receptacle with antheridia (A) of a liverwort, *Marchantia*, in longitudinal section. (X40) The antheridial head (B) showing a developing antheridium. (X200)

1. Antheridia
2. Spermatogenous tissue
3. Antheridium

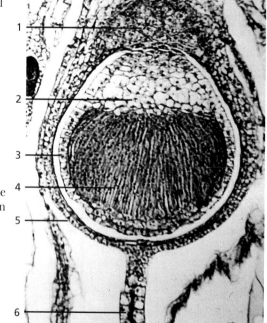

Figure 6.8. Young sporophyte of a liverwort, *Marchantia*, in longitudinal section. (X240)

1. Foot
2. Seta (stalk)
3. Young sporophyte
4. Sporogenous tissue ($2n$)
5. Enlarged archegonium (calyptra)
6. Neck of archegonium

DIVISION BRYOPHYTA ━━━━━━━━   # Class Hepaticae (liverworts)

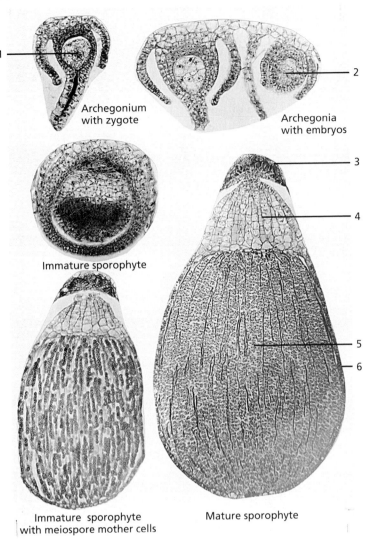

Archegonium
with zygote

Archegonia
with embryos

Immature sporophyte

Immature sporophyte
with meiospore mother cells

Mature sporophyte

Figure 6.10.   Developmental stages of the sporophyte of *Marchantia*.
(X65)

1. Zygote   3. Foot   5. Spores and elaters
2. Embryo   4. Seta (stalk) 6. Sporangium (capsule)

Figure 6.11.   **A** sporophyte of the liverwort *Porella*.

1. Sporophyte (2*n*)   3. Seta (stalk)
2. Capsule   4. Gametophyte (*n*)

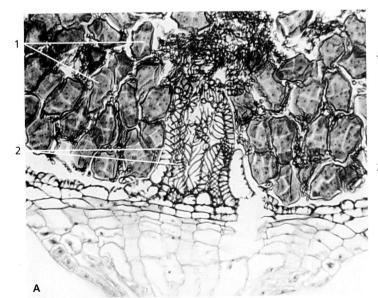

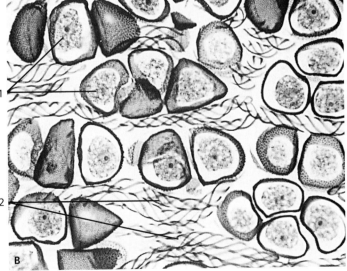

Figure 6.12.   A capsule from the leafy liverwort *Pelia;* (A) in longitudinal view (X180) and (B) in cross sectional view. (X430)

1. Spores   2. Elaters

DIVISION BRYOPHYTA ━━━━━━━━━━━━━━━━━━━━━━━━━━━━━━ **Class Antherocerotae (hornworts)**

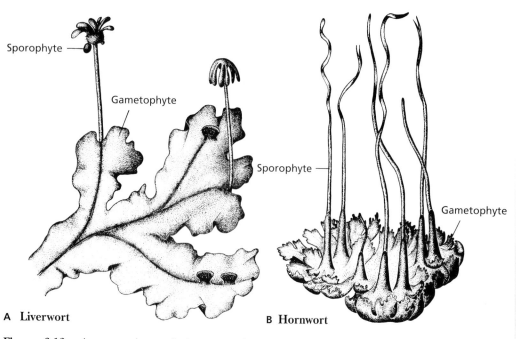

A Liverwort

B Hornwort

Figure 6.13. A comparison of the sporophytes and gametophytes of (A) the liverwort *Marchantia*, and (B) the hornwort *Anthoceros*.

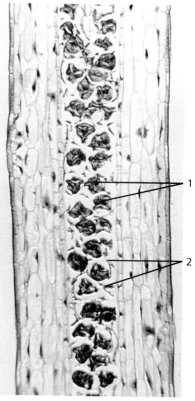

Figure 6.14. Hornwort, *Anthoceros*, a longitudinal section of the sporangium of a sporophyte. (X100)

1. Spores   2. Elater-like structures

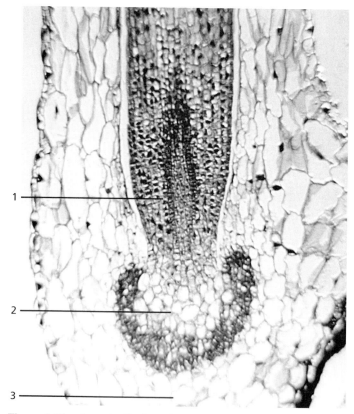

Figure 6.15. Longitudinal section of a portion of the sporophyte plant of the hornwort *Anthoceros*. (X100)

1. Meristematic region of sporophyte
2. Foot
3. Gametophyte

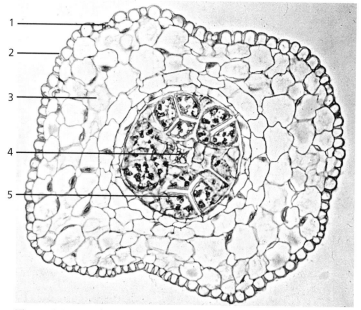

Figure 6.16. Hornwort, *Anthoceros*, cross section through the capsule of a sporophyte. (X100)

1. Stoma   3. Photosynthetic tissue   5. Spore
2. Epidermis   4. Columella

DIVISION BRYOPHYTA ━━━━━━━━━━━━━━━ Class Musci (mosses)

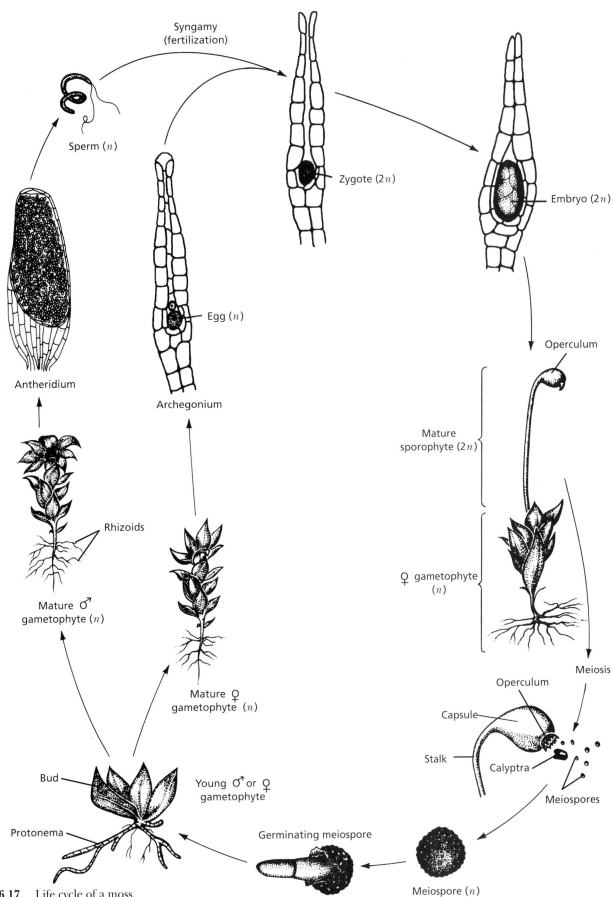

Figure 6.17. Life cycle of a moss.

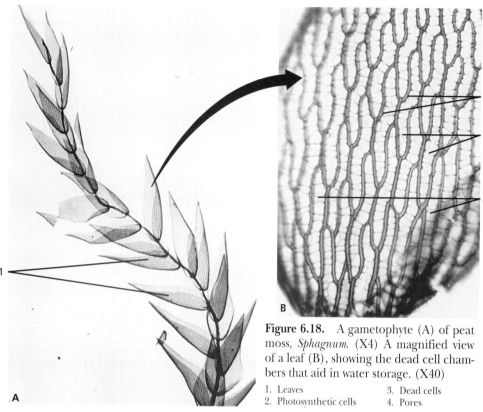

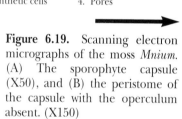

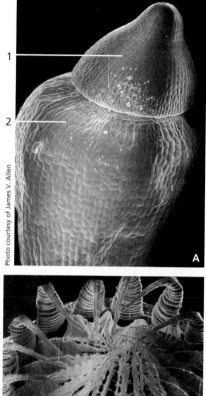

Photo courtesy of James V. Allen

Photo courtesy of James V. Allen

Figure 6.18. A gametophyte (A) of peat moss, *Sphagnum*. (X4) A magnified view of a leaf (B), showing the dead cell chambers that aid in water storage. (X40)

1. Leaves   3. Dead cells
2. Photosynthetic cells 4. Pores

Figure 6.19. Scanning electron micrographs of the moss *Mnium*. (A) The sporophyte capsule (X50), and (B) the peristome of the capsule with the operculum absent. (X150)

1. Operculum
2. Capsule
3. Outer teeth of peristome
4. Inner teeth of peristome
5. Capsule

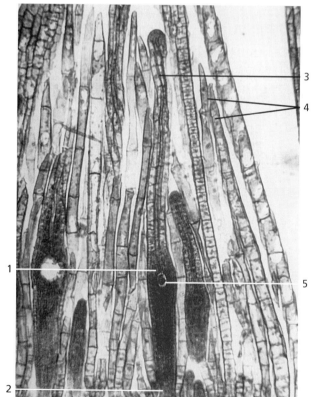

Figure 6.20. The archegonal head of the moss, *Mnium*, in longitudinal section. The paraphyses are non-reproductive filaments that support the archegonia. (X180)

1. Venter   2. Stalk   3. Neck   4. Paraphyses   5. Egg

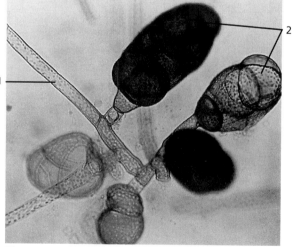

Figure 6.21. The gametophyte of a moss develops from buds along the protonema. Several buds attached to a protonema are illustrated here. (X430)

1. Protonema   2. Buds

DIVISION BRYOPHYTA ━━━━━━━━━━━━━━━━━━━━━━━━━━━━━━━ Class Musci (mosses)

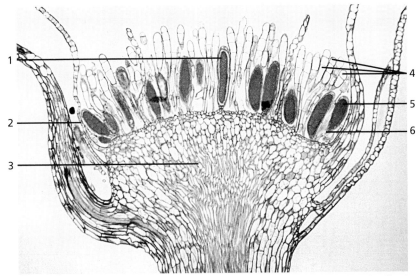

Figure 6.22. The antheridial head of the moss, *Mnium*, in longitudinal section. (X65)

1. Spermatogenous tissue
2. Sterile jacket layer
3. Male gametophyte (*n*)
4. Paraphyses (sterile filaments)
5. Antheridium (*n*)
6. Stalk

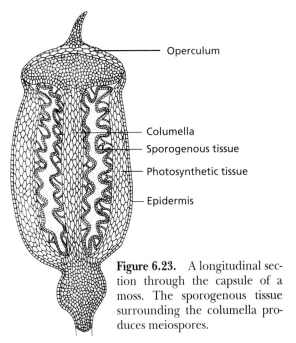

Figure 6.23. A longitudinal section through the capsule of a moss. The sporogenous tissue surrounding the columella produces meiospores.

Operculum
Columella
Sporogenous tissue
Photosynthetic tissue
Epidermis

DIVISION PSILOTOPHYTA (whisk ferns) ━━━━━━━━━━━━━━━━━━━━━━━

Figure 6.24. The whisk fern, *Psilotum nudum*, is a simple vascular plant lacking true leaves and roots.

Photo courtesy of James V. Allen

Figure 6.25. Scanning electron micrograph of a ruptured synangium (3 fused sporangia) of *Psilotum*, which is spilling spores. (X75)

1. Sporangium (often called a synangium)   2. Branch   3. Spores

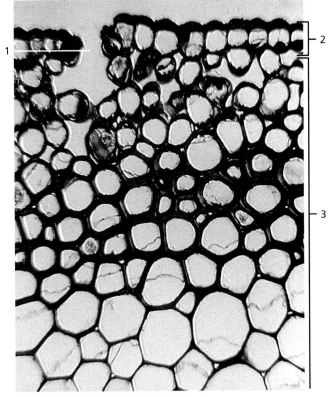

Figure 6.26. A photomicrograph of a scalelike outgrowth from the branch of the wisk fern *Psilotum nudum*. (X430)

1. Stoma   2. Epidermis   3. Ground tissue

DIVISION PSILOTOPHYTA (whisk ferns)

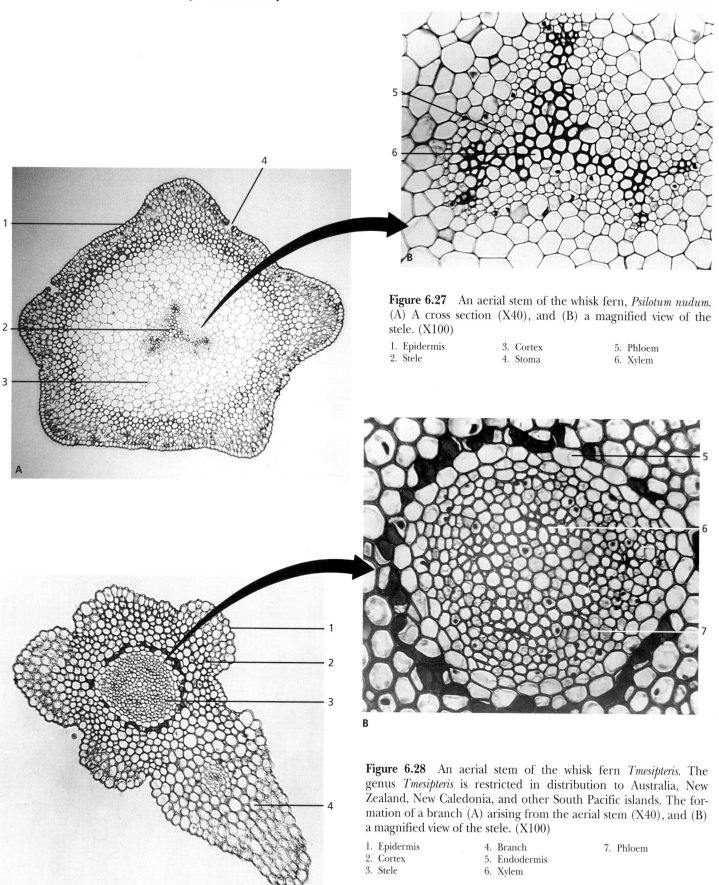

Figure 6.27   An aerial stem of the whisk fern, *Psilotum nudum.* (A) A cross section (X40), and (B) a magnified view of the stele. (X100)

1. Epidermis	3. Cortex	5. Phloem
2. Stele	4. Stoma	6. Xylem

Figure 6.28   An aerial stem of the whisk fern *Tmesipteris.* The genus *Tmesipteris* is restricted in distribution to Australia, New Zealand, New Caledonia, and other South Pacific islands. The formation of a branch (A) arising from the aerial stem (X40), and (B) a magnified view of the stele. (X100)

1. Epidermis	4. Branch	7. Phloem
2. Cortex	5. Endodermis	
3. Stele	6. Xylem	

DIVISION LYCOPHYTA (club mosses and quillworts)

Figure 6.29. A club moss *Lycopodium*. Club mosses occur from the arctic to the tropics. Being evergreen, they are most obvious during winter.

1. Strobilus
2. Leaves (microphylls)
3. Aerial stem

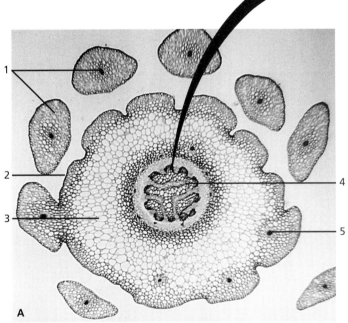

Figure 6.30 An aerial stem of the club moss *Lycopodium*. (A) A cross section (X100), and (B) a magnified view of the stele. (X200)

1. Leaves (microphylls)
2. Epidermis
3. Cortex
4. Stele
5. Leaf trace
6. Xylem
7. Phloem
8. Pericycle
9. Endodermis

Figure 6.31. An enlargement of a branch tip of *Lycopodium*, showing sporangia. (X10)

1. Sporangia   2. Sporophylls (leaves with attached sporangia)

Figure 6.32. *Lycopodium*, rhizome. A rhizome of *Lycopodium* is similar to an aerial stem but lacks microphylls. (X40)

1. Xylem   2. Epidermis   3. Phloem   4. Cortex

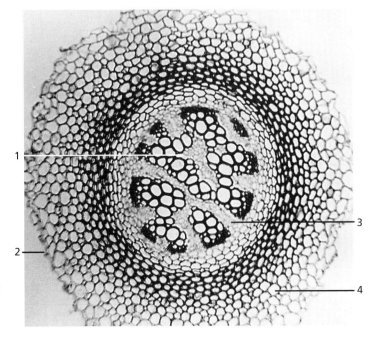

DIVISION LYCOPHYTA (club mosses and quillworts)

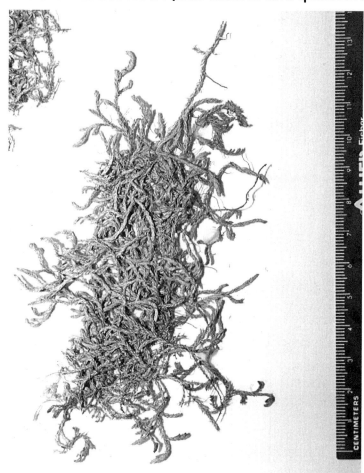

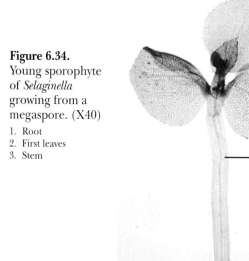

Figure 6.34.
Young sporophyte
of *Selaginella*
growing from a
megaspore. (X40)

1. Root
2. First leaves
3. Stem

Figure 6.33. Herbarium specimen of the lycopod *Selaginella*. These lycopods are mainly tropical in distribution. Some are found in arid regions, however, where they are dormant during dry seasons.

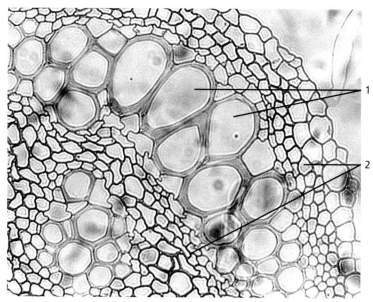

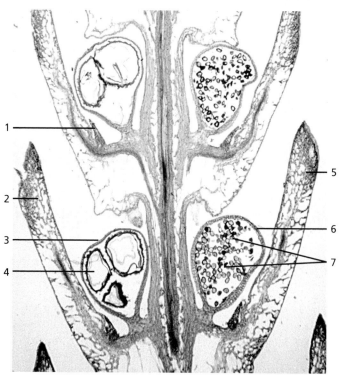

Figure 6.35. Vascular tissue in the stele of *Selaginella*. (X150)

1. Xylem   2. Phloem

Figure 6.36 Longitudinal view of a strobilus of *Selaginella*. (X40)

1. Ligule   4. Megaspore (*n*)   7. Microspores (*n*)
2. Megasporophyll   5. Microsporophyll
3. Megasporangium   6. Microsporangium

DIVISION LYCOPHYTA (club mosses and quillworts)

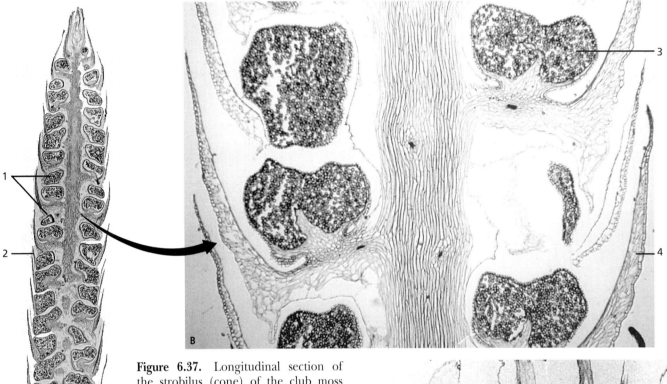

Figure 6.37. Longitudinal section of the strobilus (cone) of the club moss *Lycopodium*. (A) A homosporus lycopod (X15), and (B) a magnified view of the sporangia. (X100)

1. Sporangia
2. Sporophyll
3. Sporangium
4. Sporophyll

Figure 6.38. Herbarium specimen of a quillwort, *Isoetes melanopoda*. The quillwort is an aquatic plant with a small, underground stem and quill-like leaves. They are hetero-sporous and the megasporangia and microsporangia are located at the bases of different leaves.

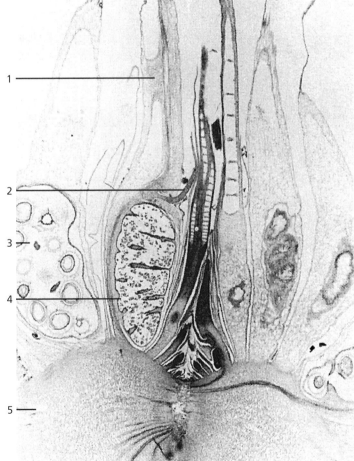

Figure 6.39 A longitudinal section of *Isoetes*, an aquatic, grass-like lycopod.

1. Leaf base
2. Ligule
3. Megasporangium
4. Microsporangium
5. Corm

DIVISION SPHENOPHYTA (horsetails)

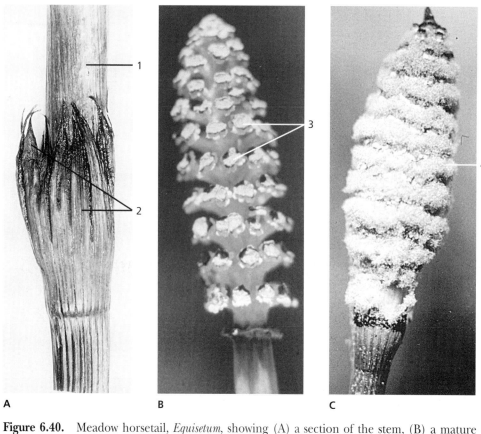

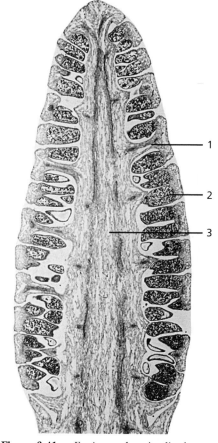

Figure 6.40.   Meadow horsetail, *Equisetum*, showing (A) a section of the stem, (B) a mature strobilus, and (C) a strobilus shedding its spores. (X2)

1. Stem
2. Whorl of leaves
3. Separated sporangiophores revealing sporangia
4. Sporangia shedding spores

Figure 6.41.   *Equisetum*, longitudinal section of the strobilus. (X2)

1. Sporangiophore
2. Sporangium
3. Strobilus axis

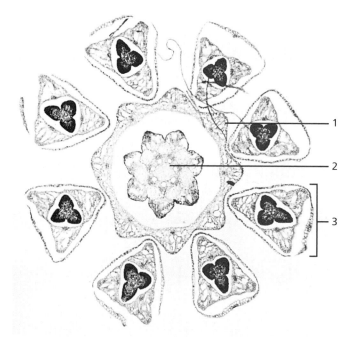

Figure 6.42.   Cross section of the stem of *Equisetum* just above a node. (X10)

1. Leaf sheath
2. Main stem
3. Branch

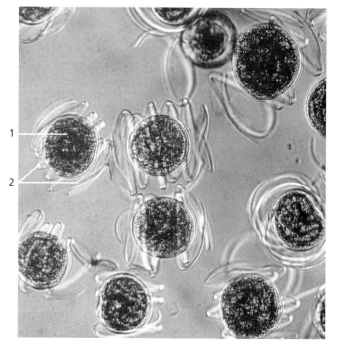

Figure 6.43.   Meiospores of *Equisetum* showing the elaters coiled about them. (X430)

1. Meiospore
2. Elater

DIVISION PTEROPHYTA (ferns)

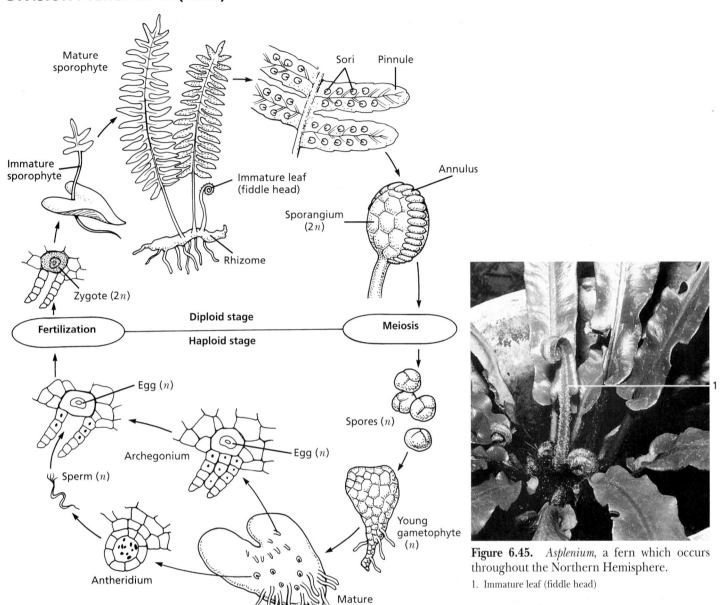

Figure 6.44. Life cycle of a fern.

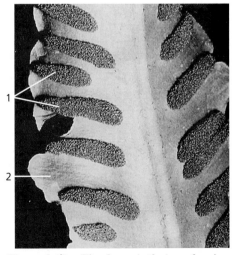

Figure 6.45. *Asplenium*, a fern which occurs throughout the Northern Hemisphere.

1. Immature leaf (fiddle head)

Figure 6.46. The fern, *Cyrtomium*, showing sori on the underside of the pinnae.

1. Pinna (pinnule)   2. Sori

Figure 6.47. The fern, *Asplenium*, showing the sori on the undersurface of the pinna.

1. Sori   2. Pinna (pinnule)

DIVISION PTEROPHYTA (ferns)

Figure 6.48. The fern *Polypodium*. (A) Sori on the undersurface of the pinnae, and (B) a scanning electron micrograph of a sorus. (X100)

1. Pinna (pinnule)
2. Sori
3. Annulus
4. Sporangium

Figure 6.49. Sori of the fern *Polystichum*.

1. Sori
2. Sporangia
3. Veins of pinna (pinnule)

Figure 6.50. The maidenhair fern *Adiantum*. (A) Pinnae and sori, and (B) A magnified view (X100) of the tip of a pinna folded under to form a false indusium that encloses the sorus.

1. Sori   2. False indusium   3. Pinna (pinnule)   4. Sporangia with spores   5. False indusium enclosing a sorus   6. Vascular tissue of the pinna

DIVISION PTEROPHYTA (ferns)

Figure 6.51. Sorus of a homosporous fern, *Cyrtomium falcatum*. The indusium protects the sporangia. (X100)

1. Spores (*n*)
2. Indusium (2*n*)
3. Epidermis of pinna
4. Pinna tissue
5. Sporangia (2*n*)

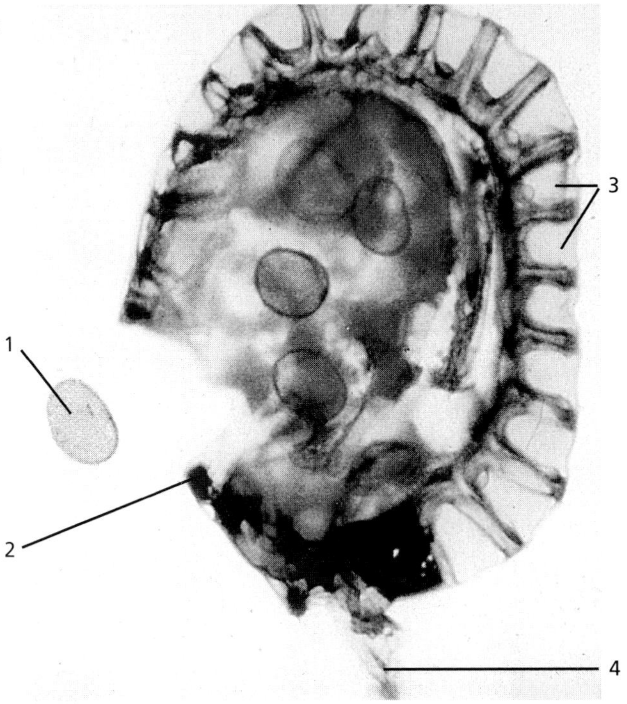

Figure 6.52. A sporangium of the fern *Cyrtomium* discharging a spore. (X430)

1. Spore　　2. Lip cell　　3. Annulus　　4. Stalk

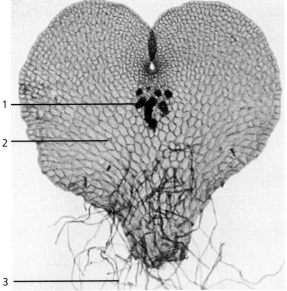

Figure 6.53. Fern gametophyte showing archegonia. (X40)

1. Archegonia
2. Gametophyte (prothallus)
3. Rhizoids

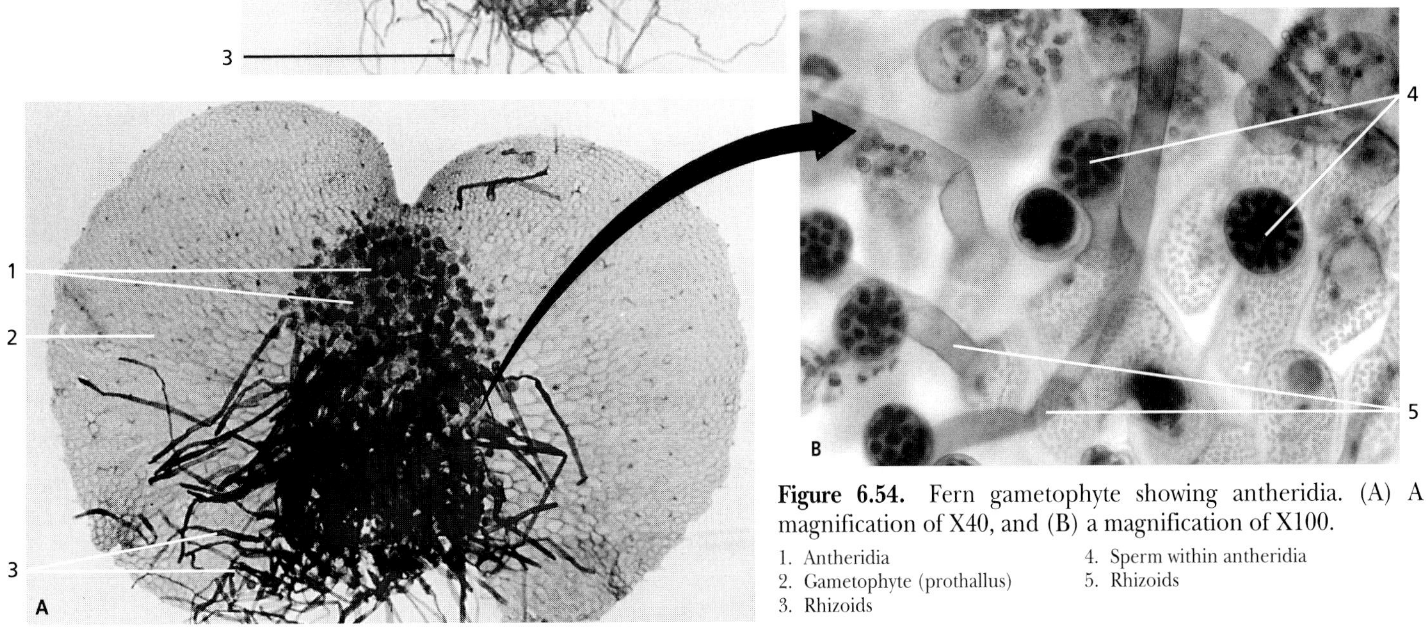

Figure 6.54. Fern gametophyte showing antheridia. (A) A magnification of X40, and (B) a magnification of X100.

1. Antheridia
2. Gametophyte (prothallus)
3. Rhizoids
4. Sperm within antheridia
5. Rhizoids

TABLE 6.3
Gymnosperms Within the Kingdom Plantae

Divisions and Representative Kinds	Characteristics
Division Cycadophyta (cycads)	Gymnosperms with pollen and seed cones borne on different plants; plants mostly shrubby; leaves large, palm-like
Division Ginkgophyta (ginkgo)	Gymnosperm with deciduous, fan-shaped leaves; large tree
Division Coniferophyta (conifers)	Woody gymnosperms producing seeds in cones; most with needlelike leaves that lack air spaces, and the stomata are sunken
Division Gnetophyta (gnetophytes)	Gymnosperms that contain vessels; motile sperm are absent

DIVISION CYCADOPHYTA (cycads) ━━━

Figure 6.55   Male cones of *Cycas revoluta*. Cycads are a group of gymnosperms that were very abundant during the Mesozoic Era. Currently, there are 10 living genera with about 100 species that are mainly found in tropical and subtropical areas. The trunk of many cycads is densely covered with petioles of shed leaves.

1. Cones

Figure 6.57.   Microsporangiate cones of the cycad, *Zamia*. The cone on the left (A) is longitudinally sectioned.

1. Microsporangia   2. Microsporophyll

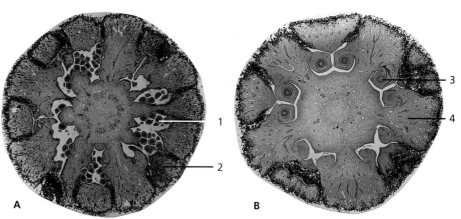

Figure 6.56.   Cross sections of a microsporangiate cone (A) and a megasporangiate cone (B) of the cycad *Zamia*.

1. Microsporangia   2. Microsporophyll   3. Ovule   4. Megasporophyll

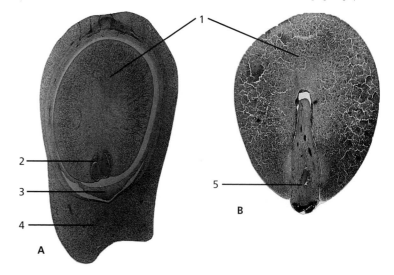

Figure 6.58.   The ovule of the cycad *Zamia*. In (A) the ovule has two archegonia and is ready to be fertilized. In (B) the ovule has been fertilized and contains an embryo, but the seed coat has been removed. (X5)

1. Female gametophyte   4. Integument
2. Archegonium   5. Embryo
3. Nucellus

DIVISION GINKGOPHYTA (ginkgo)

Figure 6.59. The ginkgo, or maidenhair tree, *Ginkgo biloba*. Consisting of a central trunk with lateral branches, a mature ginkgo grows to 80 to 100 feet tall. Native to China, *Ginkgo biloba* has been introduced in temperate climates throughout the world as an interesting ornamental tree.

Figure 6.60. A leaf from the ginkgo tree, *Ginkgo biloba*. Note the characteristic pattern of venation and fan-shaped leaf.

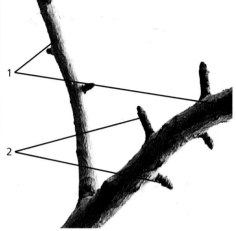

Figure 6.61. A branch of a ginkgo tree, *Ginkgo biloba*.

1. Long shoots   2. Short shoots (spurs)

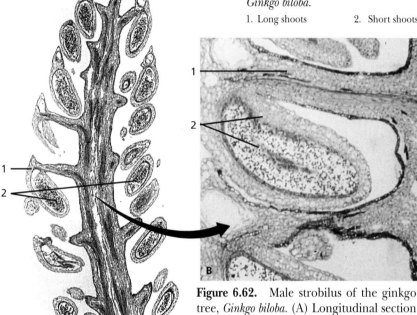

Figure 6.62. Male strobilus of the ginkgo tree, *Ginkgo biloba*. (A) Longitudinal section (X10), and (B) longitudinal section (X40).

1. Sporophyll   2. Microsporangia

Figure 6.63. Male strobili of the ginkgo tree, *Ginkgo biloba*.

1. Leaf   2. Male strobili

Figure 6.64. A mature seed of *Ginkgo* with the outer layer of the seed coat still intact.

Figure 6.65. A mature seed of *Ginkgo* with the outer layer of the seed coat removed.

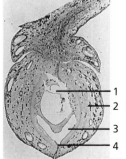

Figure 6.66. A longitudinal section of an ovule of *Ginkgo* prior to fertilization.

1. Nucellus
2. Integument
3. Pollen chamber
4. Micropyle

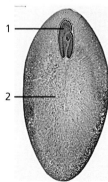

Figure 6.67. A longitudinal section of a seed of *Ginkgo*.

1. Embryo
2. Female gametophyte

DIVISION CONIFEROPHYTA (conifers)

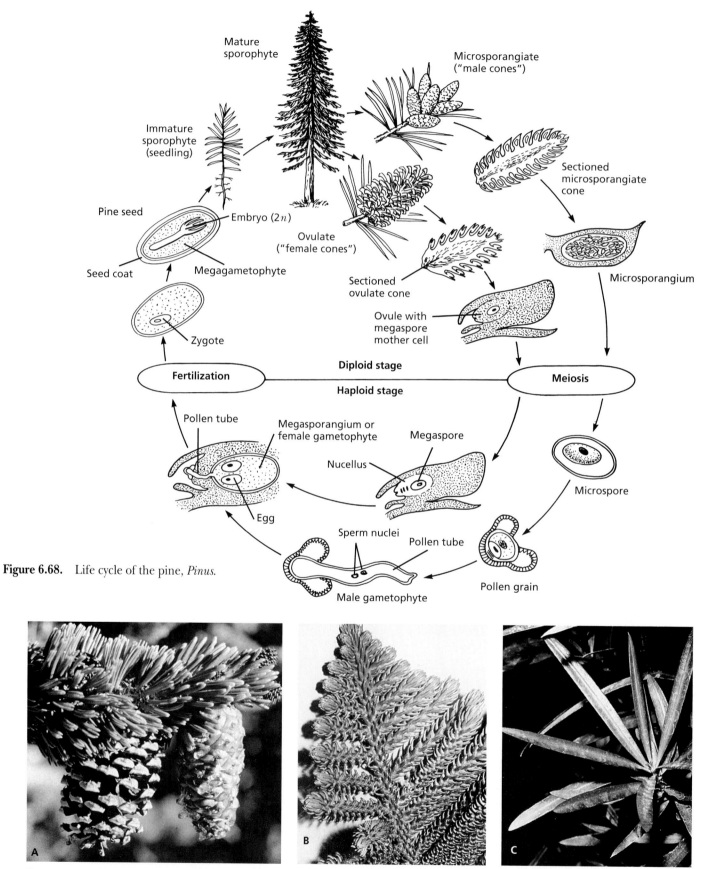

Figure 6.68. Life cycle of the pine, *Pinus*.

Figure 6.69. The leaves of most species of conifers are needle-shaped (A) such as these of the blue spruce, *Picea pungens. Araucaria* (B), however, has awl-shaped leaves, and *Podocarpus* (C) has strap-shaped leaves.

DIVISION CONIFEROPHYTA (conifers)

Figure 6.70. A young sporophyte (seedling) of a pine, *Pinus.*

1. Seedling needles   2. Seed coat   3. Cotyledons

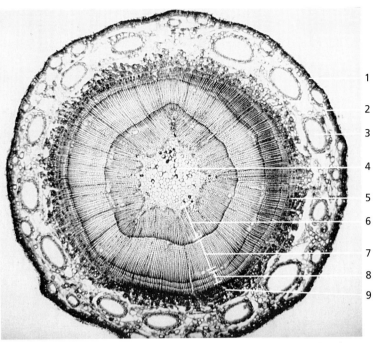

Figure 6.71. Cross section through the stem of a young conifer showing the arrangement of the tissue layers. (X20)

1. Periderm
2. Cortex
3. Resin duct
4. Pith
5. Cambium
6. First year's xylem
7. Second year's xylem
8. Third year's xylem
9. Phloem

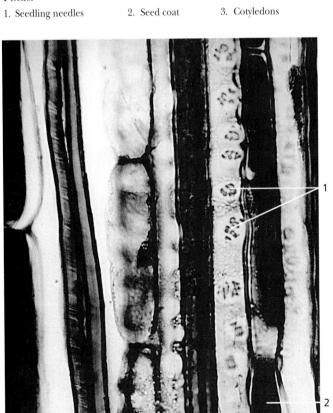

Figure 6.72. Longitudinal section through the phloem of *Pinus.* (X100)

1. Sieve areas on a sieve cell   2. Storage parenchyma

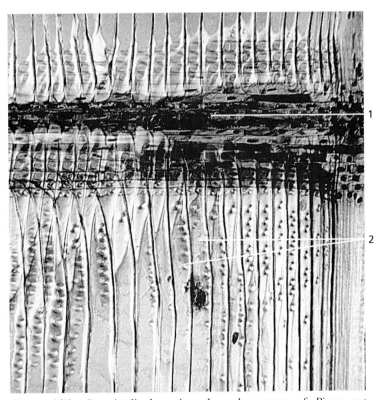

Figure 6.73. Longitudinal section through a stem of *Pinus,* cut through the xylem tissue. (X100)

1. Ray parenchyma   2. Tracheids

DIVISION CONIFEROPHYTA (conifers)

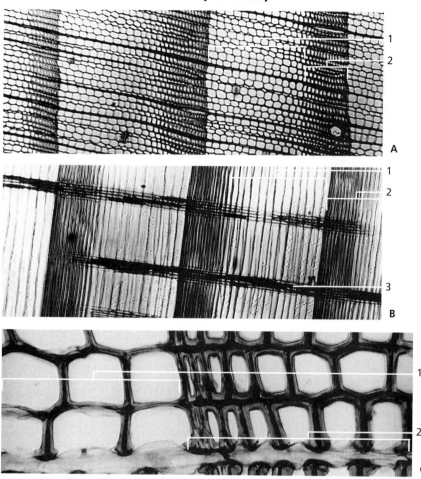

Figure 6.74. Growth rings in *Pinus,* (A) a cross section through a *Pinus* stem (X20); (B) a longitudinal section through a *Pinus* stem (X20); and (C) a magnified view of the xylem cells of spring and summer wood. (X100)

1. Spring wood   2. Summer wood   3. Vascular ray

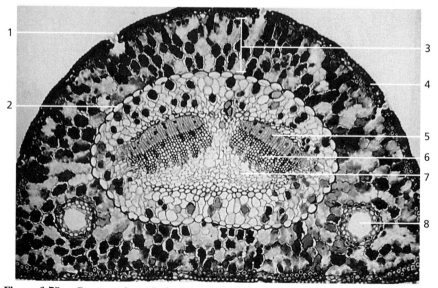

Figure 6.75. Cross section of a leaf (needle) of *Pinus.* (X40)

1. Stoma
2. Endodermis
3. Photosynthetic mesophyll
4. Epidermis
5. Phloem
6. Xylem
7. Transfusion tissue
8. Resin duct

Figure 6.76. Megasporangiate (female) cones from various species of conifers. (A) Southern hemisphere pine, *Aruacaria;* (B) larch, *Larix;* (C) pinion pine, *Pinus edulis;* and (D) sugar pine, *Pinus lambertiana.*

DIVISION CONIFEROPHYTA (conifers)

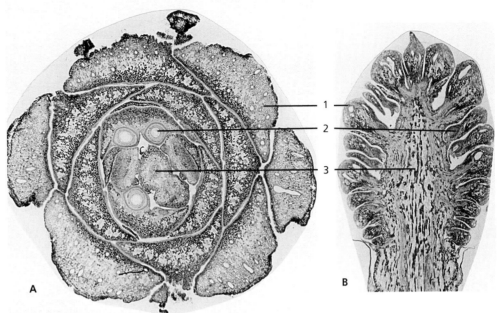

Figure 6.77. Relative positions of the male and female cones in *Pinus*.

1. Microsporangiate cone (male)
2. Ovulate cone (female)

Figure 6.78 Female cones of a conifer. (A) cross section (X20); and (B) longitudinal section (X10).

1. Ovuliferous scale (megasporophyll)   2. Ovule   3. Cone axis

Figure 6.79. Microsporangiate (male) cones of a conifer. (A) Staminate cones at end of branch; (B) a longitudinal section through a branch tip; (C) a longitudinal section through a single cone; and (D) a cross section through a single cone.

1. Pine needles (leaves)
2. Sporophylls
3. Staminate cone
4. Microsporangium
5. Sporophyll
6. Cone axis
7. Branch tip

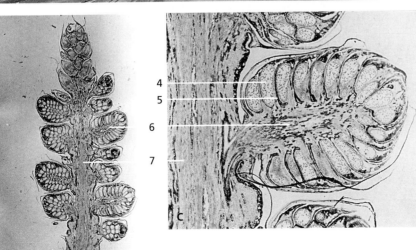

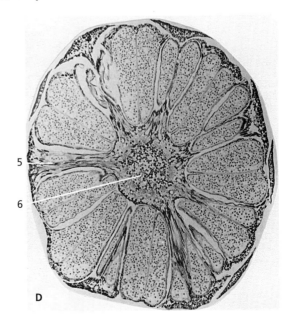

DIVISION ANTHOPHYTA (angiosperms: monocots and dicots)

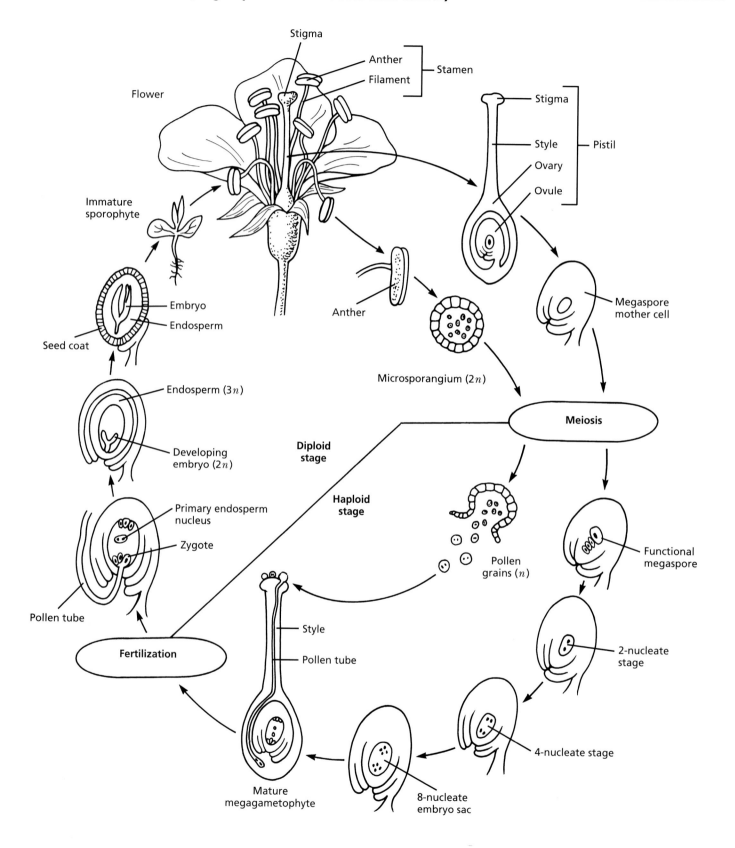

Figure 6.80. Life cycle of an angiosperm.

DIVISION ANTHOPHYTA (angiosperms: monocots and dicots)

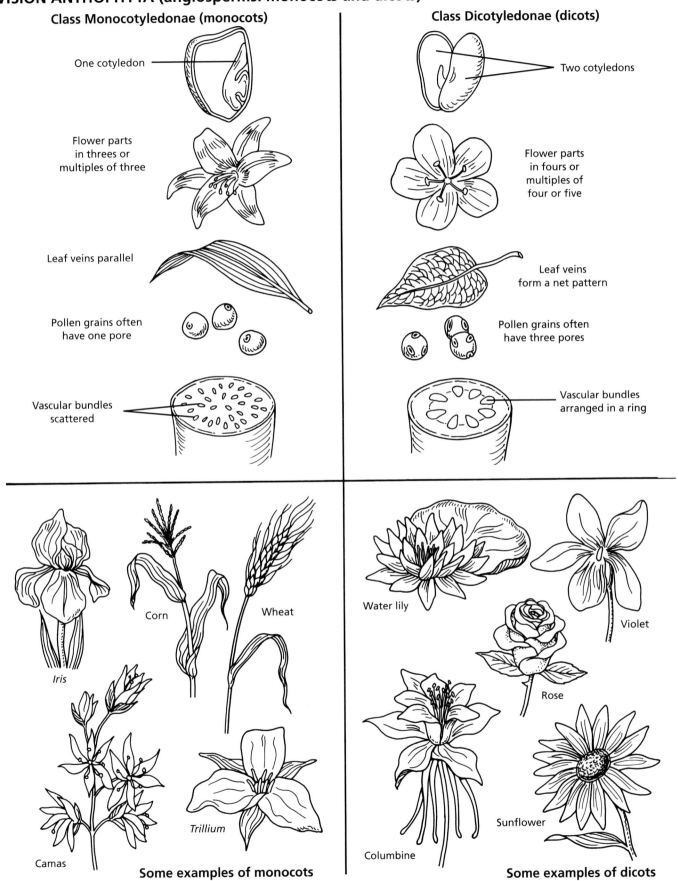

Figure 6.81. Comparison and examples of monocots and dicots.

DIVISION ANTHOPHYTA (angiosperms: monocots and dicots)

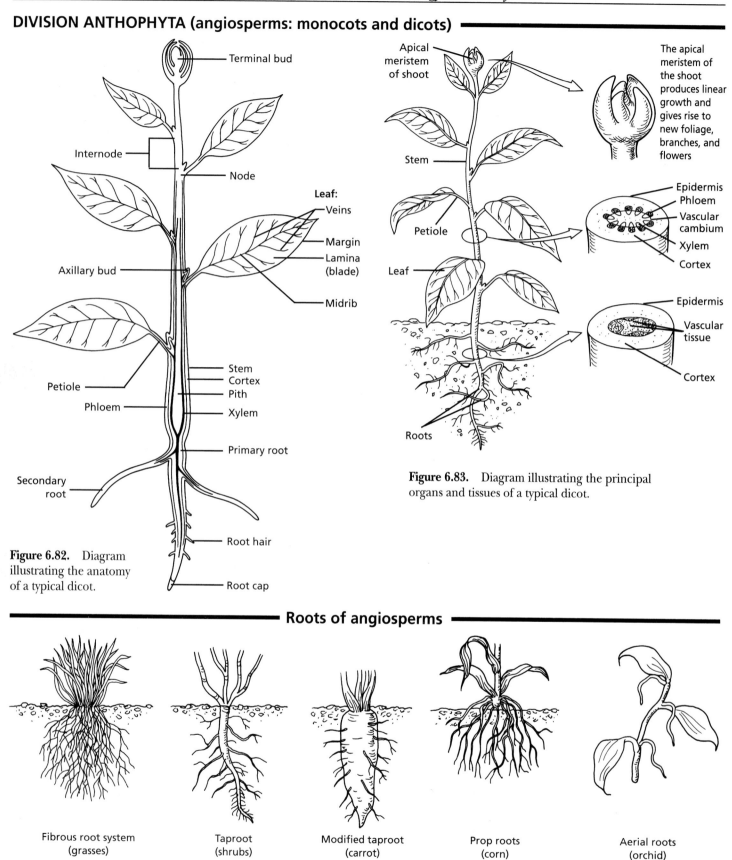

Terminal bud

Internode

Node

Leaf:
Veins

Margin
Lamina (blade)

Axillary bud

Midrib

Stem
Cortex
Pith
Xylem

Petiole

Phloem

Primary root

Secondary root

Root hair

Figure 6.82. Diagram illustrating the anatomy of a typical dicot.

Root cap

Apical meristem of shoot

Stem

Petiole

Leaf

Roots

The apical meristem of the shoot produces linear growth and gives rise to new foliage, branches, and flowers

Epidermis
Phloem
Vascular cambium
Xylem
Cortex

Epidermis
Vascular tissue

Cortex

Figure 6.83. Diagram illustrating the principal organs and tissues of a typical dicot.

Roots of angiosperms

Fibrous root system (grasses)

Taproot (shrubs)

Modified taproot (carrot)

Prop roots (corn)

Aerial roots (orchid)

Figure 6.84. Diagrams of typical root systems of angiosperms. The root system of an angiosperm is the descending portion of the plant specialized for anchorage and absorption, storage, and conduction of water and nutrients. Monocots, such as grasses, typically have fibrous root systems. Dicots, such as shrubs and most woody plants, typically have taproot systems. Specialized supporting root systems include prop roots and aerial roots. Taproots, such as found in carrots and turnips, are capable of storing large amounts of food.

Roots of Angiosperms

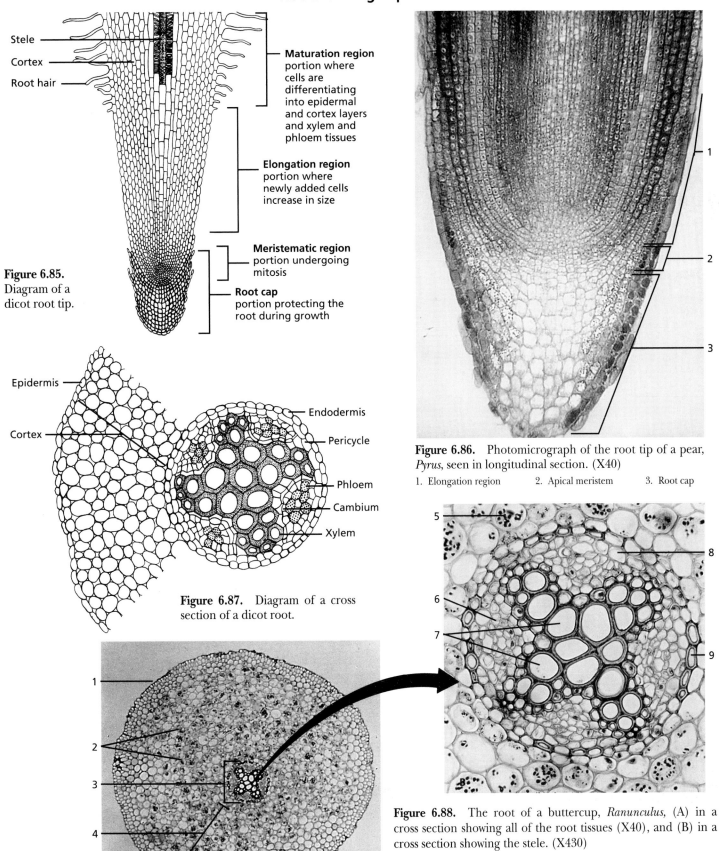

Stele

Cortex

Root hair

Maturation region
portion where
cells are
differentiating
into epidermal
and cortex layers
and xylem and
phloem tissues

Elongation region
portion where
newly added cells
increase in size

Meristematic region
portion undergoing
mitosis

Root cap
portion protecting the
root during growth

Figure 6.85.
Diagram of a
dicot root tip.

Epidermis

Cortex

Endodermis

Pericycle

Phloem

Cambium

Xylem

Figure 6.87. Diagram of a cross
section of a dicot root.

Figure 6.86. Photomicrograph of the root tip of a pear,
Pyrus, seen in longitudinal section. (X40)

1. Elongation region   2. Apical meristem   3. Root cap

Figure 6.88. The root of a buttercup, *Ranunculus*, (A) in a
cross section showing all of the root tissues (X40), and (B) in a
cross section showing the stele. (X430)

1. Epidermis
2. Parenchyma cells of
    cortex
3. Stele

4. Cortex
5. Starch grains within
    parenchyma cells
6. Phloem

7. Xylem
8. Pericycle
9. Endodermis

Roots of Angiosperms

Photo courtesy of James V. Allen

Figure 6.89. A scanning electron micrograph of a young root from a corn plant, *Zea mays.*

1. Root hairs
2. Epidermis
3. Stele
4. Cortex
5. Endodermis
6. Xylem
7. Phloem

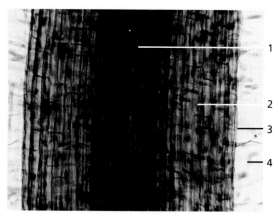

Figure 6.90. Photomicrograph of a young root of wheat, *Triticum,* showing root hairs. (X40)

1. Stele   2. Cortex   3. Epidermis   4. Root hair

Stems of Angiosperms

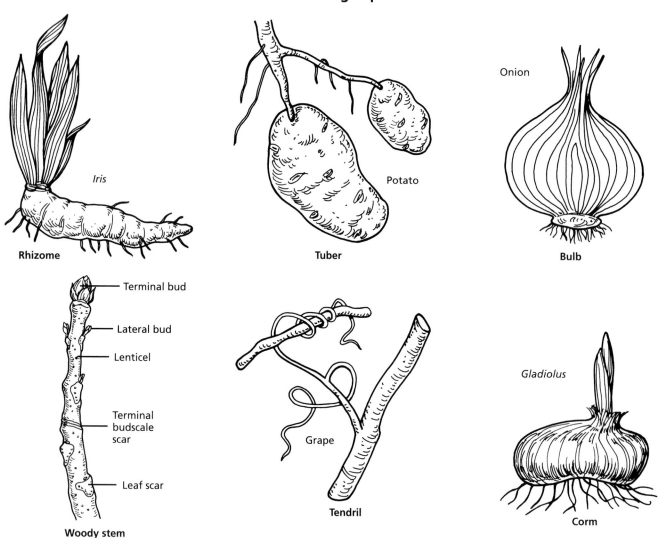

Iris

Rhizome

Potato

Tuber

Onion

Bulb

Terminal bud

Lateral bud

Lenticel

Terminal budscale scar

Leaf scar

Woody stem

Grape

Tendril

Gladiolus

Corm

Figure 6.91. Examples of the variety and specialization of angiosperm stems. The stem of an angiosperm is the ascending portion of the plant specialized to produce and support leaves and flowers, transport and store water and nutrients, and provide growth through cell division. Stems of plants are utilized extensively by humans in products including paper, building materials, furniture, and fuel. In addition, the stems of potatoes, onions, celery, cabbage, and other plants are important food crops.

Stems of Angiosperms

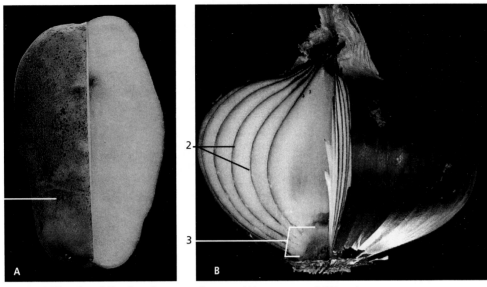

Figure 6.92. Specialized underground stems, (A) potato, and (B) onion.

1. Node (eye) bearing a minute scale leaf   2. Bulb scales (modified leaves)   3. Short stem

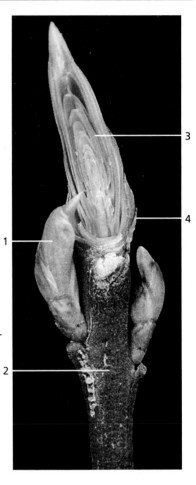

Figure 6.93. A terminal bud of a woody stem that has been longitudinally sectioned to show a developing leaf.

1. Lateral bud
2. Stem
3. Leaf primordium
4. Scale

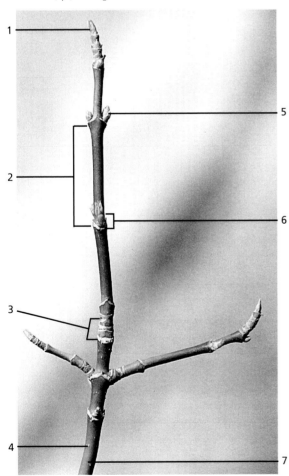

Figure 6.94 A woody branch of a dicot seen in early spring just as the buds are beginning to swell. Branches and twigs are small extensions of the stems of certain angiosperms that directly support leaves and flowers.

1. Terminal bud
2. Internode
3. Terminal budscale scars
4. Lenticel
5. Lateral bud
6. Node area
7. Stem

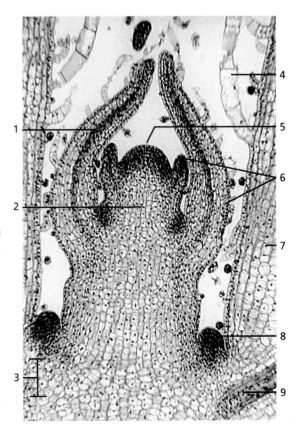

Figure 6.95. Longitudinal section of the stem tip of the common houseplant *Coleus*. (X40)

1. Procambium
2. Ground meristem
3. Leaf gap
4. Trichome
5. Apical meristem
6. Developing leaf primorda
7. Leaf primordium
8. Axillary bud
9. Developing vascular tissue

Stems of Angiosperms

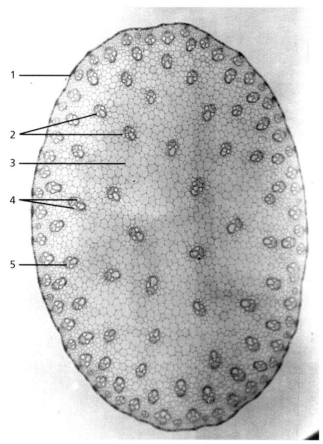

Figure 6.96.   A cross section from the stem of a monocot, *Zea mays*, corn. (X40)

1. Epidermis
2. Vascular bundles
3. Parenchyma cells (ground tissue)
4. Xylem
5. Phloem

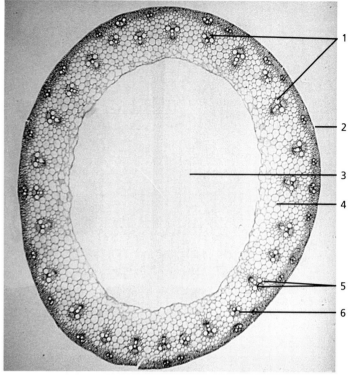

Figure 6.97.   Cross section through the stem of a monocot, *Triticum*, wheat. (X40)

1. Vascular bundles
2. Epidermis
3. Ground tissue cavity
4. Parenchyma cells of ground tissue
5. Xylem
6. Phloem

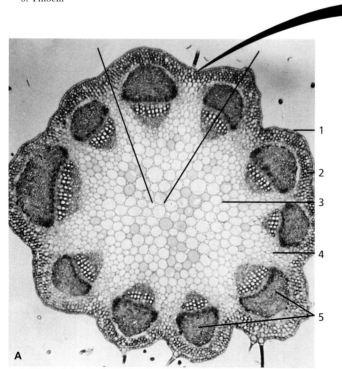

Interfascicular region

Epidermis
Cortex
Phloem fibers
Phloem
Fascicular cambium
xylem

Pith

B

Figure 6.98.   A cross section from the stem of a dicot. (A) The stem of a clover magnified at X40, and (B) a diagram.

1. Epidermis
2. Cortex
3. Pith
4. Interfascicular cambium
5. Vascular bundles with caps

Stems of Angiosperms

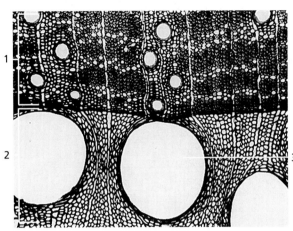

Figure 6.99. A cross section through the secondary xylem (wood) of the stem of an oak, *Quercus.* (X100)

1. Summer wood
2. Spring wood
3. Vessel element

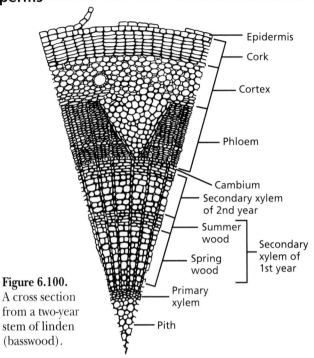

Figure 6.100. A cross section from a two-year stem of linden (basswood).

Epidermis
Cork
Cortex
Phloem
Cambium
Secondary xylem of 2nd year
Summer wood
Spring wood
Secondary xylem of 1st year
Primary xylem
Pith

Leaves of Angiosperms

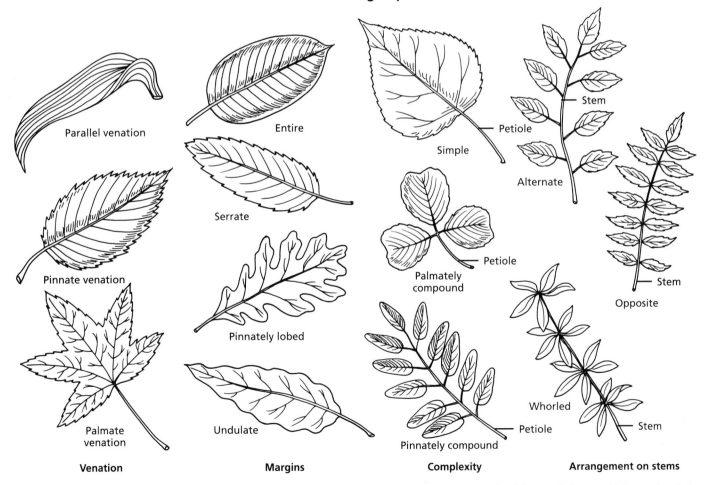

Parallel venation

Entire

Serrate

Simple

Petiole

Stem

Alternate

Pinnate venation

Pinnately lobed

Palmately compound

Petiole

Opposite

Stem

Palmate venation

Undulate

Pinnately compound

Petiole

Whorled

Stem

Venation   **Margins**   **Complexity**   **Arrangement on stems**

Figure 6.101. Diagram showing several representative angiosperm leaf types. Leaves comprise the foliage of plants which provide habitat and a food source for many animals including humans. Leaves also provide protective ground cover and are the portion of the plant most responsible for oxygen replenishment into the atmosphere.

Leaves of Angiosperms

Figure 6.102. Typical angiosperm leaf showing characteristic surface features. Leaves are organs modified to carry out photosynthesis. Photosynthesis is the manufacture of food (sugar) from carbon dioxide and water, in the presence of chlorophyll, with sunlight as the source of energy.

1. Lamina (blade)
2. Petiole
3. Serrate margin
4. Midrib
5. Veins

Figure 6.103. The organic decomposition of a leaf is a gradual process beginning with the softer tissues of the lamina, leaving only the vascular tissues of the midrib and the veins, as seen in this photograph. With time, these will also decompose. A deciduous leaf is one that is shed during the autumn season as the petiole detaches from the stem.

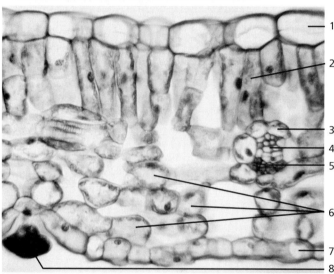

1. Upper epidermis
2. Palisade mesophyll
3. Bundle sheath
4. Xylem
5. Phloem
6. Spongy mesophyll
7. Lower epidermis
8. Trichome (leaf hair)

Figure 6.104. A cross section through the leaf of the common hedge privet, *Ligustrum,* magnified X100. The typical tissue arrangement of a leaf includes an upper epidermis, a lower epidermis and the centrally located mesophyll. Containing chloroplasts, the cells of the mesophyll are often divided into palisade mesophyll and spongy mesophyll. Veins within the mesophyll conduct material through the leaf.

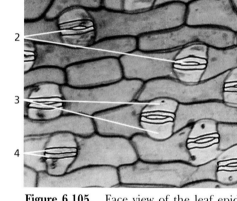

Figure 6.105. Face view of the leaf epidermis of *Tradescantia.* (X430)

1. Epidermal cells   3. Subsidiary cells
2. Stomata   4. Guard cells

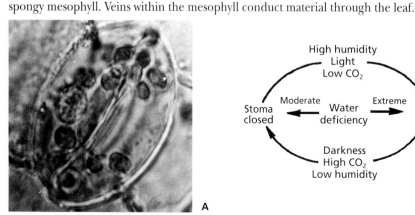

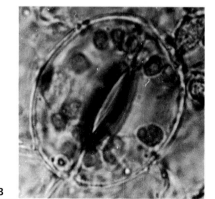

High humidity
Light
Low CO_2

Moderate   Water   Extreme

Stoma closed ←   deficiency   → Stoma open

Darkness
High CO_2
Low humidity

A   **B**

Figure 6.106. Guard cells regulate the opening of the stomata according to the environmental factors as indicated in this diagram. (A) A face view of a closed stoma of a geranium, and (B) an open stoma. (X1500)

Flowers of Angiosperms

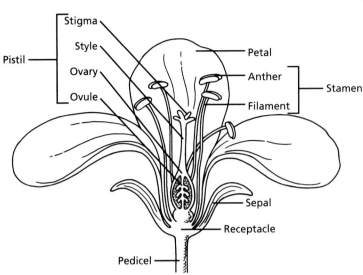

Figure 6.107. Diagram of angiosperm flower. The angiosperm flower is typically composed of sepals, petals, stamens, and one or more pistils. The sepals are the outermost circle of protective leaf-like structures. They are usually green and are collectively called the calyx. The petals generally form a whorl to the inside of the calyx and are collectively called the corolla. The stamens and the pistils are the reproductive parts of the flower. A stamen consists of the filament (stalk) and the anther, where pollen is produced. The pistil consists of a sticky stigma at the tip that receives pollen and a style that leads to the ovary.

Figure 6.108. Floral parts of a tulip.

1. Stigma   2. Anther   3. Petal   4. Ovary

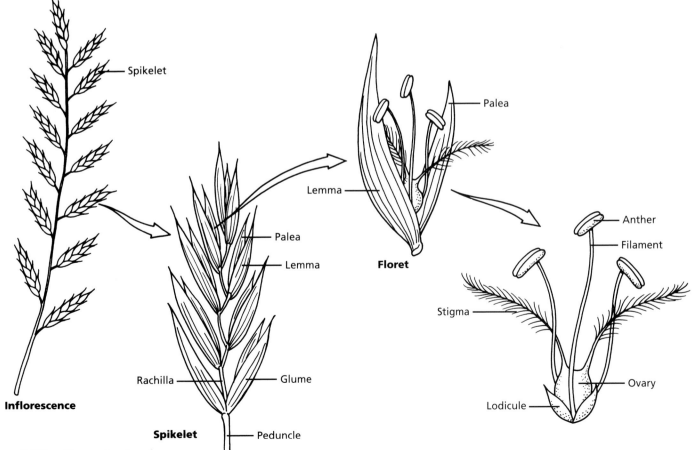

Figure 6.109. Flower structure in grasses.

Flowers of Angiosperms

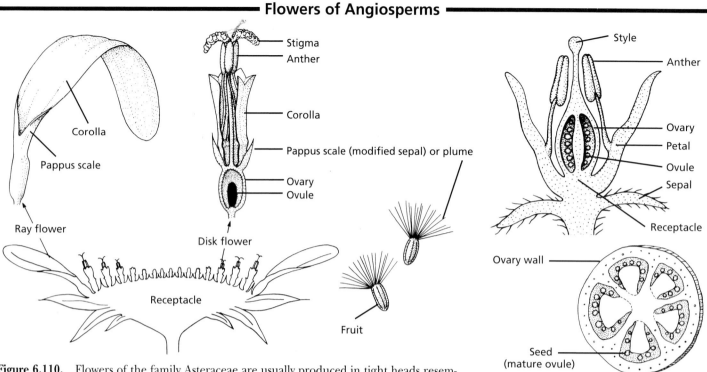

Figure 6.110.   Flowers of the family Asteraceae are usually produced in tight heads resembling single large flowers. One of these inflorescences can contain hundreds of individual flowers. Examples of this family include dandelions, sunflowers, asters, and marigolds.

Figure 6.111.   Flower and fruit of a tomato, *Lycopersicon esculentum.*

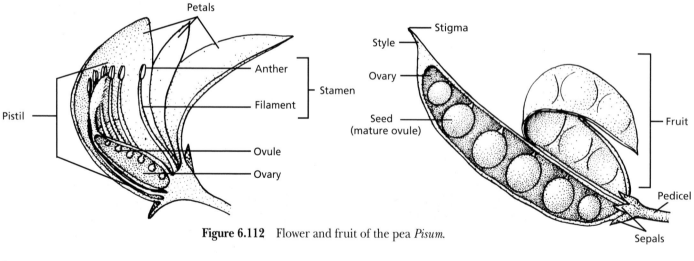

Figure 6.112   Flower and fruit of the pea *Pisum.*

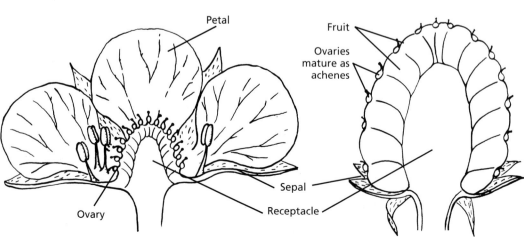

Figure 6.113. Flower and fruit of a strawberry, *Fragaria.* The strawberry has an aggregate fruit.

Flowers of Angiosperms

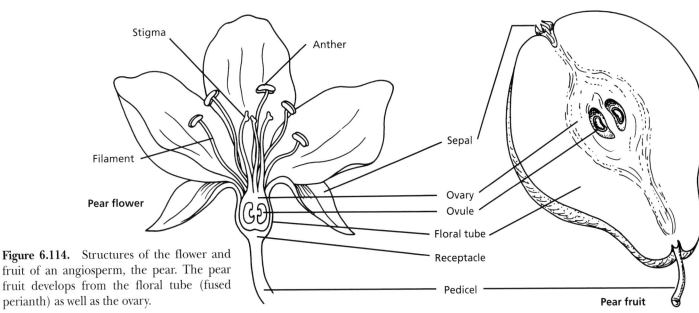

Figure 6.114. Structures of the flower and fruit of an angiosperm, the pear. The pear fruit develops from the floral tube (fused perianth) as well as the ovary.

Figure 6.115. *Gladiolus* flower with petals and sepals removed. The stamen is comprised of the anther and filament whereas the pistil is made up of the stigma, style, and ovary.

1. Anther   5. Stigma
2. Filament   6. Style
3. Ovules   7. Ovary
4. Receptacle

Figure 6.116.
Rose flower.

1. Petals
2. Stigma
3. Anther
4. Sepal
5. Pistils
6. Stamens
7. Ovaries

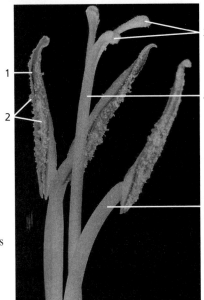

Figure 6.117.

Gladiolus anthers and stigma.

1. Anther
2. Pollen (*n*)
3. Stigma
4. Style
5. Filament

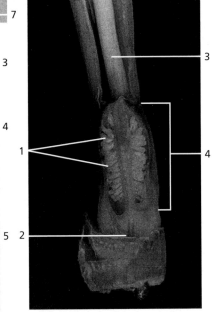

Figure 6.118. *Gladiolus* ovary.

1. Ovules (unfertilized seeds)   3. Style
2. Receptacle   4. Ovary

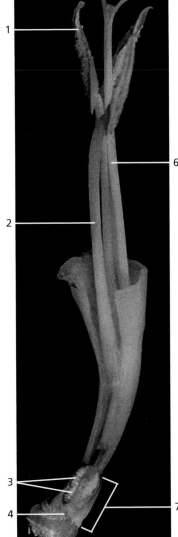

Seeds, Fruits, and Seed Germination of Angiosperms

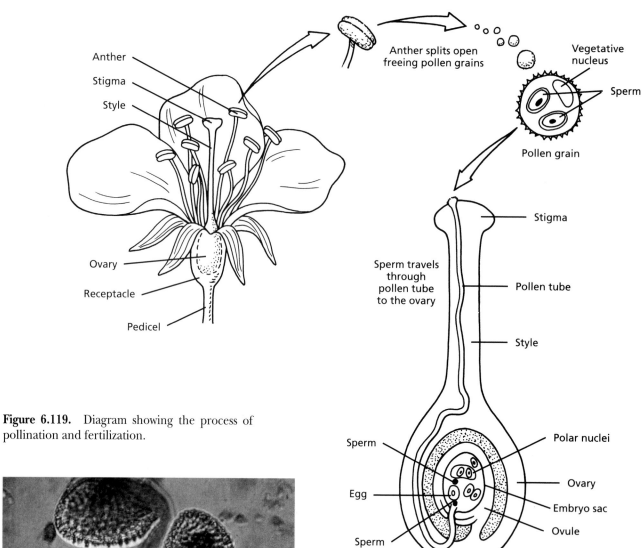

Anther

Stigma

Style

Ovary

Receptacle

Pedicel

Anther splits open freeing pollen grains

Vegetative nucleus

Sperm

Pollen grain

Stigma

Sperm travels through pollen tube to the ovary

Pollen tube

Style

Sperm

Polar nuclei

Egg

Ovary

Embryo sac

Sperm

Ovule

Figure 6.119. Diagram showing the process of pollination and fertilization.

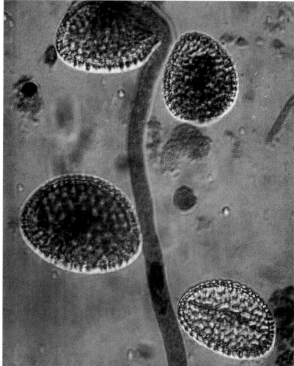

Figure 6.120. Pollen grains of a lily. The pollen grain at the top of the photo has germinated to produce a pollen tube. (about X500)

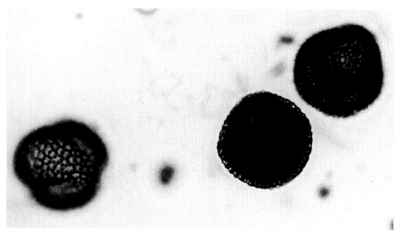

Figure 6.121. Pollen grains of the dicot, *Lilac.* (X430)

Seeds, Fruits, and Seed Germination of Angiosperms

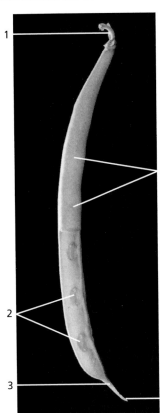

Figure 6.122. A legume, string bean.

1. Pedicel
2. Seeds
3. Style
4. Fruit
5. Stigma

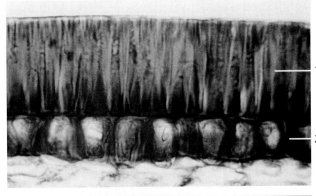

Figure 6.123. Closeup of the seed coat of the garden bean, *Phaseolus*, showing sclerified epidermis. (X100)

1. Macrosclerids
2. Subepidermal sclerids

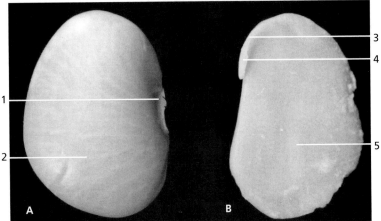

Figure 6.124. Lima bean. (A) Entire bean seed; (B) longitudinally sectioned seed.

1. Hilum
2. Integument (seed coat)
3. Hypocotyl
4. Radicle
5. Cotyledon

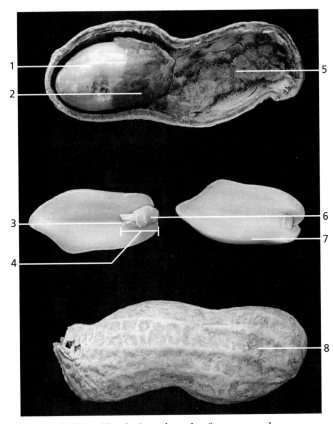

Figure 6.126. The fruit and seeds of a peanut plant.

1. Cotyledon
2. Integument (seed coat)
3. Plumule
4. Embryo
5. Interior of fruit
6. Radicle
7. Cotyledon
8. Fruit wall

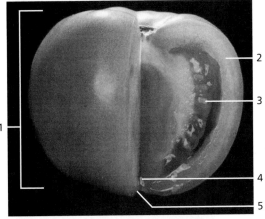

Figure 6.125. Longitudinal section of a tomato fruit (a berry).

1. Fruit (mature ovary)
2. Ovary wall
3. Seed (mature ovule)
4. Remnant of style
5. Remnant of stigma

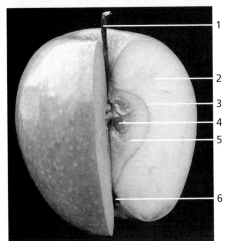

Figure 6.127. Longitudinal section of an apple fruit (a pome).

1. Pedicel
2. Floral tube
3. Ovary wall
4. Seed (mature ovule)
5. Mature ovary
6. Remnants of floral parts

Seeds, Fruits, and Seed Germination of Angiosperms

Grape

Berry

Acorn   Filbert

Nuts

Peach

Apple

Drupe

Fruit   Lima bean

Seed

Legumes

Corn   Grass

Caryopses
(grains)

Pome   **Simple fruits**

Blackberry   Strawberry

Aggregate fruits

Fig

Pineapple

Multiple fruits

Figure 6.128.   Drawings of several types of fruit.

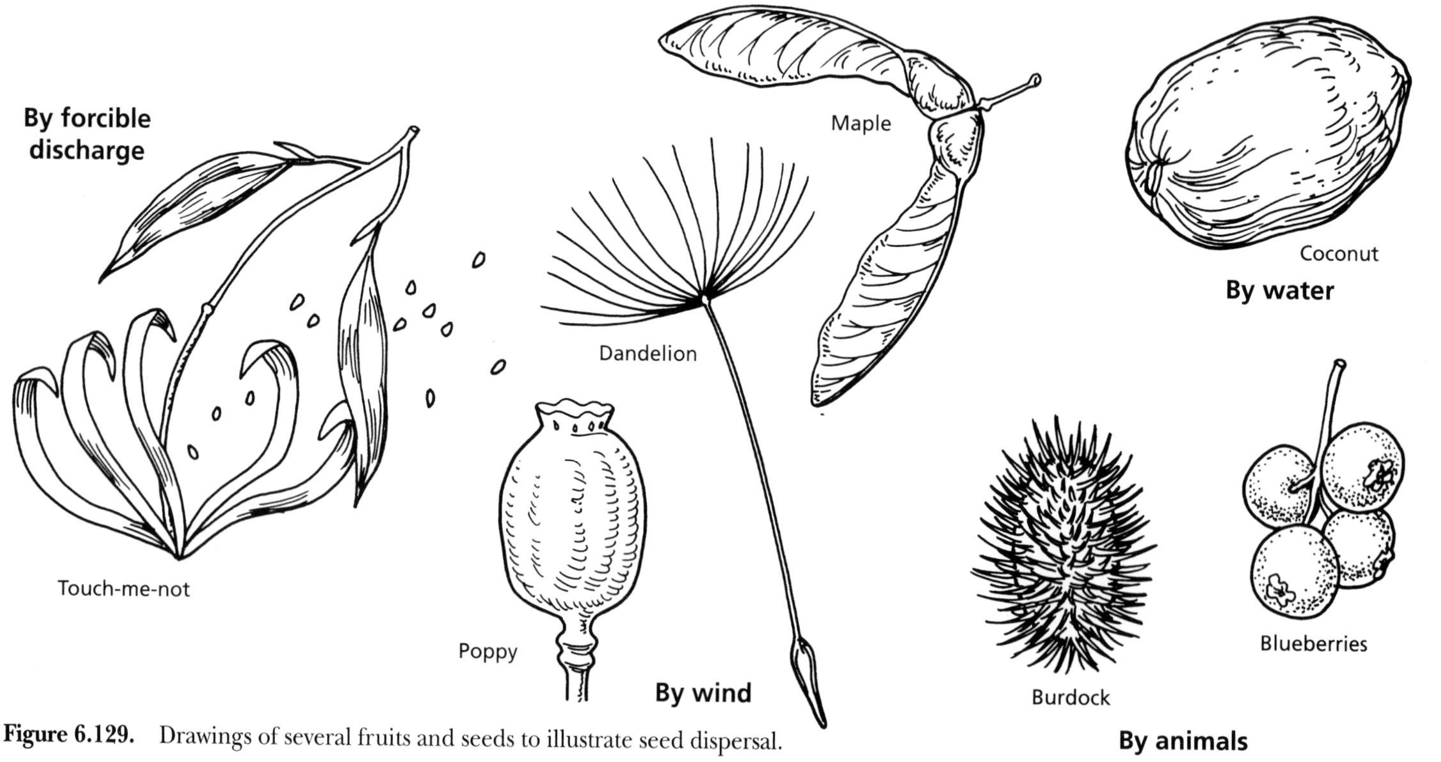

By forcible discharge

Maple

Touch-me-not

Dandelion

Poppy

By wind

Coconut

By water

Burdock   Blueberries

By animals

Figure 6.129.   Drawings of several fruits and seeds to illustrate seed dispersal.

Seeds, Fruits, and Seed Germination of Angiosperms

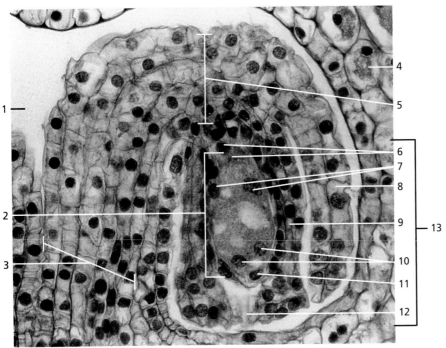

Figure 6.130. A cross section of an eight-nucleate embryo sac of an ovule from a lily, *Lilium*. (X430)

1. Locule
2. Embryo sac (*n*)
3. Funiculus
4. Wall of ovary
5. Chalaza
6. Antipodal cells (*n*)
7. Polar nuclei (*n*)
8. Outer integument (2*n*)
9. Inner integument (2*n*)
10. Synergid cells (*n*)
11. Egg (*n*)
12. Micropyle (pollen tube entrance)
13. Components of ovule

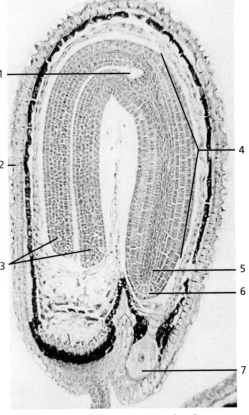

Figure 6.131. Photomicrograph of a mature dicot embryo and seed from a shepherd's purse, *Capsella bursapastoris*. (X100)

1. Apical meristem
2. Seed coat
3. Cotyledons
4. Hypocotyl axis
5. Developing vascular tissue (procambium)
6. Root tip
7. Basal cell

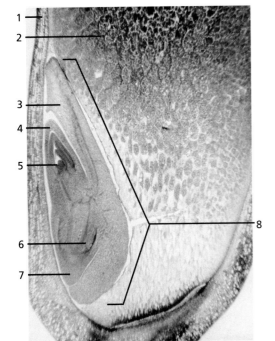

Figure 6.132. Photomicrograph of a mature grain, or caryopsis, of wheat, *Triticum aestivum.* (X100)

1. Pericarp
2. Starchy endosperm
3. Scutellum
4. Coleoptile
5. Shoot apex
6. Radicle
7. Coleorhiza
8. Embryo

Figure 6.133. Diagram of bean, *Phaseolus*, germination.

1. Hypocotyl
2. Seed coat
3. Primary root
4. Foliage leaves
5. Epicotyl
6. Cotyledon
7. Hypocotyl
8. Primary roots
9. Lateral roots

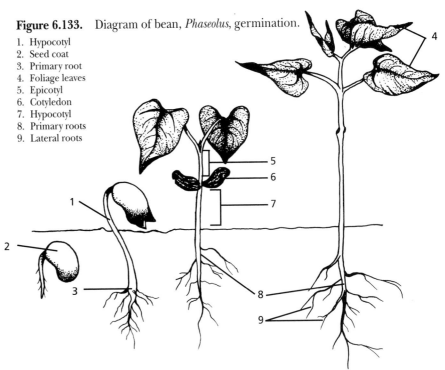

Kingdom Animalia

Animals are multicellular, heterotrophic eukaryotes that ingest food materials and store carbohydrate reserves as glycogen or fat. The cells of animals lack cell walls, but do contain intercellular connections including desmosomes, gap junctions, and tight junctions. Animal cells are also highly specialized into the specific kinds of tissues described in chapter 1. Most animals are motile through the contraction of muscle fibers containing actin and myosin proteins. The complex body systems of animals include elaborate sensory and neuromotor specializations that accommodate dynamic behavioral mechanisms.

Reproduction in animals is primarily sexual, with the diploid stage generally dominating the life cycle. Primary sex organs, or *gonads*, produce the haploid gametes called *sperm* and *ova*. Propagation begins as a small flagellated sperm fertilizes a larger, nonmotile egg forming a diploid *zygote* that has genetic traits of both parents. The zygote then undergoes a succession of mitotic divisions called *cleavage*. In most animals, cleavage is followed by the formation of a multicellular stage called a *blastula*. With further development, the *germ layers* form which eventually give rise to each of the body organs. The developmental cycle of many animals includes *larval forms*, which are still developing, free-living, and sexually immature. Larvae usually have different food and habitat requirements from those of the adults. Larvae eventually undergo metamorphosis that transforms them into sexually mature adults.

Animals inhabit nearly all aquatic and terrestrial habitats of the biosphere. The greatest number of animals are marine, where the first animals probably evolved. Depending on the classification scheme, animals may be grouped into as many as 35 phyla. The most commonly known phylum is *Chordata* that includes the subphylum *Vertebrata*, or the backboned animals. Chordates, however, comprise only about 5% of all the animal species. All other animals are frequently referred to as *invertebrates*, and they account for approximately 95% of the animal species.

<table>
<tr><td colspan="2" align="center">TABLE 7.1.
Some Representatives of the Kingdom Animalia</td></tr>
<tr><td>Phyla and Representative Kinds</td><td>Characteristics</td></tr>
<tr><td>Porifera (sponges)</td><td>Multicellular, aquatic animals, with stiff or fibrous bodies perforated by pores</td></tr>
<tr><td>Cnidaria (corals, hydra, jellyfish)</td><td>Aquatic animals, radially symmetrical, mouth surrounded by tentacles bearing cnidocytes (stinging cells); body composed of epidermis and gastrodermis, separated by mesoglea</td></tr>
<tr><td>Platyhelminthes (flatworms)</td><td>Elongated, flattened, and bilaterally symmetrical; distinct head containing ganglia; nerve cords; protonephridia of flame cells</td></tr>
<tr><td>Nematoda (roundworms)</td><td>Mostly microscopic, unsegmented worms; body enclosed in cuticle; whiplike body movement</td></tr>
<tr><td>Mollusca (mollusks: clams, snails, squids)</td><td>Bilaterally symmetrical with true coelom; mantle; may have muscular foot and protective shell</td></tr>
<tr><td>Annelida (segmented worms)</td><td>Body segmented by septa; a series of hearts; hydrostatic skeleton and circular and longitudinal muscles</td></tr>
<tr><td>Arthropoda (crustaceans, insects)</td><td>Body segmented; paired and jointed appendages; chitinous exoskeleton; hemocoel for blood flow</td></tr>
<tr><td>Echinodermata (echinoderms: sea stars, sand dollars, sea cucumbers, sea urchins)</td><td>Larvae have bilateral symmetry; adults have pentaradial symmetry; coelom with complete digestive tract; regeneration of body parts</td></tr>
<tr><td>Chordata (amphioxus, amphibians, fishes, reptiles, birds, and mammals)</td><td>Fibrous notochord, pharyngeal gill pouches, and dorsal hollow nerve cord present at some stage in their life cycle</td></tr>
</table>

PHYLUM PORIFERA (sponges)

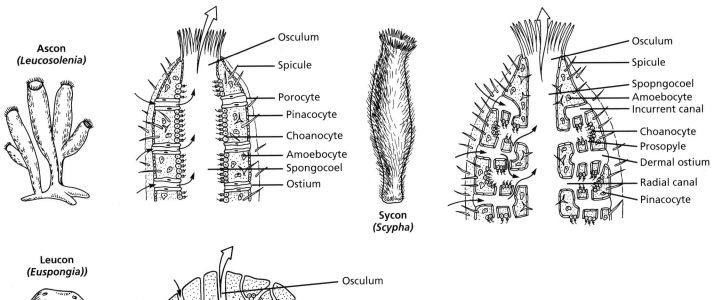

Ascon (*Leucosolenia*)

- Osculum
- Spicule
- Porocyte
- Pinacocyte
- Choanocyte
- Amoebocyte
- Spongocoel
- Ostium

Sycon (*Scypha*)

- Osculum
- Spicule
- Spopngocoel
- Amoebocyte
- Incurrent canal
- Choanocyte
- Prosopyle
- Dermal ostium
- Radial canal
- Pinacocyte

Leucon (*Euspongia*))

- Osculum
- Excurrent canal
- Ostium
- Incurrent canal
- Flagellated chamber

Figure 7.1. Examples of sponge body types. A diagrammatic representative of each of the three types depicts with arrows the flow of water through the body of the sponge.

PHYLUM PORIFERA ——————————————— Class Calcarea

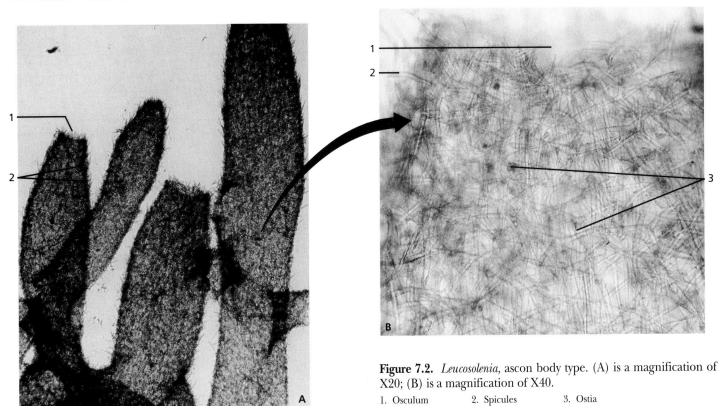

Figure 7.2. *Leucosolenia*, ascon body type. (A) is a magnification of X20; (B) is a magnification of X40.

1. Osculum   2. Spicules   3. Ostia

PHYLUM PORIFERA ━━━━━━━━━━━━━━━━━━━━━━━━━━━━━━━━━━━━━━ Class Calcarea

Figure 7.3. Cross section of sponge, *Scypha.* (A) is a magnification of X40; (B) is a magnification of X100.

1. Spongocoel	5. Choanocytes (collar cells)	8. Blastulas
2. Ostium	6. Apopyle	9. Ostium
3. Incurrent canal	7. Incurrent canals	10. Pinacocytes
4. Radial canal		11. Radial canal

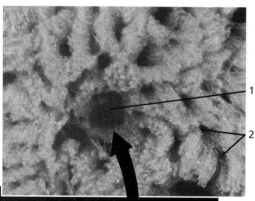

PHYLUM PORIFERA ━━━━━━━━━━━━━━━━━━━━━━━━━━━━━━━━━━━ Class Desmospongiae

Figure 7.4. Bath sponge, class Desmospongiae. This sponge exhibits a leuconoid body structure.

1. Osculum
2. Ostia

Figure 7.5. Spicules of a freshwater sponge. (X430)

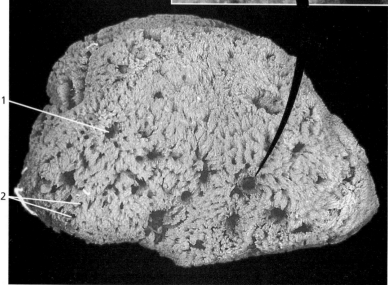

Figure 7.6. A leucon type sponge occurring in a natural habitat.

PHYLUM CNIDARIA (corals, hydra, and jellyfish)

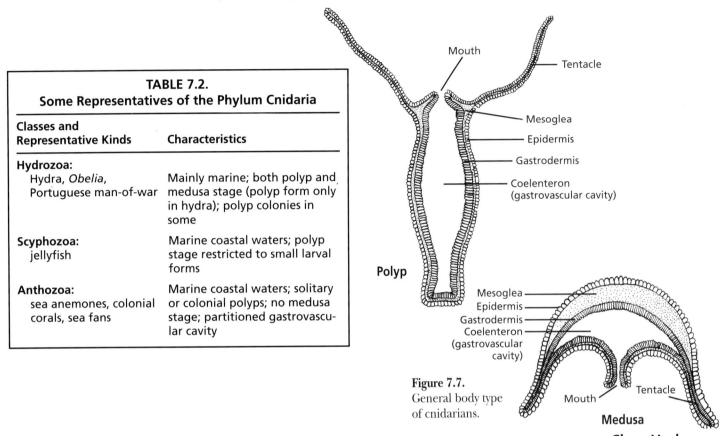

TABLE 7.2. Some Representatives of the Phylum Cnidaria	
Classes and Representative Kinds	**Characteristics**
Hydrozoa: Hydra, *Obelia*, Portuguese man-of-war	Mainly marine; both polyp and medusa stage (polyp form only in hydra); polyp colonies in some
Scyphozoa: jellyfish	Marine coastal waters; polyp stage restricted to small larval forms
Anthozoa: sea anemones, colonial corals, sea fans	Marine coastal waters; solitary or colonial polyps; no medusa stage; partitioned gastrovascular cavity

Figure 7.7.
General body type of cnidarians.

PHYLUM CNIDARIA ━━━━━━━━━━━━━━━━━━━━━━━━━━━━━ Class Hydrozoa

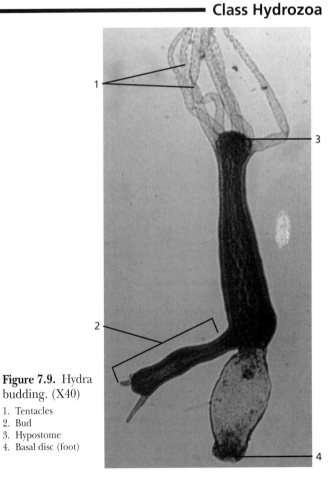

Figure 7.8.
Hydra, vertical section.

Figure 7.9. Hydra budding. (X40)

1. Tentacles
2. Bud
3. Hypostome
4. Basal disc (foot)

PHYLUM CNIDARIA ▬▬▬▬▬▬▬▬▬▬▬▬▬▬▬▬▬▬▬▬▬▬▬▬▬▬▬▬▬▬▬ **Class Hydrozoa**

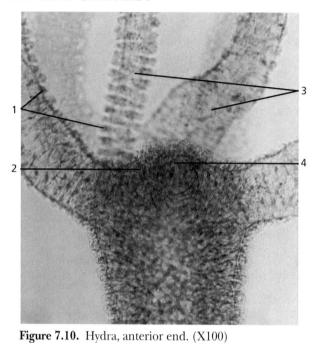

Figure 7.10. Hydra, anterior end. (X100)

1. Cnidocysts   3. Tentacles
2. Hypostome   4. Mouth

Figure 7.11. Hydra, cross section. (X430)

1. Epidermis (ectoderm)   3. Mesoglea
2. Coelenteron   4. Gastrodermis (endoderm)

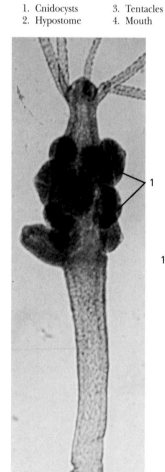

Figure 7.12. Hydra, male. (X40)

1. Testes

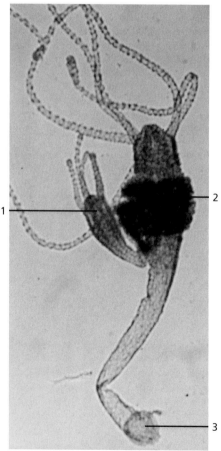

Figure 7.13. Hydra, female. (X40)

1. Bud
2. Ovary
3. Basal disc (foot)

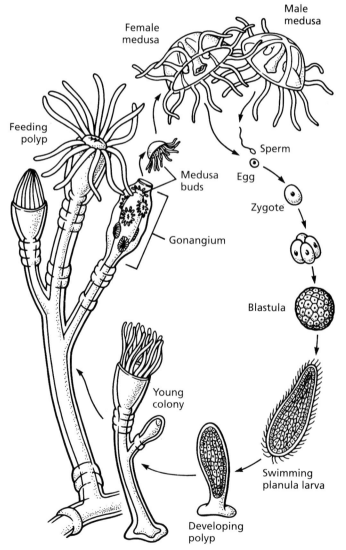

Figure 7.14. Life cycle of *Obelia*.

PHYLUM CNIDARIA ━━━━━━━━━━━━━━━━━━━━━━━━━━━ Class Hydrozoa

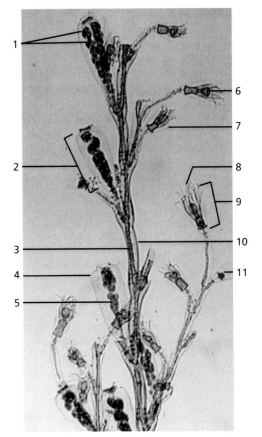

Figure 7.15. *Obelia* colony. (X40)

1. Medusa buds	7. Hydrotheca
2. Gonangium	8. Tentacles
3. Coenosarc	9. Hydranth
4. Gonotheca	10. Perisarc
5. Blastostyle	11. Developing hydranth
6. Hypostome	

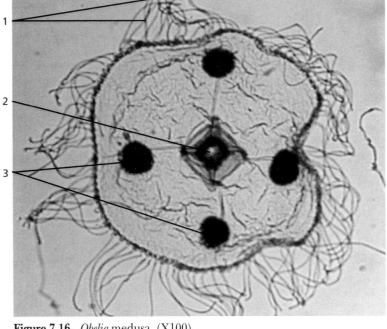

Figure 7.16. *Obelia* medusa. (X100)

1. Tentacles   2. Manubrium   3. Gonads

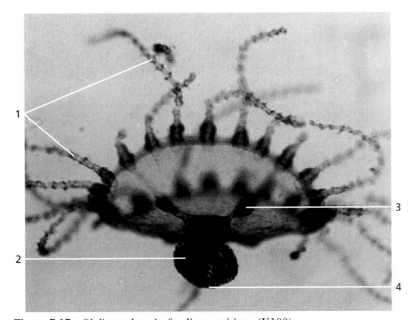

Figure 7.17. *Obelia* medusa in feeding position. (X100)

1. Tentacles	3. Gonad
2. Manubrium	4. Mouth

Figure 7.18. *Physalia*, Portugese man-of-war. This is actually a colony of medusae and polyps acting as a single organism. The tentacles are comprised of three types of polyps: the gastrozooids (feeding polyps), the dactylozooids (stinging polyps), and the gonozooids (reproductive polyps).

1. Pneumatophore (float)   2. Tentacles

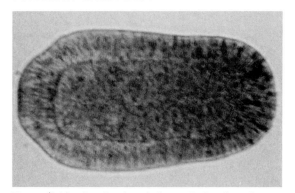

Figure 7.19. *Aurelia* planula larva. It develops from a fertilized egg that may be retained on the oral arm of the medusa. (X40)

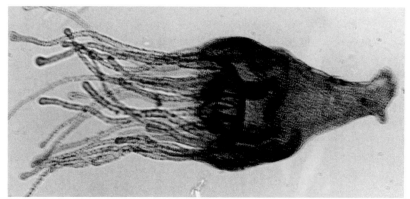

Figure 7.20. *Aurelia* scyphistoma. The polyp is a developmental stage in the life cycle of the jelly fish. (X40)

Figure 7.21. *Aurelia* strobilia. Under favorable conditions the scyphistoma develops into the strobilia. (X40)

1. Developing ephyrae

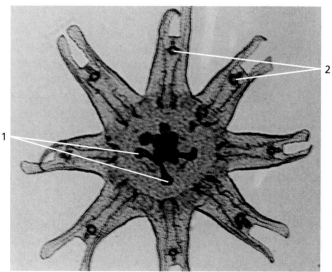

Figure 7.22. *Aurelia*, ephyra larva. These gradually develop into adult jelly fish. (X40)

1. Gonads   2. Sense organs

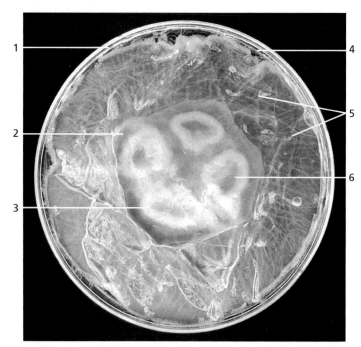

Figure 7.23. *Aurelia* medusa, dorsal (aboral) view.

1. Circular canal	3. Gonad	5. Radial canals
2. Gastric pouch	4. Marginal tentacles	6. Subgenital pit

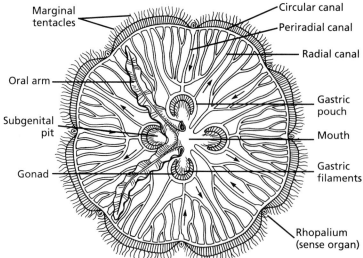

Figure 7.24. *Aurelia* medusa, oral view. In this diagram, the right oral arms have been removed. The arrows depict circulation through the canal system.

PHYLUM CNIDARIA ━━━━━━━━━━━━━━━━━━━━━━━━━━━━━━━━━━━━━━━ **Class Anthozoa**

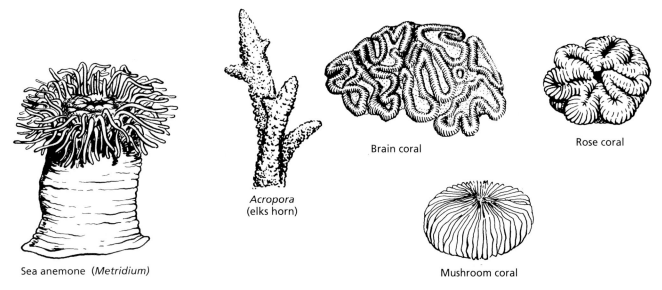

Brain coral

Rose coral

Acropora
(elks horn)

Mushroom coral

Figure 7.25. Representatives of the class Anthozoa.

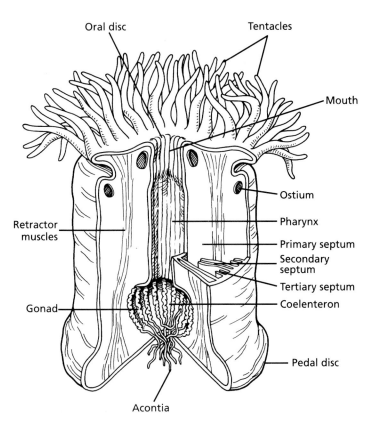

Oral disc

Tentacles

Mouth

Ostium

Retractor
muscles

Pharynx

Primary septum

Secondary
septum

Tertiary septum

Coelenteron

Gonad

Pedal disc

Acontia

Figure 7.26. Sea anemone, *Metridium,* partly dissected.

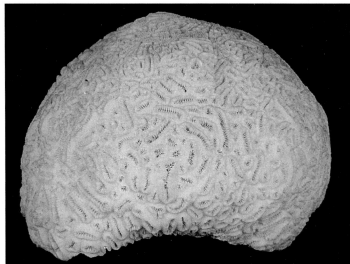

Figure 7.27. Dried skeleton of a brain coral, class Anthozoa.

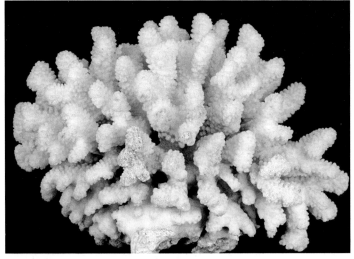

Figure 7.28. Coral, class Anthozoa. Coral like this comprises coral reefs.

PHYLUM PLATYHELMINTHES

TABLE 7.3.
Some Representatives of the Phylum Platyhelminthes

Classes and Representative Kinds	Characteristics
Turbellaria: planarians	Mostly free-living, carnivorous, freshwater forms; body covered by ciliated epidermis
Trematoda: flukes (e.g., schistosomes)	Parasitic with wide range of invertebrate and vertebrate hosts; suckers for attachment to host
Cestoda: tapeworms	Parasitic on many vertebrate hosts; complex life cycle with intermediate hosts; suckers or hooks on scolex for attachment to host; eggs are produced and shed within proglottids

PHYLUM PLATYHELMINTHES — Class Turbellaria

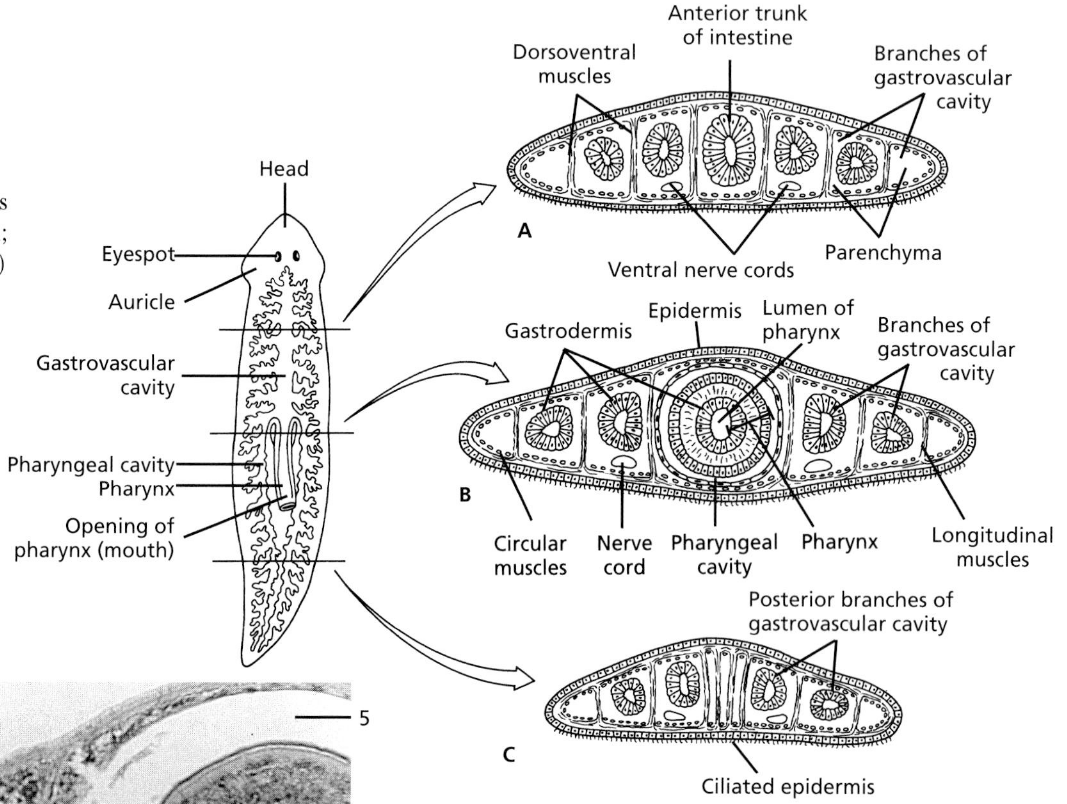

Figure 7.29. *Planaria,* cross sections through the: (A) anterior region; (B) pharyngeal region; and, (C) posterior region.

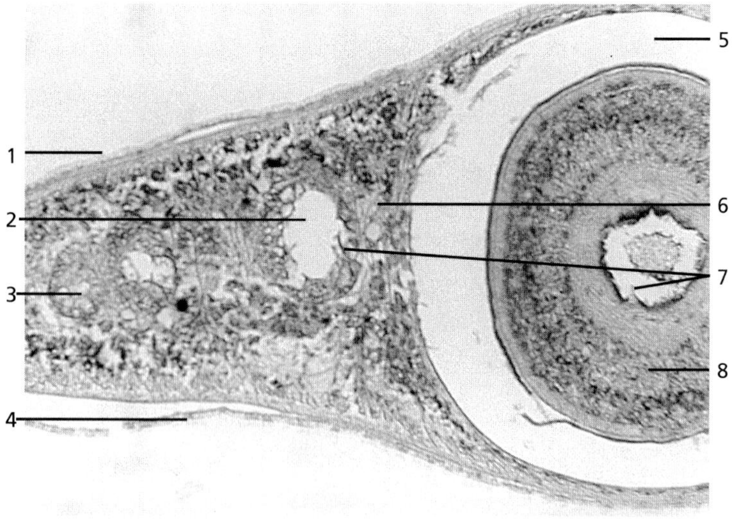

Figure 7.30. *Planaria,* cross section through pharyngeal region. (X100)

1. Epidermis
2. Gastrovascular cavity
3. Testis
4. Cilia
5. Pharyngeal cavity
6. Dorsoventral muscles
7. Gastrodermis (endoderm)
8. Pharynx

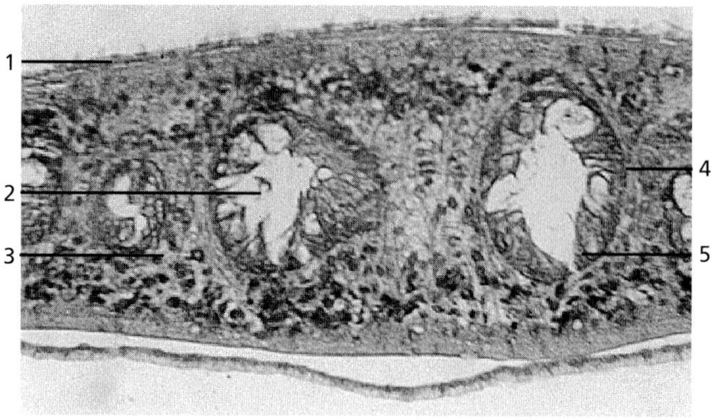

Figure 7.31. *Planaria,* cross section through posterior region. (X100)

1. Epidermis
2. Gastrovascular cavity
3. Mesenchyme
4. Dorsoventral muscles
5. Intestine (endoderm)

PHYLUM PLATYHELMINTHES ═══ **Class Trematoda**

Figure 7.32.   Cow liver fluke, *Fasciola magna*. This is one of the largest flukes, measuring about 3 inches long.

1. Yolk gland   2. Ventral sucker   3. Oral sucker

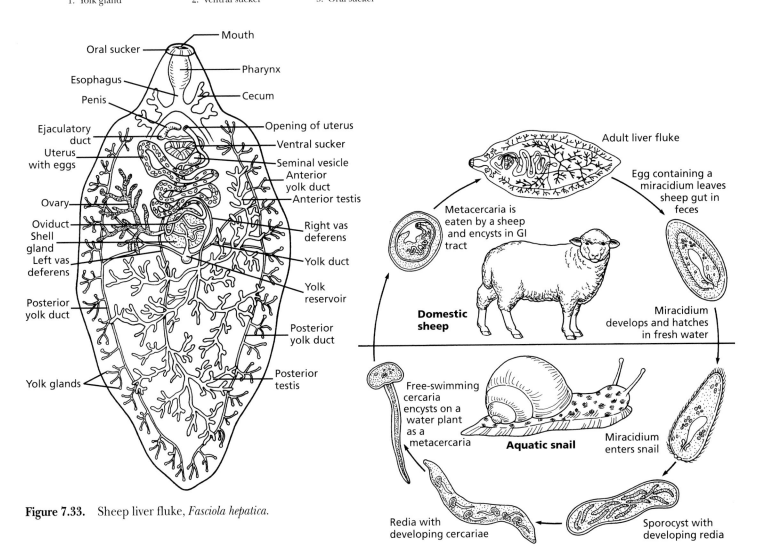

Figure 7.33.   Sheep liver fluke, *Fasciola hepatica*.

Figure 7.34.   Life cycle of sheep liver fluke, *Fasciola hepatica*.

PHYLUM PLATYHELMINTHES — Class Trematoda

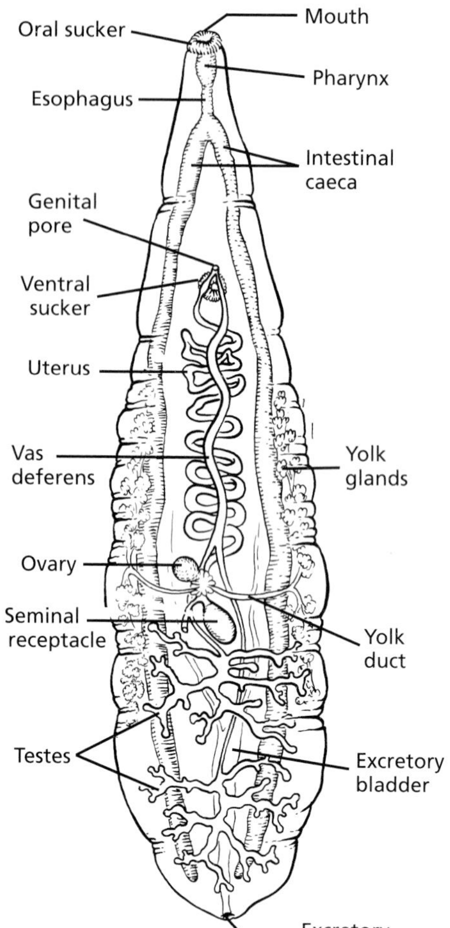
Mouth
Oral sucker
Pharynx
Esophagus
Intestinal caeca
Genital pore
Ventral sucker
Uterus
Vas deferens
Yolk glands
Ovary
Seminal receptacle
Yolk duct
Testes
Excretory bladder
Excretory pore

Figure 7.35. *Clonorchis sinensis,* Chinese liver fluke.

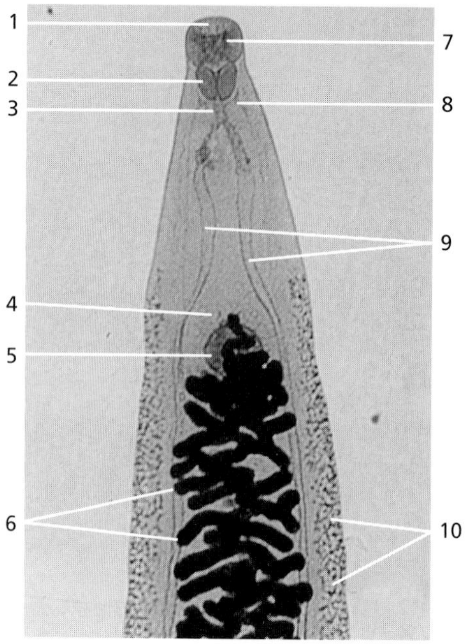
1
2
3
4
5
6
7
8
9
10

Figure 7.36. *Clonorchis,* anterior end. (X100)

1. Mouth
2. Pharynx
3. Esophagus
4. Genital pore
5. Ventral sucker
6. Uterus
7. Oral sucker
8. Cerebral ganglion
9. Intestinal caeca
10. Yolk glands

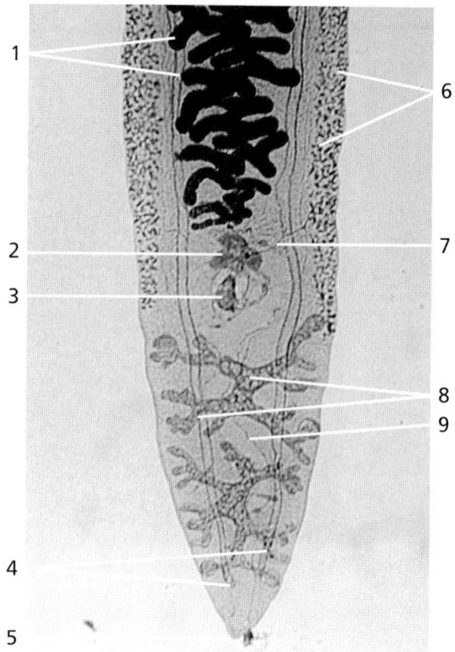
1
2
3
4
5
6
7
8
9

Figure 7.37. *Clonorchis,* posterior end. (X100)

1. Uterus
2. Ovary
3. Seminal receptacle
4. Intestinal caeca
5. Excretory pore
6. Yolk glands
7. Yolk duct
8. Testis
9. Excretory bladder

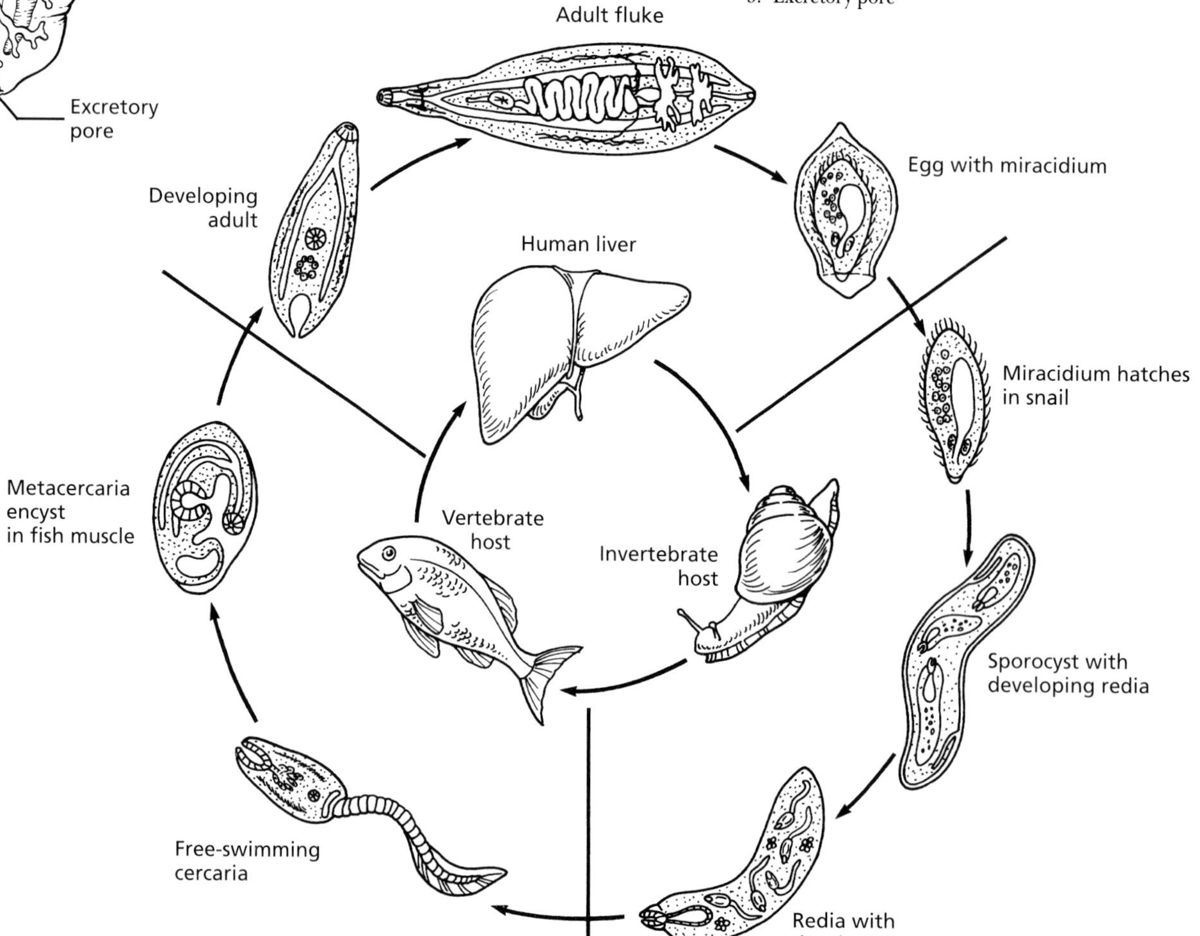
Adult fluke
Egg with miracidium
Miracidium hatches in snail
Sporocyst with developing redia
Redia with developing cercaria
Free-swimming cercaria
Metacercaria encyst in fish muscle
Developing adult
Human liver
Vertebrate host
Invertebrate host

Figure 7.38. Life cycle of the human liver fluke, *Clonorchis sinesis.*

PHYLUM PLATYHELMINTHES ━━━━━━━━━━━━━━━━ Class Cestoda

Figure 7.39. *Taenia pisiformis*, a parasitic tapeworm.

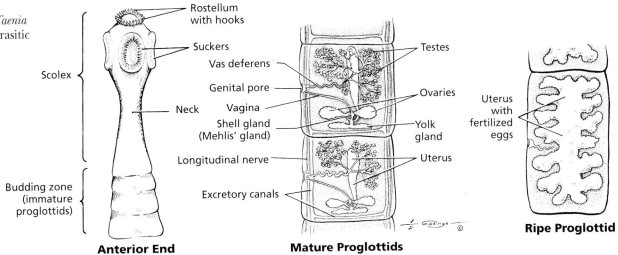

Rostellum with hooks
Suckers
Scolex
Neck
Budding zone (immature proglottids)
Anterior End

Vas deferens
Genital pore
Vagina
Shell gland (Mehlis' gland)
Longitudinal nerve
Excretory canals
Mature Proglottids

Testes
Ovaries
Yolk gland
Uterus

Uterus with fertilized eggs
Ripe Proglottid

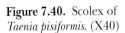

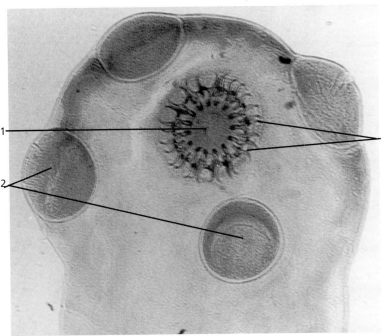

Figure 7.40. Scolex of *Taenia pisiformis.* (X40)

1. Rostellum
2. Suckers
3. Hooks

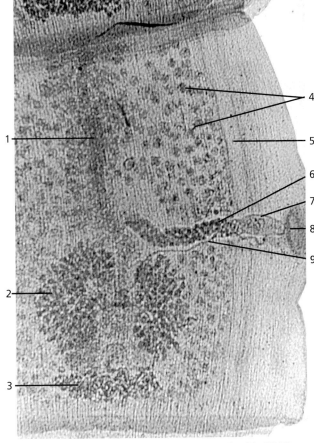

Figure 7.41. *Taenia pisiformis*, immature proglottids. (X40)

1. Early ovary
2. Early testes
3. Excretory canal
4. Immature vagina and ductus deferens

Figure 7.42. *Taenia pisiformis*, mature proglottid. (X40)

1. Uterus   4. Testes   7. Cirrus
2. Ovary   5. Excretory canal   8. Genital pore
3. Yolk gland   6. Ductus deferens   9. Vagina

PHYLUM NEMATODA

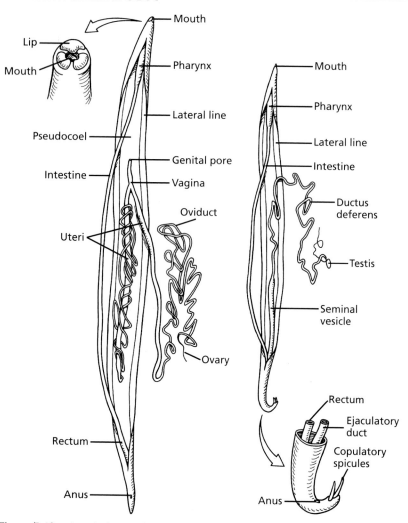

Figure 7.43. *Ascaris,* internal anatomy.

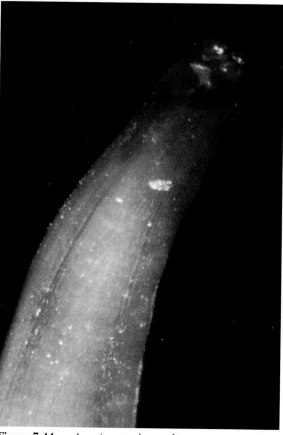

Figure 7.44. *Ascaris,* anterior end.

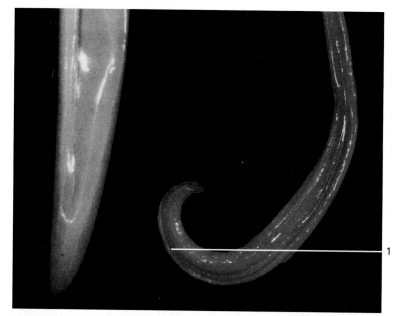

Figure 7.45. *Ascaris,* surface anatomy, posterior region, female (left) and male (right).

1. Ejaculatory duct

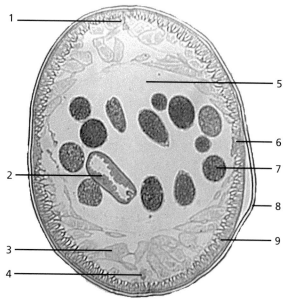

Figure 7.46. *Ascaris,* male, cross section. (X40)

1. Dorsal nerve cord	6. Lateral line
2. Intestine	7. Testis
3. Longitudinal muscle cell body	8. Cuticle
4. Ventral nerve	9. Contractile sheath of muscle cell
5. Pseudocoel	

PHYLUM NEMATODA

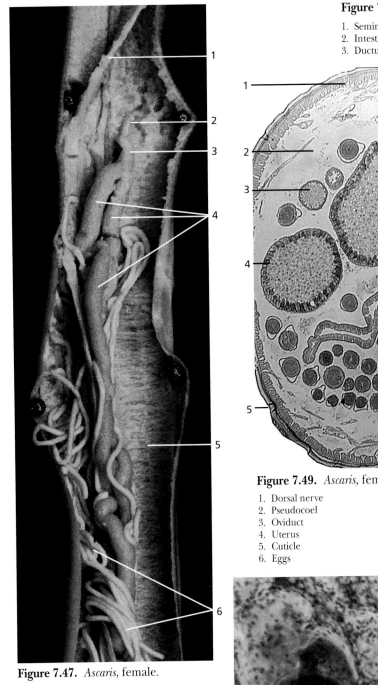

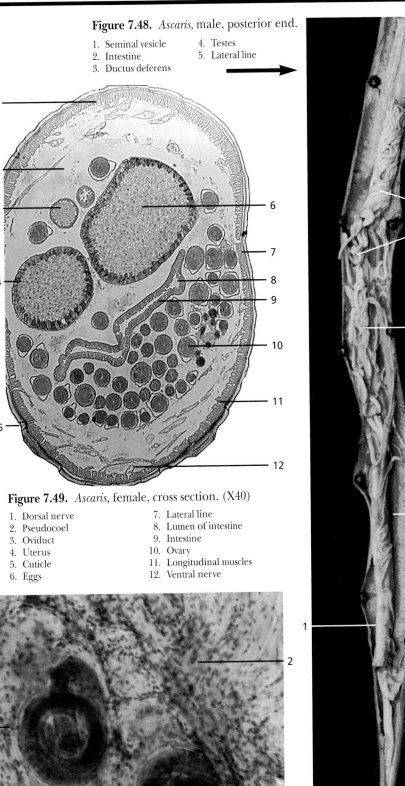

Figure 7.48. *Ascaris*, male, posterior end.

1. Seminal vesicle
2. Intestine
3. Ductus deferens
4. Testes
5. Lateral line

Figure 7.49. *Ascaris*, female, cross section. (X40)

1. Dorsal nerve
2. Pseudocoel
3. Oviduct
4. Uterus
5. Cuticle
6. Eggs
7. Lateral line
8. Lumen of intestine
9. Intestine
10. Ovary
11. Longitudinal muscles
12. Ventral nerve

Figure 7.47. *Ascaris*, female.

1. Intestine
2. Genital pore
3. Vagina
4. Uterus (Y-shaped)
5. Lateral line
6. Oviducts

Figure 7.50. *Trichinella spiralis* encysted in muscle. (X100)

1. Cyst
2. Muscle
3. Larva

PHYLUM ROTIFERA

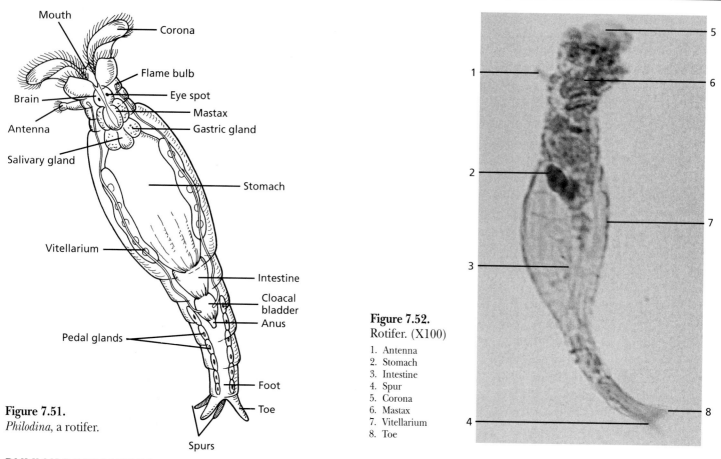

Figure 7.51.
Philodina, a rotifer.

Figure 7.52.
Rotifer. (X100)

1. Antenna
2. Stomach
3. Intestine
4. Spur
5. Corona
6. Mastax
7. Vitellarium
8. Toe

PHYLUM MOLLUSCA

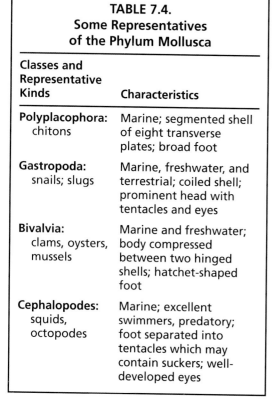

TABLE 7.4. Some Representatives of the Phylum Mollusca	
Classes and Representative Kinds	**Characteristics**
Polyplacophora: chitons	Marine; segmented shell of eight transverse plates; broad foot
Gastropoda: snails; slugs	Marine, freshwater, and terrestrial; coiled shell; prominent head with tentacles and eyes
Bivalvia: clams, oysters, mussels	Marine and freshwater; body compressed between two hinged shells; hatchet-shaped foot
Cephalopodes: squids, octopodes	Marine; excellent swimmers, predatory; foot separated into tentacles which may contain suckers; well-developed eyes

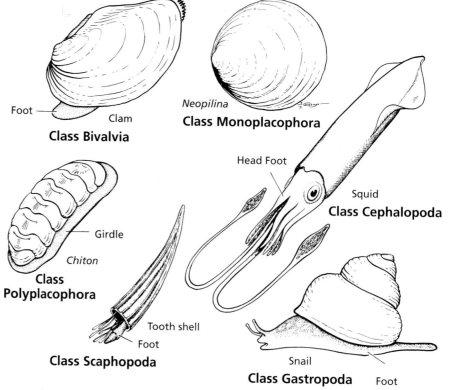

Figure 7.53. Specimens representing several classes of molluscs.

PHYLUM MOLLUSCA ━━━━━━━━━━━━━━━━━ Class Polyplacophora

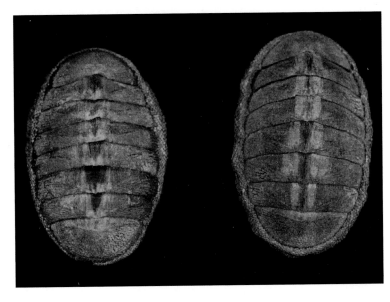

Figure 7.54. Chitons, *Chiton stoke-sii*, are easily recognized by their eight dorsal plates.

PHYLUM MOLLUSCA ━━━━━━━━━━━━━━━━━ Class Gastropoda

Figure 7.55. Slug, class Gastropoda. Slug locomotion requires the production of mucus. Slugs differ from snails in that a shell is absent.

1. Head
2. Antenna (stalked eye)
3. Foot
4. Mucus

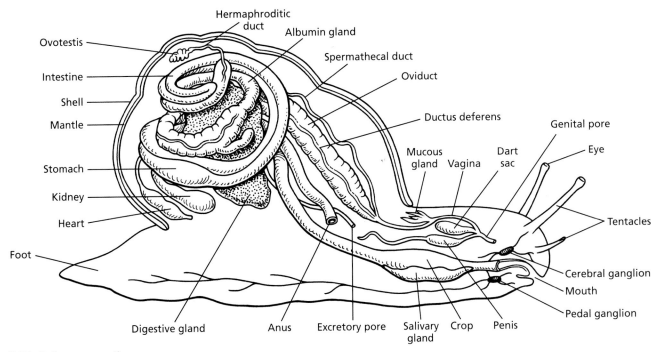

Figure 7.56. Pulmonate snail anatomy.

PHYLUM MOLLUSCA ━━━━━━━━━━━━━━━━━━━━━━━━━━━━━━━━━ Class Bivalvia

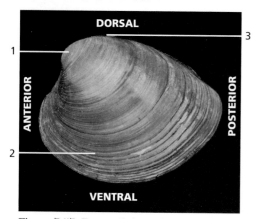

Figure 7.57. External view of clamshell, left valve.

1. Umbo
2. Growth lines
3. Hinge ligament

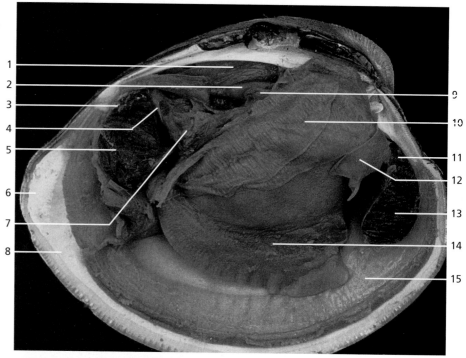

Figure 7.58. Lateral view of a clam.

1. Pericardium
2. Ventricle of heart
3. Anus
4. Posterior retractor muscle
5. Posterior adductor muscle
6. Excurrent siphon
7. Nephridium (kidney)
8. Incurrent siphon
9. Atrium of heart
10. Gills
11. Anterior retractor muscle
12. Labial palps
13. Anterior adductor muscle
14. Foot
15. Mantle

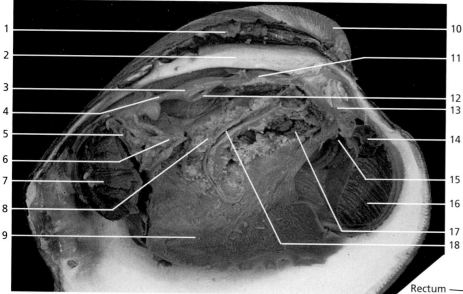

Figure 7.59. Lateral view of a clam, foot cut.

1. Hinge ligament
2. Hinge
3. Ventricle of heart
4. Posterior aorta
5. Posterior retractor muscle
6. Nephridium (kidney)
7. Posterior adductor muscle
8. Gonad
9. Foot
10. Umbo
11. Anterior aorta
12. Opening between atrium and ventricle
13. Esophagus
14. Anterior retractor muscle
15. Mouth
16. Anterior adductor muscle
17. Digestive gland
18. Intestine

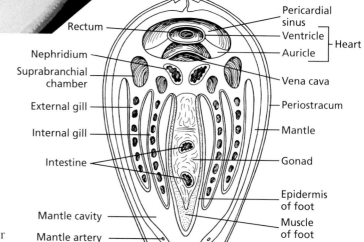

Figure 7.60. Freshwater clam, cross section through heart region.

Rectum

Nephridium

Suprabranchial chamber

External gill

Internal gill

Intestine

Mantle cavity

Mantle artery

Pericardial sinus

Ventricle ⎫
Auricle ⎬ Heart

Vena cava

Periostracum

Mantle

Gonad

Epidermis of foot

Muscle of foot

Shell

PHYLUM MOLLUSCA ━━━━━━━━━━━━━━━━━━━━━━━━━━ **Class Cephalopoda**

Figure 7.61. *Nautilus*, a cephalopod, has gas-filled chambers within its shell, as seen in this dissected specimen. These chambers regulate buoyancy.

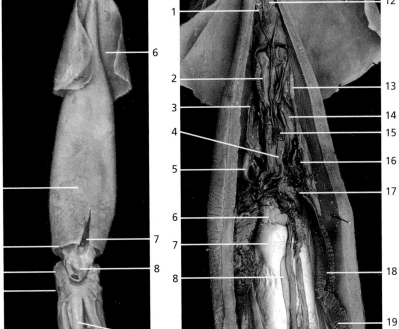

Figure 7.62.
Squid, *Loligo*,
posterior view.
1. Body tube
2. Collar
3. Olfactory crest
4. Eye
5. Tentacles
6. Fin
7. Mantle cavity
8. Siphon
9. Arms

Figure 7.63.
Squid internal
anatomy, female,
posterior view.
1. Lateral mantle artery
2. Ovary
3. Posterior vena cava
4. Nidamental gland
5. Oviductal opening
6. Kidney
7. Liver
8. Ink sac
9. Siphon retractor
muscle
10. Rectum
11. Anus
12. Caecum
13. Oviduct
14. Stomach
15. Median mantle artery
16. Branchial heart
17. Systemic heart
18. Branchial vein
19. Gill
20. Anterior vena cava
21. Mantle

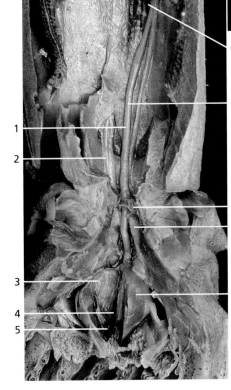

Figure 7.64.
Squid, head
region.
1. Esophagus
2. Pleural nerve
3. Radula
4. Ligula
5. Mandible
6. Quill (pen)
7. Cephalic aorta
8. Visceral ganglion
9. Pedal ganglion
10. Buccal bulb

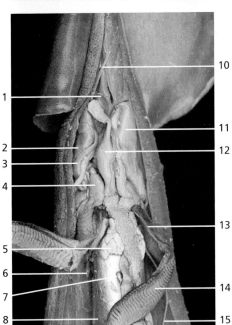
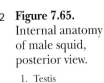

Figure 7.65.
Internal anatomy
of male squid,
posterior view.
1. Testis
2. Spermatophoric gland
3. Vas deferens
4. Branchial heart
5. Kidney
6. Penis
7. Ink sac
8. Liver
9. Rectum
10. Caecum
11. Stomach
12. Posterior vena cava
13. Branchial vein
14. Gill
15. Siphon retractor
muscle
16. Anterior vena cava
17. Anus

PHYLUM ANNELIDA ━━━ **Class Polychaeta**

TABLE 7.5.
Some Representatives of the Phylum Annelida

Classes and Representative Kinds	Characteristics
Polychaeta: tubeworms, sandworms	Mostly marine; segments with parapodia
Oligochaeta: earthworms	Freshwater and burrowing terrestrial forms; small setae; poorly developed head
Hirudinea: leeches	Freshwater; most are blood-sucking parasites; lack setae; prominent muscular suckers

Figure 7.66. Tubeworms (also called sea feathers), *Sabellastarte indico,* shown here feeding.

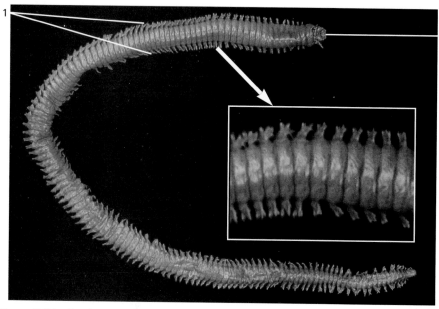

Figure 7.67. Sandworm, *Neanthes,* dorsal view.

1. Parapodia   2. Mouth

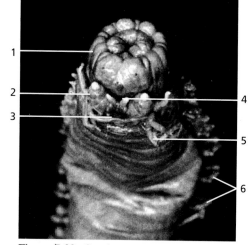

Figure 7.68. Sandworm, *Neanthes,* head region, dorsal view.

1. Everted pharynx   4. Prostomial tentacle
2. Palp   5. Peristomial tentacle
3. Prostomium   6. Parapodia

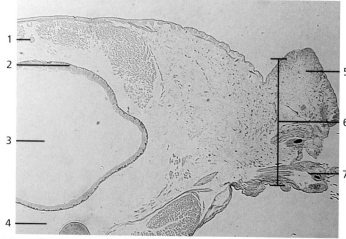

Figure 7.69. *Neanthes,* cross section. (X40)

1. Dorsal blood vessel   4. Coelom   7. Neuropodium
2. Intestine   5. Notopodium
3. Lumen of gut   6. Parapodium

Figure 7.70. *Neanthes,* parapodium. (X100)

1. Dorsal cirrus   2. Notopodium   3. Setae   4. Neuropodium

PHYLUM ANNELIDA ━━━━━━━━━━━━━━ Class Oligochaeta

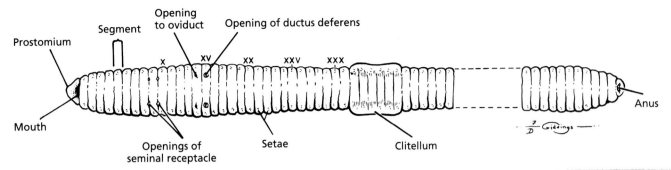

Figure 7.71. Earthworm, *Lumbricus*, surface anatomy.

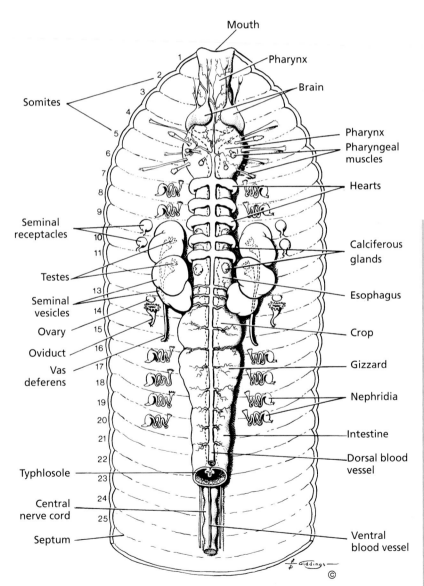

Figure 7.73 Earthworm, *Lumbricus*, anterior end.

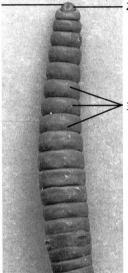

Figure 7.72.
Earthworm,
Lumbricus,
anterior end.

1. Prostomium
2. Mouth
3. Setae

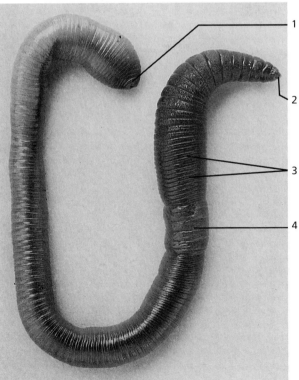

Figure 7.74. Earthworm, *Lumbricus*, surface anatomy.

1. Anus   3. Segments
2. Prostomium   4. Clitellum

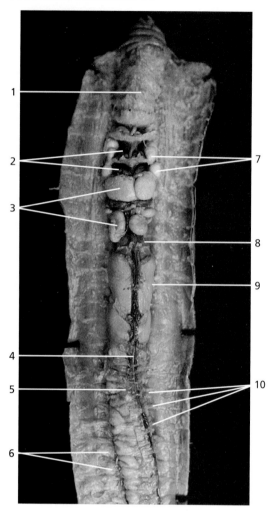

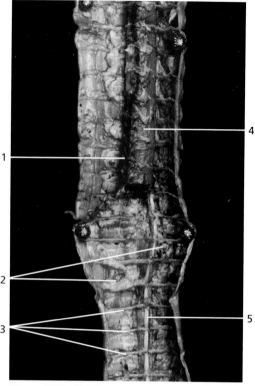

Figure 7.75. Earthworm, *Lumbricus*, anterior end.

1. Pharynx
2. Hearts
3. Seminal vesicles
4. Dorsal blood vessel
5. Intestine
6. Nephridia
7. Seminal receptacles
8. Crop
9. Gizzard
10. Septa

Figure 7.76. Earthworm, *Lumbricus*, posterior region. Part of the intestine has been removed.

1. Dorsal blood vessel
2. Nephridia
3. Septa
4. Intestine
5. Ventral nerve cord

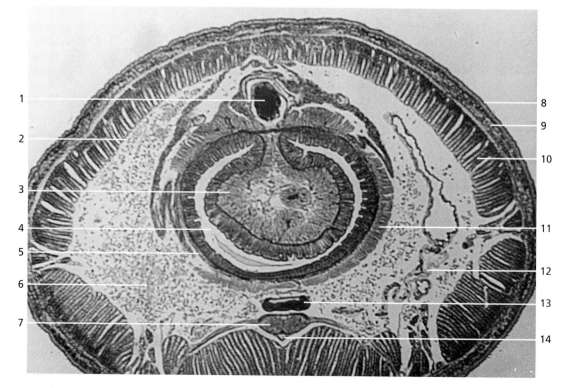

Figure 7.77. Earthworm, *Lumbricus*, cross section, posterior to clitellum.

1. Dorsal blood vessel
2. Peritoneum
3. Typhlosole
4. Lumen of intestine
5. Intestine
6. Coelom
7. Ventral nerve cord
8. Epidermis
9. Circular muscles
10. Longitudinal muscles
11. Chloragogue cells
12. Nephridium
13. Ventral blood vessel
14. Subneural blood vessel

PHYLUM ANNELIDA ══════════════════════════════ Class Oligochaeta

Figure 7.78. Earthworm cocoons (each line represents 1 mm).

PHYLUM ANNELIDA ══════════════════════════════ Class Hirudinea

Figure 7.79. Leeches are more specialized than other annelids. They have lost their setae and developed suckers for attachment while sucking blood.

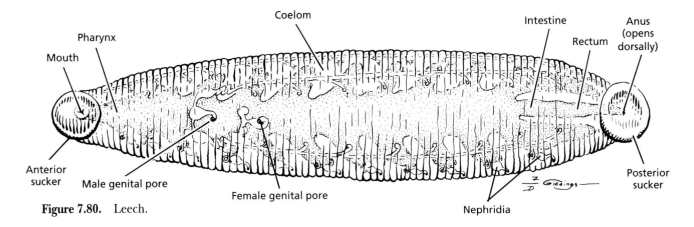

Figure 7.80. Leech.

PHYLUM ARTHROPODA

TABLE 7.6.
Some Representatives of the Phylum Arthropoda

Classes and Representative Kinds	Characteristics
Merostomata: horseshoe crab	Cephalothorax and abdomen; specialized front appendages into chelicerae; lack antennae and mandibles
Arachnida: spiders, mites, ticks, scorpions	Cephalothorax and abdomen; four pairs of legs; book lungs or trachea; lack antennae and mandibles
Malacostraca lobsters, crabs, shrimp	Cephalothorax and abdomen; two pair of antennae, pair of mandibles and maxillae; biramous appendages; gills
Insecta: beetles, butterflies, ants	Head, thorax, and abdomen; three pairs of legs; well-developed mouth parts; usually two pairs of wings; trachea
Chilopoda: centipedes	Head with segmented body; one pair of legs per segment; trachea; one pair of antennae
Diplopoda: millipedes	Head with segmented body; usually two pair of legs per segment; trachea

PHYLUM ARTHROPODA   Class Merostomata

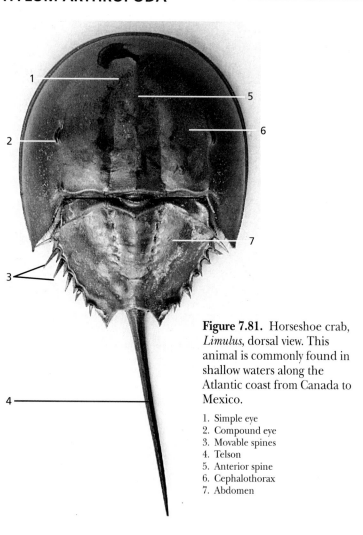

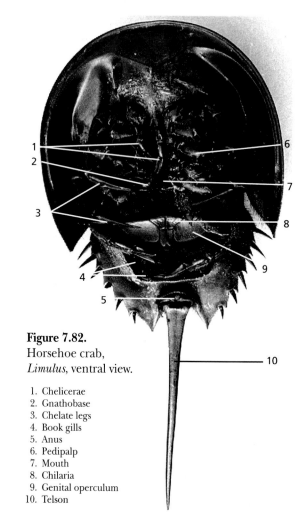

Figure 7.81. Horseshoe crab, *Limulus*, dorsal view. This animal is commonly found in shallow waters along the Atlantic coast from Canada to Mexico.

1. Simple eye
2. Compound eye
3. Movable spines
4. Telson
5. Anterior spine
6. Cephalothorax
7. Abdomen

Figure 7.82. Horsehoe crab, *Limulus*, ventral view.

1. Chelicerae
2. Gnathobase
3. Chelate legs
4. Book gills
5. Anus
6. Pedipalp
7. Mouth
8. Chilaria
9. Genital operculum
10. Telson

PHYLUM ARTHROPODA ━━━━━━━━━━━━━━━━━━━━ Class Arachnida

Figure 7.83. Black widow spider, *Latrodectus mactans*, along with the brown recluse are the two spiders in the United States which can give severe or fatal bites.

1. Tibia	4. Tarsus	7. Prosoma
2. Femur	5. Patella	8. Trochanter
3. Metatarsus	6. Opisthosoma	

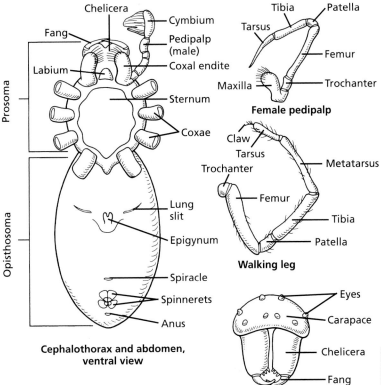

Cephalothorax and abdomen, ventral view

Female pedipalp

Walking leg

Head, anterior view

Figure 7.85. Anatomy of the garden spider, *Argiope*.

Figure 7.84. Tarantula, *Dugesiella*, in a defensive posture. The pautrons and the fangs comprise the chelicera.

1. Pautrons	3. Pedipalp	5. Prosoma
2. Opisthosoma	4. Fangs	

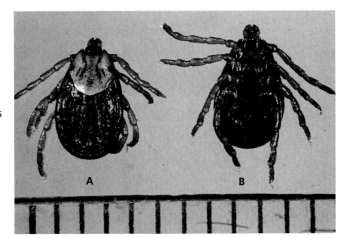

Figure 7.86. Ticks, family Ixodidae, are specialized parasitic arthropods; (A) dorsal view, (B) ventral view.

Figure 7.87. Scorpion, *Pandinus*. Scorpions are most commonly found in tropical and subtropical regions, but they are also found in temperate zones.

1. Stinging apparatus	4. Cephalothorax
2. Postabdomen (tail)	5. Preabdomen
3. Pedipalp	6. Walking legs

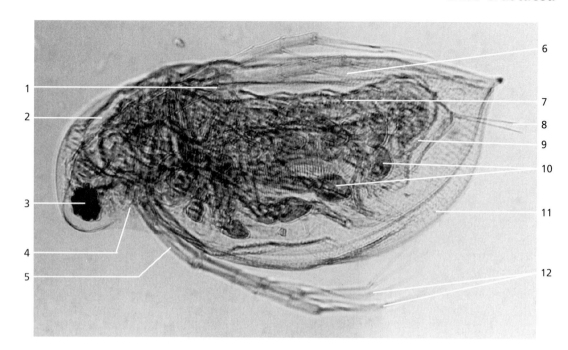

Figure 7.88. Water flea, *Daphnia*, a common microscopic crustacean.

1. Heart
2. Midgut
3. Compound eye
4. Mouth
5. 2nd Antenna
6. Brood chamber
7. Hindgut
8. Abdominal seta
9. Anus
10. Thoracic appendages
11. Carapace
12. Setae

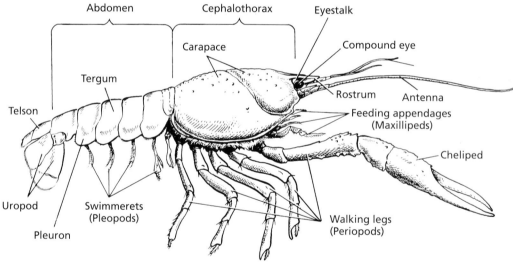

Figure 7.89. Crayfish, *Cambarus*, lateral view.

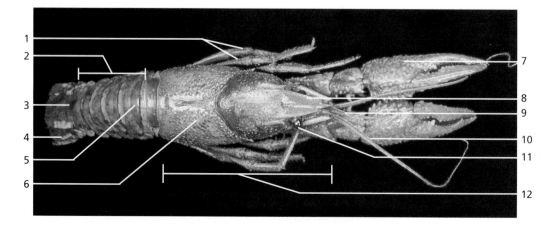

Figure 7.90. Crayfish, *Cambarus*, dorsal view.

1. Walking legs
2. Abdomen
3. Telson
4. Uropod
5. Tergum
6. Carapace
7. Cheliped
8. Rostrum
9. Antennule
10. Antenna
11. Eye
12. Cephalothorax

PHYLUM ARTHROPODA ━━━━━━━━━━━━━━━━━━━━━━━━━━━━━━━━ **Class Crustacea**

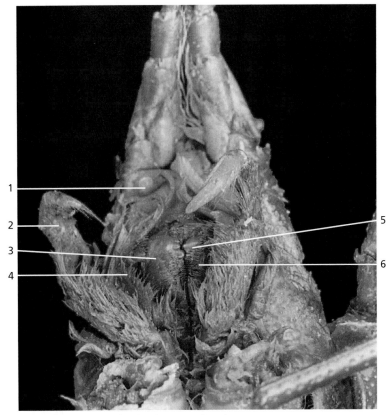

Figure 7.91. Crayfish, *Cambarus*, oral region.

1. Green gland duct
2. Third maxilliped
3. First maxilliped
4. Second maxilliped
5. Mandible
6. Maxilla

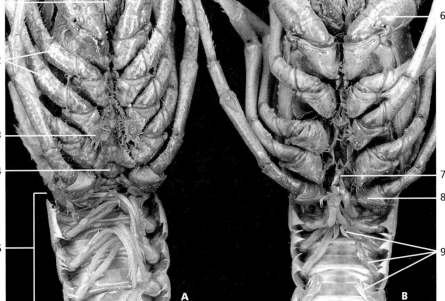

Figure 7.92. Crayfish, *Cambarus*, ventral view of a female (A) and a male (B). The first pair of swimmerets are greatly enlarged in the male for the depositing of sperm in the female's seminal receptacle.

1. Third maxilliped
2. Walking legs
3. Disc covering oviduct
4. Seminal receptacle
5. Abdomen
6. Chiliped
7. First pair of swimmerets (pleopods)
8. Sperm ducts (genital pores)
9. Swimmerets (pleopods)

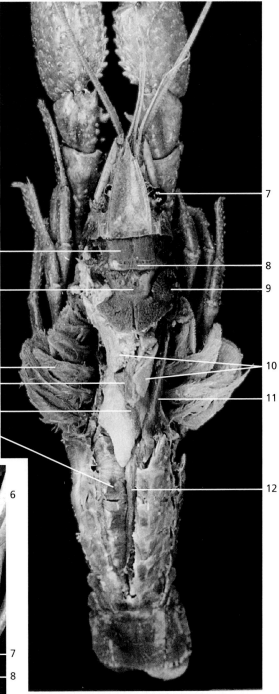

Figure 7.93. Crayfish, dorsal view of a female, internal organs.

1. Gastric muscle
2. Stomach
3. Gills
4. Heart
5. Ovary with eggs
6. Abdominal flexor muscles
7. Compound eye
8. Gastrolith
9. Mandibular muscle
10. Digestive gland
11. Abdominal extensor muscle
12. Intestine

Class Crustacea

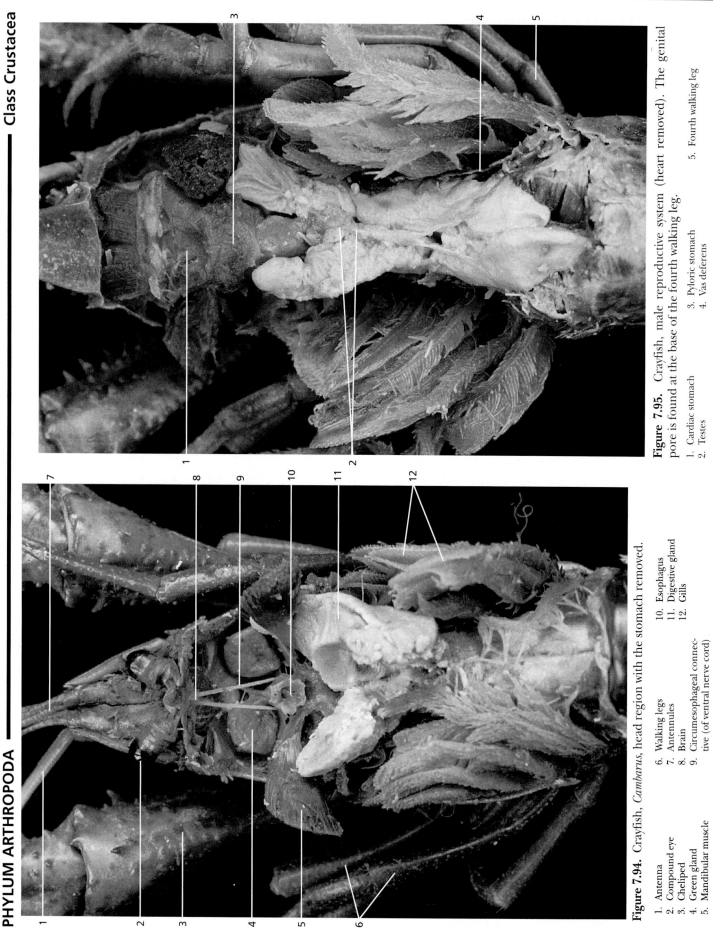

PHYLUM ARTHROPODA

Figure 7.95. Crayfish, male reproductive system (heart removed). The genital pore is found at the base of the fourth walking leg.

1. Cardiac stomach
2. Testes
3. Pyloric stomach
4. Vas deferens
5. Fourth walking leg

Figure 7.94. Crayfish, *Cambarus*, head region with the stomach removed.

1. Antenna
2. Compound eye
3. Cheliped
4. Green gland
5. Mandibular muscle
6. Walking legs
7. Antennules
8. Brain
9. Circumesophageal connective (of ventral nerve cord)
10. Esophagus
11. Digestive gland
12. Gills

PHYLUM ARTHROPODA ━━━━━━━━━━━━━━━━━━━━━━━━━━━━━━━━━━━━━━ **Class Insecta**

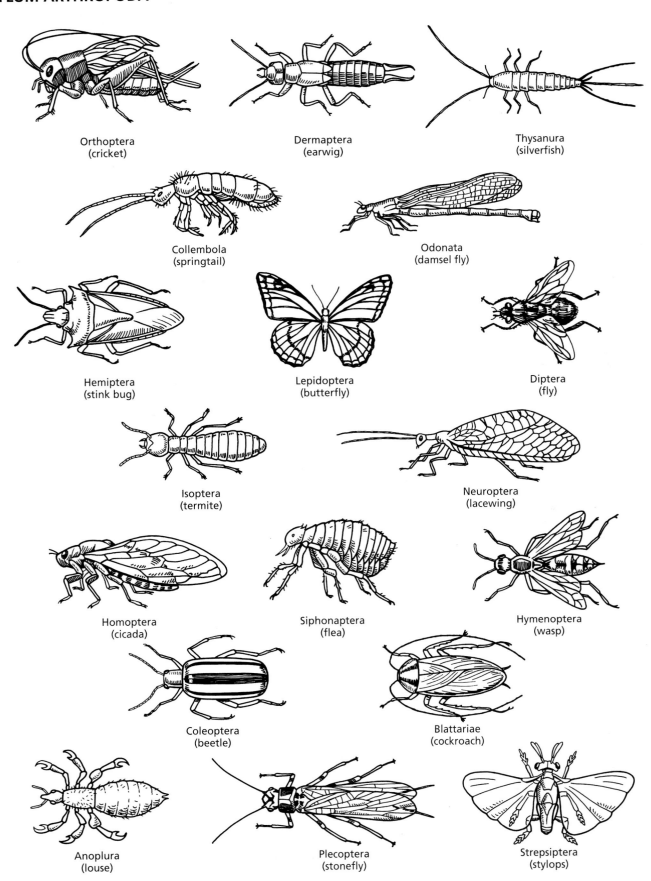

Figure 7.96. Representatives from some of the orders of insects.

PHYLUM ARTHROPODA ━━━━━━━━━━━━━━━━━━━━━━━━━━━━━━━━━ **Class Insecta**

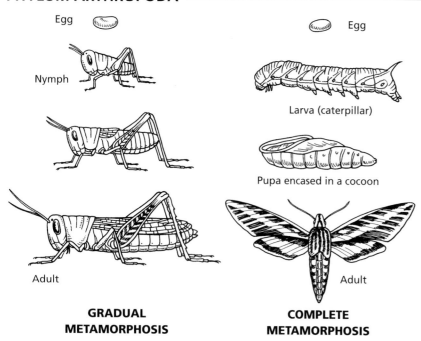

Egg

Nymph

Adult

**GRADUAL
METAMORPHOSIS**

Egg

Larva (caterpillar)

Pupa encased in a cocoon

Adult

**COMPLETE
METAMORPHOSIS**

Figure 7.97. Insect development. In gradual (incomplete) metamorphosis the young resemble the adults but they are smaller and have different body proportions. In complete metamorphosis, the larvae look different than the adult and generally have different food requirements.

Figure 7.98. The common grey cricket molting. All arthropods must periodically shed their exoskeleton in order to grow. This process is called molting, or ecdysis.

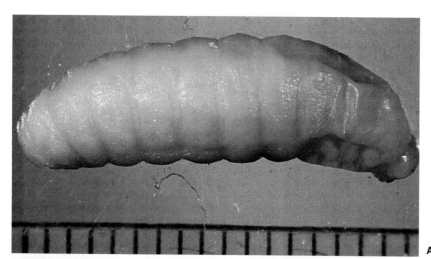

A

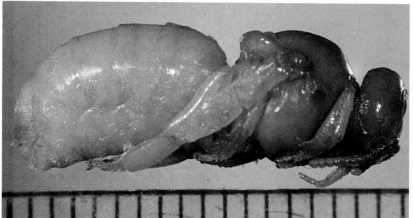

B

C

Figure 7.99. Developmental stages of the common honeybee: (A) larval stage, (B) pupa, and (C) adult.

PHYLUM ARTHROPODA ━━━━━━━━━━━━━━━━━━━━ Class Insecta

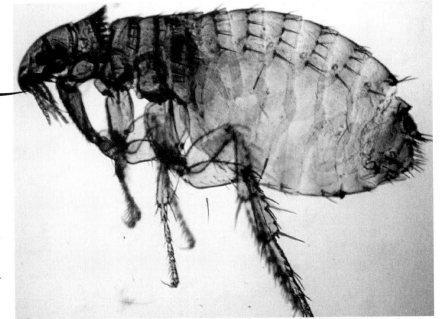

Figure 7.100. The head of a butterfly, lateral view. (A) the compound eyes and the curled tongue, which is used to siphon nectar from flowers, are the most obvious structures on the head of a butterfly. (B) a magnified view of the compound eye of an insect. It is believed that because of compound eyes insects can see in all directions (panoramically), although images are fuzzy.

1. Compound eye   2. Tongue

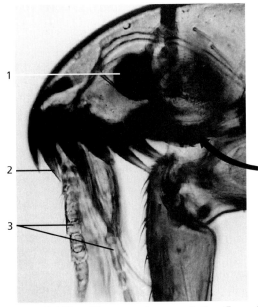

Figure 7.101. The mouthparts of the flea, *Ctenocephalides*, are specialized for parasitism. Notice the oral bristles beneath the mouth which aid the flea in penetrating between hairs to feed on the blood of mammals.

1. Eye   2. Oral bristle   3. Maxillary palps

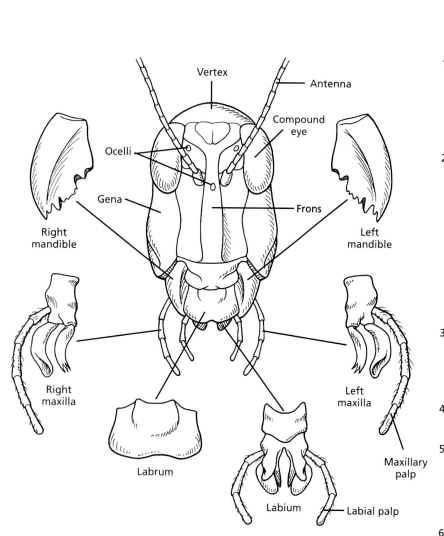

Figure 7.102. A preserved specimen of a grasshopper, order Orthoptera, surface anatomy.

1. Mesothorax
2. Prothorax
3. Vertex
4. Compound eye
5. Gena
6. Frons
7. Clypeus
8. Labrum
9. Antenna
10. Claw
11. Coxa
12. Trochanter
13. Abdomen
14. Metathorax
15. Tympanic membrane
16. Wing
17. Femur
18. Tibia
19. Spiracle
20. Tarsus

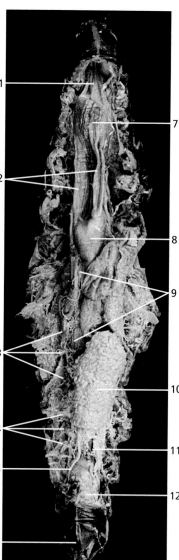

Figure 7.104. Grasshopper, female, internal anatomy.

1. Esophagus
2. Gastric caeca
3. Malpighian tubules
4. Trachea
5. Oviduct
6. Ovipositor
7. Crop
8. Gizzard
9. Stomach
10. Ovary
11. Large intestine
12. Rectum

Figure 7.103. Grasshopper, head and mouthparts.

PHYLUM ECHINODERMATA

TABLE 7.7 Some Representatives of the Phylum Echinodermata	
Classes and Representative Kinds	**Characteristics**
Asteroidea: sea stars (star fish)	Pentaradial symmetry; appendages arranged around a central disk containing the mouth;
Echinoidea: sea urchins, sand dollars	Disk-shaped with no arms; compact skeleton; movable spines; tube feet with suckers
Ophiuroidea: brittle stars	Pentaradial symmetry; appendages sharply marked off from central disk; tube feet without suckers
Holothuroidea: sea cucumbers	Cucumber-shaped with no arms; spines absent; tube feet with tentacles and suckers
Crinoidea: sea lilies and feather stars	Sessile during much of life cycle; calyx supported by elongated stalk

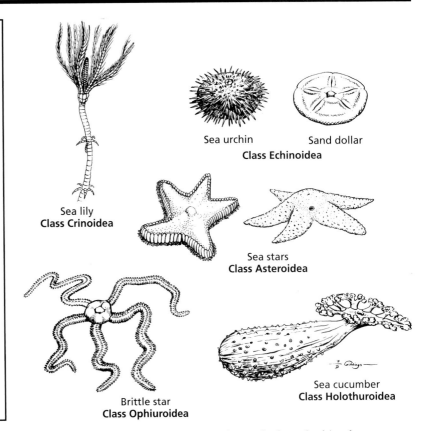

Figure 7.105. Specimens representing each class of echinoderms.

PHYLUM ECHINODERMATA ━━━━━━━━━━━━━━ Class Asteroidea

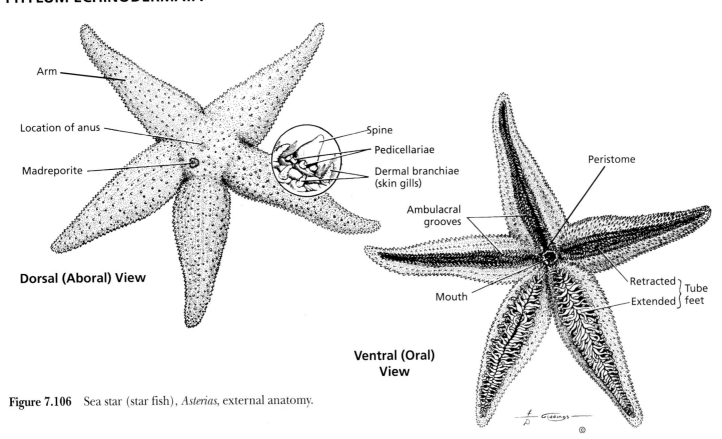

Figure 7.106 Sea star (star fish), *Asterias*, external anatomy.

PHYLUM ECHINODERMATA ━━━━━━━━━━━━━━━━━━━━━━━━━━ **Class Asteroidea**

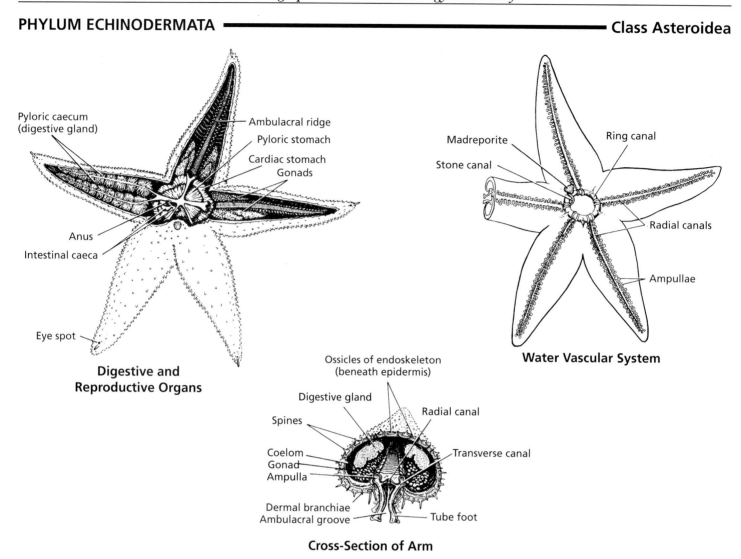

Digestive and Reproductive Organs

Pyloric caecum (digestive gland)
Ambulacral ridge
Pyloric stomach
Cardiac stomach
Gonads
Anus
Intestinal caeca
Eye spot

Water Vascular System

Madreporite
Stone canal
Ring canal
Radial canals
Ampullae

Cross-Section of Arm

Ossicles of endoskeleton (beneath epidermis)
Digestive gland
Radial canal
Spines
Coelom
Gonad
Ampulla
Transverse canal
Dermal branchiae
Ambulacral groove
Tube foot

Figure 7.107. Sea star, *Asterias*, internal anatomy.

Figure 7.108. Sea star, internal anatomy, dorsal (aboral) view.

1. Spines
2. Pyloric duct
3. Ambulacral ridge
4. Ampullae
5. Gonad
6. Pedicellariae
7. Madreporite
8. Pyloric stomach
9. Pyloric caecum (digestive gland)

PHYLUM ECHINODERMATA ━━━━━━━━━━━━━━━━━━━━━━━━━━ Class Asteroidea

Figure 7.109.
Sea star, dorsal
(aboral) view, stomach
removed.

1. Madreporite
2. Polian vesicle
3. Ampulla
4. Ambulacral ridge
5. Gonad
6. Stone canal
7. Circular canal
8. Oral spines
9. Pyloric caecum

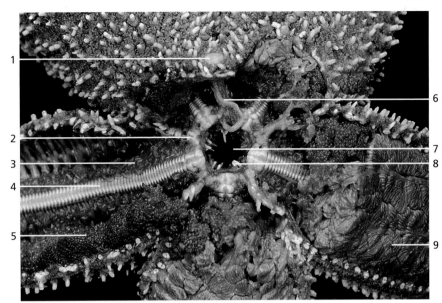

Figure 7.110 Sea star,
ventral (oral) view.

1. Tube feet
2. Pedicellariae
3. Peristome
4. Oral spines
5. Mouth
6. Ambulacral groove

Figure 7.111. Sea star arm,
cross section.

1. Coelom
2. Tube foot
3. Sucker
4. Pyloric caecum
5. Radial canal
6. Gonad
7. Ambulacral groove
8. Pedicellaria

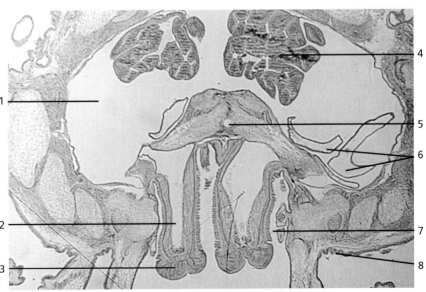

PHYLUM ECHINODERMATA ━━━━━━━━━━━━━━━━━━━━━━━━━━━━━━━━ **Class Echinoidea**

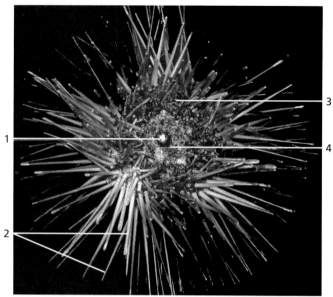

Figure 7.112. Sea urchin, *Arbacia*, ventral (oral) view.

1. Mouth   2. Spines   3. Pedicellaria   4. Peristome

Figure 7.113. Sea urchin, dorsal (aboral) view.

1. Ossicles   2. Madreporite

Figure 7.114. Sea urchin, *Arbacia*, internal anatomy.

1. Madreporite
2. Gonad
3. Intestine
4. Aristotle's lantern
5. Mouth
6. Anus
7. Esophagus
8. Calcareous tooth
9. Stomach

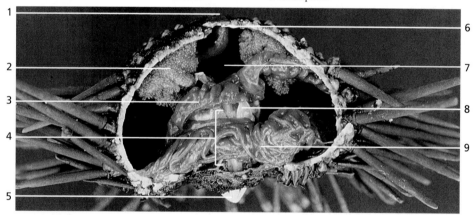

PHYLUM ECHINODERMATA ━━━━━━━━━━━━━━━━━━━━━━━━━━━━━━━━ **Class Holothuroidea**

Figure 7.115.
Sea cucumber, *Cucumaria*.

1. Tentacles
2. Tube feet

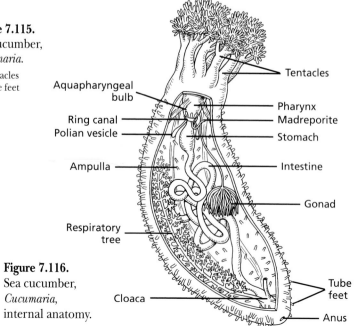

Tentacles

Aquapharyngeal bulb

Pharynx

Ring canal

Madreporite

Polian vesicle

Stomach

Ampulla

Intestine

Respiratory tree

Gonad

Figure 7.116.
Sea cucumber, *Cucumaria,* internal anatomy.

Cloaca

Tube feet

Anus

Class Holothuroidea

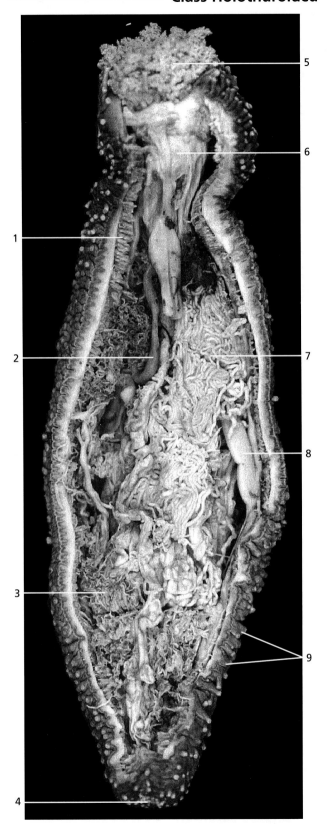

Figure 7.117. Dissection of a sea cucumber, *Cucumaria.*

1. Ampullae
2. Intestine
3. Respiratory tree
4. Anus
5. Tentacles
6. Ring canal
7. Gonad
8. Longitudinal muscle
9. Tube feet

PHYLUM CHORDATA

TABLE 7.8.
Some Representatives of the Phylum Chordate

Subphyla and Representative Kinds	Characteristics
Urochordata: tunicates	Marine, larvae are free-swimming and have notochord, gill slits, and dorsal hollow nerve cord; adults are sessile (attached), filter-feeders, saclike animals
Cephalochordata: lancelets (*Amphioxus*)	Marine, segmented, elongated body with notochord extending the length of the body; cilia surrounding the mouth for obtaining food
Vertebrata: agnathans (lampreys and hagfishes), fishes (cartilaginous and bony), amphibians, reptiles, birds, mammals	Aquatic and terrestrial forms; distinct head and trunk supported by a series of cartilaginous or bony vertebrae in the adult; closed circulatory system and ventral heart; well-developed brain and sensory organs

TABLE 7.9.
Some Representatives of the Subphylum Vertebrata

Class and Representative Kinds	Characteristics
Agnatha: hagfish, lamprey	Eel-like and aquatic; sucking mouth (some parasitic); lack paired appendages
Chondrichthyes: sharks, rays, skates	Cartilaginous skeleton; placoid scales; spiracle; spiral valve in digestive tract
Osteichthyes: bony fishes	Gills covered by bony operculum; most have swim bladder
Amphibia: salamanders, frogs, toads	Larvae have gills and adults have lungs; scaleless skin; incomplete double circulation
Reptilia: turtles, snakes, lizards	Amniotic egg; epidermal scales; three- or four-chambered heart; lungs
Aves: birds	Homeothermous (warm-blooded); feathers; toothless; air sacs; four-chambered heart with right aortic arch
Mammalia: mammals	Homeothermous; hair; mammary glands; seven cervical vertebrae; muscular diaphragm; three auditory ossicles; four-chambered heart with left aortic arch

PHYLUM CHORDATA ━━━━━━━━━━━━━━━━━━━━━━━━━━━━━ **Subphylum Cephalochordata**

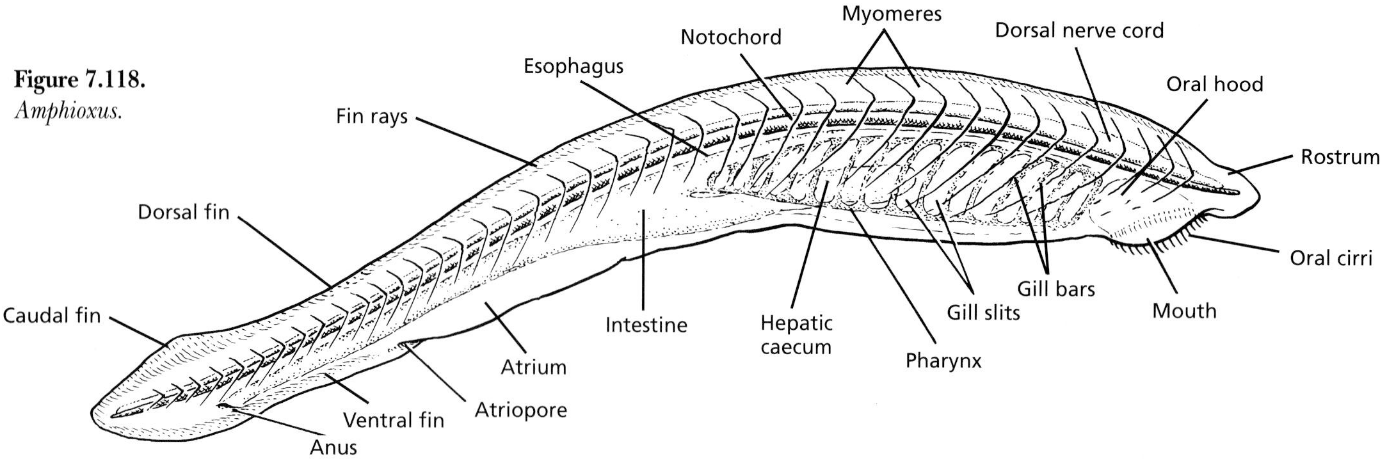

Figure 7.118.
Amphioxus.

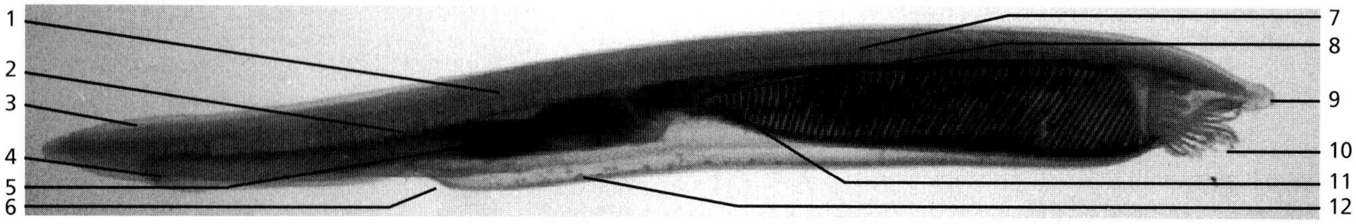

Figure 7.119. *Amphioxus*, whole mount. (X40)

1. Dorsal nerve cord
2. Notochord
3. Caudal fin
4. Anus
5. Intestine
6. Atriopore
7. Myomeres
8. Esophagus
9. Rostrum
10. Oral cirri
11. Hepatic caecum
12. Atrium

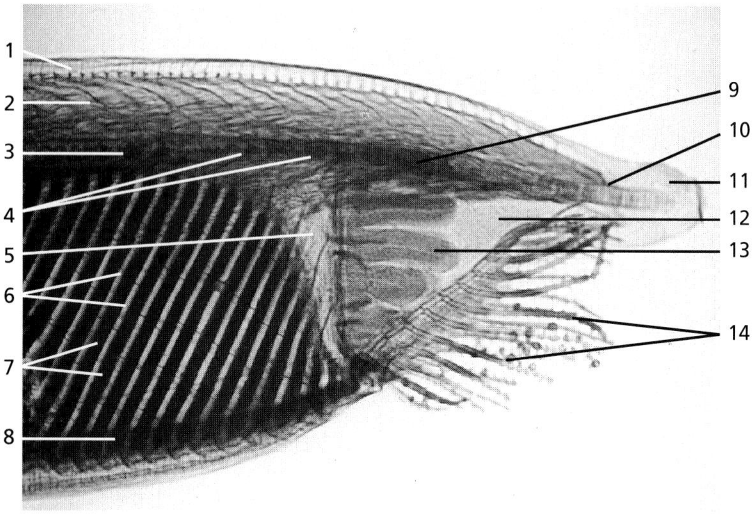

Figure 7.120
Amphioxus,
anterior
region. (X100)

1. Fin rays
2. Myomere
3. Dorsal nerve cord
4. Pigment spots
5. Velum
6. Gill slits
7. Gill bars
8. Pharynx
9. Notochord
10. Eye spot
11. Rostrum
12. Oral hood
13. Wheel organ
14. Oral cirri

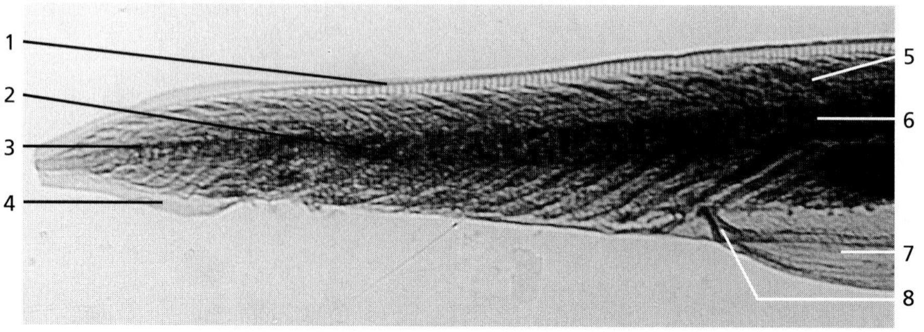

Figure 7.121. *Amphioxus*, posterior region. (X100)

1. Fin rays
2. Intestine
3. Notochord
4. Caudal fin
5. Myomeres
6. Midgut
7. Atrium
8. Atriopore

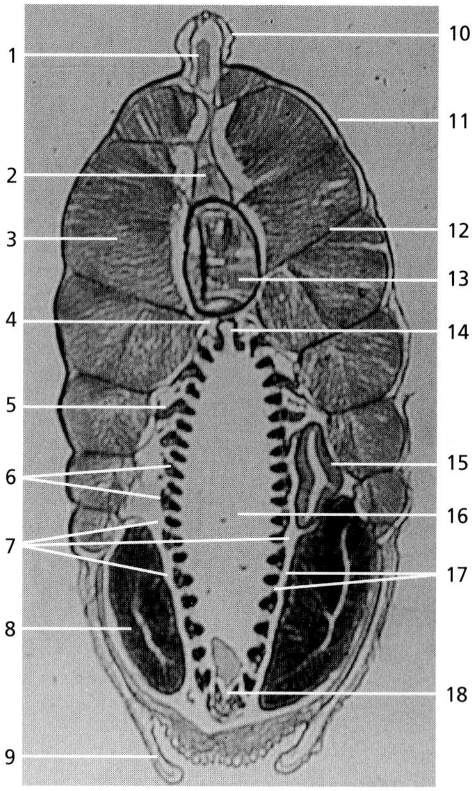

Figure 7.122. *Amphioxus*, male, cross section through pharynx. (X40)

1. Fin ray
2. Dorsal nerve cord
3. Myomere
4. Dorsal aorta
5. Nephridium
6. Gill bars
7. Atrium
8. Testis
9. Metapleural fold
10. Dorsal fin
11. Epidermis
12. Myoseptum
13. Notochord
14. Epibranchial groove
15. Hepatic caecum (liver)
16. Pharynx
17. Gill slits
18. Endostyle (hypobranchial groove)

PHYLUM CHORDATA ━━━━ **Subphylum Vertebrata** ━━━━ **Class Agnatha**

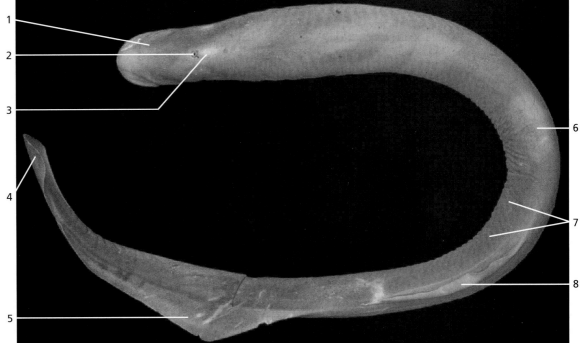

Figure 7.123.
Marine lamprey,
Petromyzon marinus,
external anatomy.

1. Head
2. Nostril
3. Pineal body
4. Caudal fin
5. Posterior dorsal fin
6. Trunk
7. Myomeres
8. Anterior dorsal fin

Figure 7.124.
Lamprey, head region.

1. Eye
2. Buccal funnel
3. External gill slits

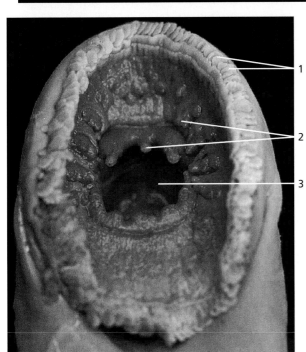

Figure 7.125.
Lamprey, oral region.

1. Buccal papillae
2. Horny teeth
3. Mouth

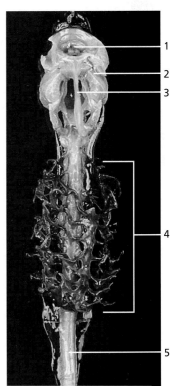

Figure 7.126.
A cartilaginous skeleton
of the marine lamprey,
Petromyzon marinus, shown
in ventral view.

1. Buccal cavity
2. Skull
3. Tongue support
4. Branchial basket
5. Notochord

PHYLUM CHORDATA ━━━━━ Subphylum Vertebrata ━━━━━ Class Agnatha

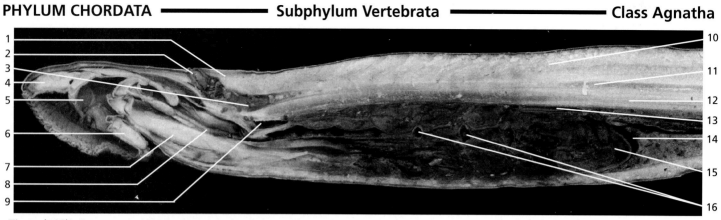

Figure 7.127. Lamprey, midsagittal section through anterior region.

1. Pineal organ	4. Annular cartilage	7. Lingual cartilage	10. Myomeres	13. Dorsal aorta	16. Internal gill slits
2. Nostril	5. Mouth	8. Pharynx	11. Spinal cord	14. Ventricle	
3. Brain	6. Annular cartilage	9. Esophagus	12. Notochord	15. Sinus venosus	

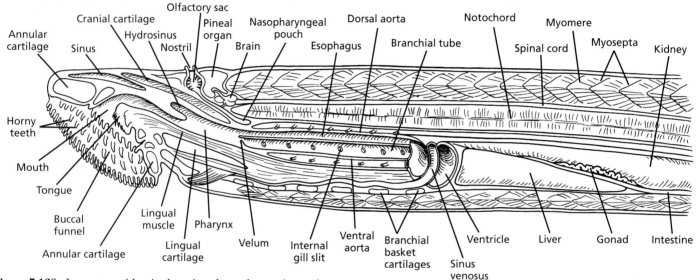

Figure 7.128. Lamprey, midsagittal section through anterior region.

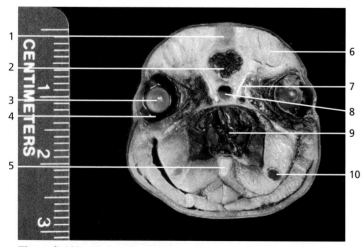

Figure 7.129. Lamprey, cross section through the eyes.

1. Pineal organ	5. Lingual cartilage	9. Pharynx
2. Brain	6. Myomere	10. Pharyngeal gland
3. Lens of eye	7. Cranial cartilage	
4. Retina of eye	8. Nasopharyngeal pouch	

Figure 7.130. Lamprey, cross section through the branchial tube anterior to the fourth gill pouch. The ventral aorta is paired at this location.

1. Spinal cord	4. Gill pouch	7. Ventral aorta
2. Notochord	5. Anterior cardinal vein	8. Lingual muscle
3. Branchial tube	6. Esophagus	

PHYLUM CHORDATA ━━━━━━━ **Subphylum Vertebrata** ━━━━━━━ **Class Amphibia**

Figure 7.131. Brown salamander, *Ambystoma gracile*. This amphibian lives in humid sites, often beneath debris along stream banks.

Figure 7.132. An adult Indonesian giant tree frog, *Litoria infrafrenata*, (A) crouched on a person's fingers; (B) suction cups on its toes aid in climbing.

PHYLUM CHORDATA ━━━━━━━ **Subphylum Vertebrata** ━━━━━━━ **Class Reptilia**

Figure 7.134. Spurred tortoise, *Geochelone sulcata*.

Figure 7.133. Hatching king snakes. Most snakes are oviparous, meaning they lay eggs such as these. Some snakes, including all American pit vipers, are ovoviviparous, giving birth to well-developed young.

Figure 7.135. Jackson's chameleon, *Chamaeleo jacksoni*. Chameleons are best known for their ability to change colors according to their surroundings.

PHYLUM CHORDATA ━━━━━━ **Subphylum Vertebrata** ━━━━━━ **Class Aves**

Coraciiformes
(kingfisher)

Apodiformes
(hummingbird)

Strigiformes
(owl)

Columbiformes
(pigeon)

Falconiformes
(falcon)

Anseriformes
(duck)

Gaviiformes
(loon)

Charadriiformes
(gull)

Pelecaniformes
(pelican)

Ciconiiformes
(heron)

Podicipediformes
(grebe)

Struthioniformes
(ostrich)

Galliformes
(pheasant)

Piciformes
(toucan)

Passeriformes
(thrush)

Sphenisciformes
(penguin)

Phoenicopteriformes
(flamingo)

Psittaciformes
(parrot)

Figure 7.136. Representatives from some of the orders of birds.

PHYLUM CHORDATA ——————— Subphylum Vertebrata ——————————— Class Aves

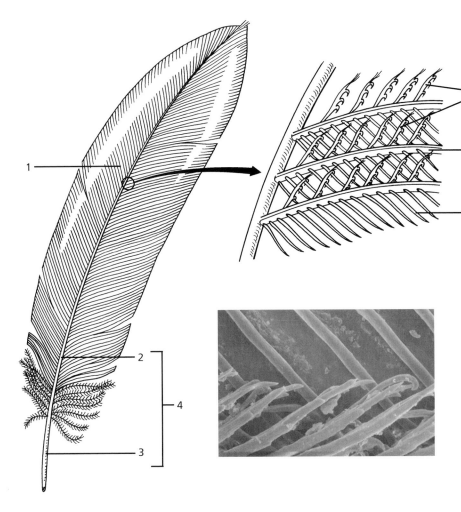

Figure 7.137. Structure of a contour (pluma) feather. The barbules and hooklets are shown in a photomicrograph.

1. Vane
2. Rachis
3. Calamus
4. Shaft
5. Hooklets
6. Barb
7. Barbule

Figure 7.138. Museum prepared bird skins are important in avian taxonomy.

PHYLUM CHORDATA ━━━━━ **Subphylum Vertebrata** ━━━━━ **Class Mammalia**

Figure 7.139. Representatives from the orders of mammals.

PHYLUM CHORDATA ━━━━━ **Subphylum Vertebrata** ━━━━━ **Class Mammalia**

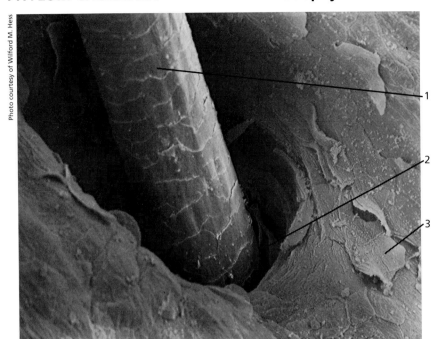

Figure 7.140. An electron micrograph of a hair emerging from a hair follicle.

1. Shaft of hair (note the scale-like pattern)
2. Hair follicle
3. Epithelial cell from stratum cornium

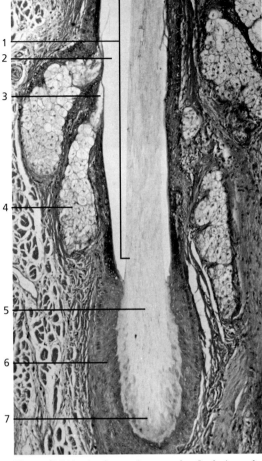

Figure 7.141. A photomicrograph of a hair and a sebaceous gland.

1. Shaft of hair
2. Ductule of sebaceous gland
3. Hair follicle
4. Sebaceous gland
5. Root of hair
6. Germinal cells of hair
7. Bulb of hair

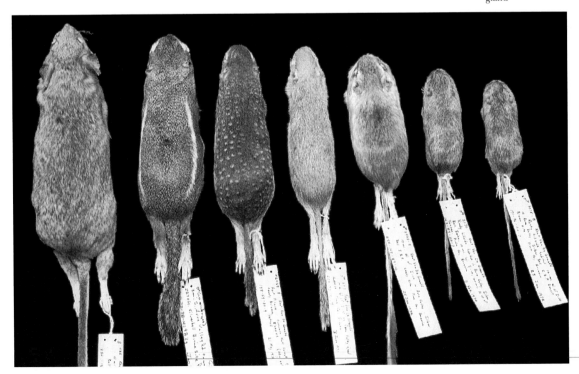

Figure 7.142. Museum prepared mammal skins are important in mammalian taxonomy.

Human Biology

The study of human biology requires learning about the anatomy and physiology of the human body. *Human anatomy* is the scientific discipline that investigates the structure of the body, and *human physiology* is the scientific discipline that investigates how body structures function. The purpose of this chapter is to present a visual overview of the principal anatomical structures of the human body.

Since both the *skeletal system* and the *muscular system* are both concerned with body movement, they are frequently discussed together as the *skeletomuscular system*. In a functional sense, the internal framework, or *bones* of the skeleton, support and provide movement at the *joints* where the muscles attached to the bones produce their actions as they are stimulated to contract.

The *nervous system* is anatomically divided into the *central nervous system* (CNS), which includes the *brain* and *spinal cord*, and the *peripheral nervous system* (PNS), which includes the *cranial nerves*, arising from the brain, and the *spinal nerves*, arising from the spinal cord. The *autonomic nervous system* (ANS) is a functional division of the nervous system devoted to regulation of involuntary activities of the body. The brain and spinal cord are the centers for integration and coordination of information. *Nerves*, composed of *neurons*, convey nerve impulse to and from the brain or spinal cord. *Sensory organs*, such as the eyes and ears, respond to impulses in the environment and convey sensations to the CNS. The nervous system functions with the *endocrine system* in coordinating body activities.

The *cardiovascular system* consists of the *heart*, *vessels* (both blood and lymphatic vessels), *blood*, and the tissues that produce the blood. The four-chambered human heart is enclosed in a *pericardial sac* within the thoracic cavity. *Arteries* and *arterioles* transport blood away from the heart, *capillaries* permeate the tissues and are the functional units for product exchange with the cells, and *venules* and *veins* transport blood toward the heart. *Lymphatic vessels* return interstitial fluid back to the circulatory system after first passing it through *lymph nodes* for cleansing. Blood cells are produced in the bone marrow. Blood cells are broken-down in the liver after they are old and worn.

The *respiratory system* consists of the *conducting division* that transports air to and from the *respiratory division* within the *lungs*. The *alveoli* of the lungs are in contact the capillaries of the cardiovascular system and are the sites for exchange of respiratory gases.

The *digestive system* consists of a *gastrointestinal tract* (GI tract) and *accessory digestive organs*. Food traveling through the GI tract is processed such that it is suitable for absorption through the intestinal wall into the blood. The *liver* and *pancreas* are the principal digestive organs that process nutrients for body utilization.

Because of commonality of prenatal development and dual functions of some of the organs, the *urinary system* and *reproductive system* may be considered together as the *urogenital system*. The urinary system, consisting of the *kidneys*, *ureters*, *urinary bladder*, and *urethra*, extracts and processes metabolic wastes from the blood in the form of urine. The male and female reproductive systems produce regulatory hormones and gametes (sperm and ova, respectively) within the *gonads* (*testes* and *ovaries*). Sexual reproduction is the mechanism for producing offspring that have traits from both parents. The process of prenatal development is made possible by the formation of *extraembryonic membranes* (placenta, umbilical cord, allantois, amnion, and yolk sac) inside the *uterus* of the mother.

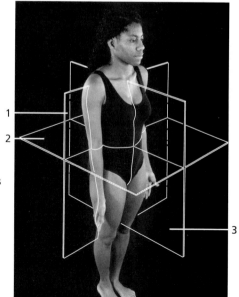

Figure 8.1. Planes of reference in a human.

1. Coronal plane (frontal plane)
2. Transverse plane (cross sectional plane)
3. Midsagittal plane (median plane)

Figure 8.2. Major body regions in humans (bipedal vertebrate).

1. Upper extremity
2. Lower extremity
3. Head
4. Neck, anterior aspect
5. Thorax (chest)
6. Abdomen
7. Cubital fossa
8. Pubic region
9. Palmar region (palm)
10. Patellar region (patella)
11. Cervical region
12. Shoulder
13. Axilla (armpit)
14. Brachium (upper arm)
15. Lumbar region
16. Elbow
17. Antebrachium (forearm)
18. Gluteal region (buttock)
19. Dorsum of hand
20. Thigh
21. Popliteal fossa
22. Calf
23. Plantar surface (sole)

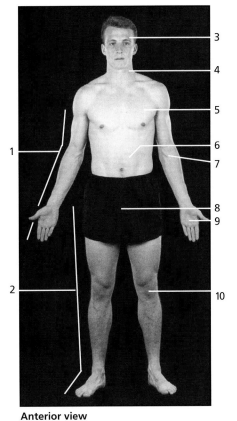

Anterior view   Posterior view

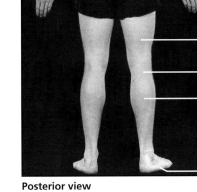

Figure 8.3. The skin and associated structures.

1. Epidermis
2. Dermis
3. Hypodermis
4. Shaft of hair
5. Stratum corneum
6. Stratum basale
7. Sweat duct
8. Sensory receptor
9. Sweat duct
10. Sebaceous gland
11. Arrector pili muscle
12. Hair follicle
13. Apocrine sweat gland
14. Eccrine sweat gland
15. Bulb of hair
16. Adipose tissue
17. Cutaneous blood vessels

Figure 8.4. Bone tissue. (X55)

1. Interstitial lamellae
2. Canaliculi
3. Osteocytes in lacunae
4. Lamellae of osteon
5. Osteons (Haversian systems)
6. Central canal (Haversian canal)

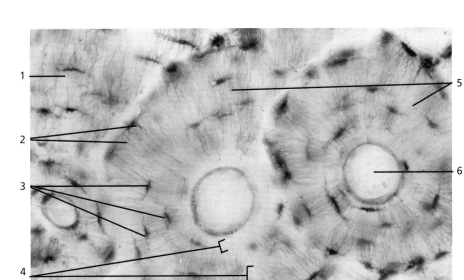

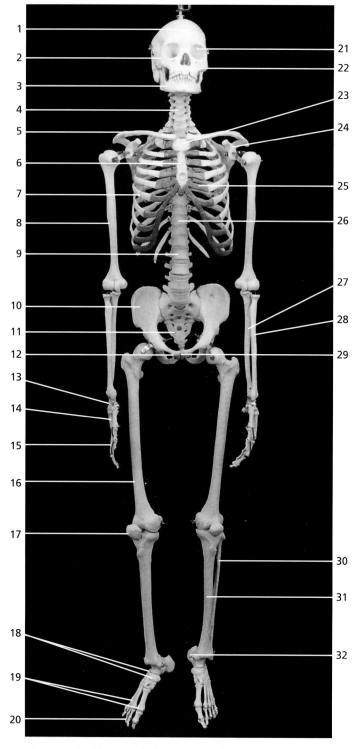

Figure 8.5. Human skeleton, anterior view.

1. Frontal bone	13. Carpal bones	23. Manubrium
2. Zygomatic bone	14. Metacarpal	24. Scapula
3. Mandible	bones	25. Costal cartilage
4. Cervical vertebra	15. Phalanges	26. Thoracic vertebra
5. Clavicle	16. Femur	27. Ulna
6. Body of sternum	17. Patella	28. Radius
7. Rib	18. Tarsal bones	29. Symphysis pubis
8. Humerus	19. Metatarsal	30. Fibula
9. Lumbar vertebra	bones	31. Tibia
10. Ilium	20. Phalanges	32. Calcaneus
11. Sacrum	21. Orbit	
12. Pubis	22. Maxilla	

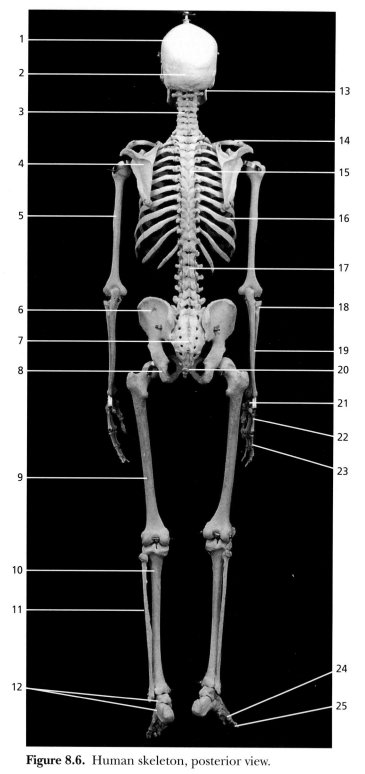

Figure 8.6. Human skeleton, posterior view.

1. Parietal bone	10. Tibia	19. Ulna
2. Occipital bone	11. Fibula	20. Coccyx
3. Cervical vertebra	12. Tarsal bones	21. Carpal bones
4. Scapula	13. Mandible	22. Metacarpal
5. Humerus	14. Clavicle	bones
6. Ilium	15. Thoracic vertebra	23. Phalanges
7. Sacrum	16. Rib	24. Metatarsal bones
8. Ischium	17. Lumbar vertebra	25. Phalanges
9. Femur	18. Radius	

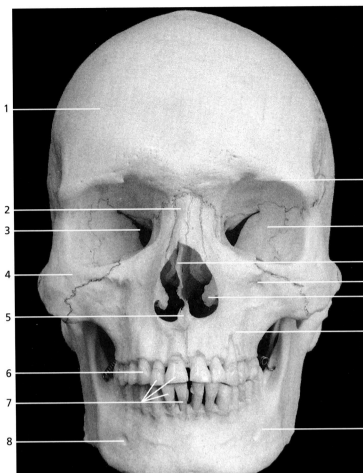

Figure 8.7. Human skull, anterior view.

1. Frontal bone
2. Nasal bone
3. Superior orbital fissure
4. Zygomatic bone
5. Vomer
6. Canine tooth
7. Incisor teeth
8. Mental foramen
9. Supraorbital margin
10. Sphenoid bone
11. Perpendicular plate of ethmoid bone
12. Infraorbital foramen
13. Inferior nasal concha
14. Maxilla
15. Mandible

Figure 8.8. Human skull, lateral view.

1. Coronal suture
2. Frontal bone
3. Lacrimal bone
4. Nasal bone
5. Zygomatic bone
6. Maxilla
7. Premolar teeth
8. Molar teeth
9. Mandible
10. Parietal bone
11. Squamosal suture
12. Temporal bone
13. Lambdoidal suture
14. Occipital bone
15. External acoustic meatus
16. Mastoid process of temporal bone
17. Condyloid process of mandible
18. Mandibular notch
19. Coronoid process of mandible
20. Angle of mandible

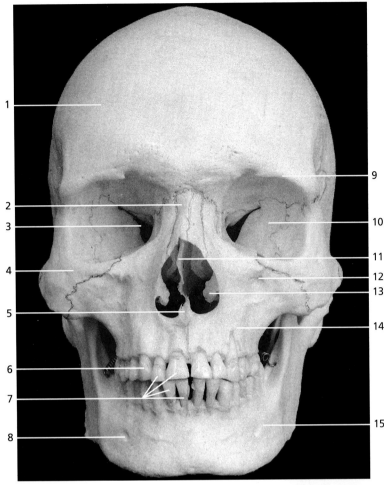

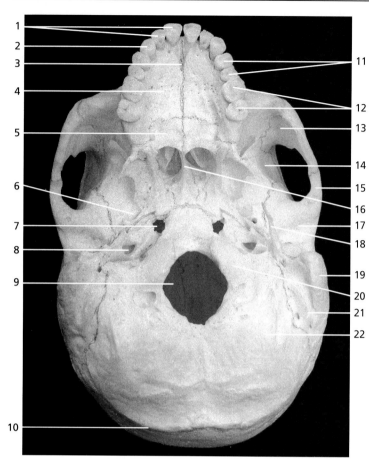

Figure 8.9. Human skull, inferior view.

1. Incisor teeth
2. Canine tooth
3. Incisive foramen
4. Maxilla
5. Palatine bone
6. Foramen ovale
7. Foramen lacerum
8. Carotid canal
9. Foramen magnum
10. Superior nuchal line
11. Premolar teeth
12. Molar teeth
13. Zygomatic bone
14. Sphenoid bone
15. Zygomatic arch
16. Vomer
17. Mandibular fossa
18. Styloid process of temporal bone
19. Mastoid process of temporal bone
20. Occipital condyle
21. Temporal bone
22. Occipital bone

Figure 8.10. Human skull, sagittal view.

1. Frontal bone
2. Frontal sinus
3. Crista galli of ethmoid bone
4. Cribriform plate of ethmoid bone
5. Nasal bone
6. Nasal concha
7. Maxilla
8. Mandible
9. Parietal bone
10. Occipital bone
11. Internal acoustic meatus
12. Sella turcica
13. Hypoglossal canal
14. Sphenoidal sinus
15. Styloid process of temporal bone
16. Vomer

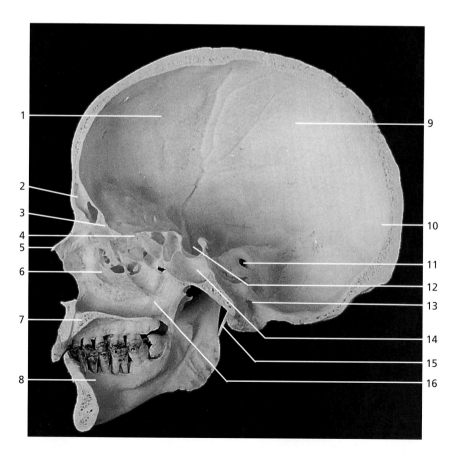

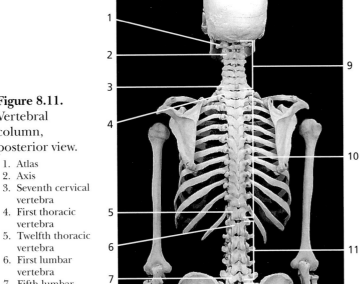

Figure 8.11.
Vertebral column, posterior view.

1. Atlas
2. Axis
3. Seventh cervical vertebra
4. First thoracic vertebra
5. Twelfth thoracic vertebra
6. First lumbar vertebra
7. Fifth lumbar vertebra
8. Sacroiliac joint
9. Cervical vertebrae
10. Thoracic vertebrae
11. Lumbar vertebrae
12. Sacrum
13. Coccyx

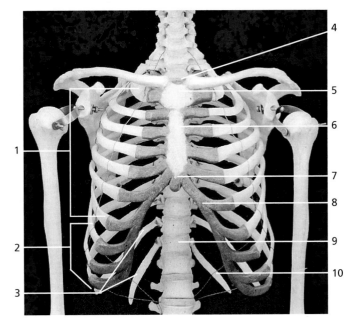

Figure 8.12. Rib cage, anterior view.

1. True ribs (seven pairs)
2. False ribs (five pairs)
3. Floating ribs (inferior two pairs of false ribs)
4. Jugular notch
5. Manubrium
6. Body of Sternum
7. Xiphoid process
8. Costal cartilage
9. Twelfth thoracic vertebra
10. Twelfth rib

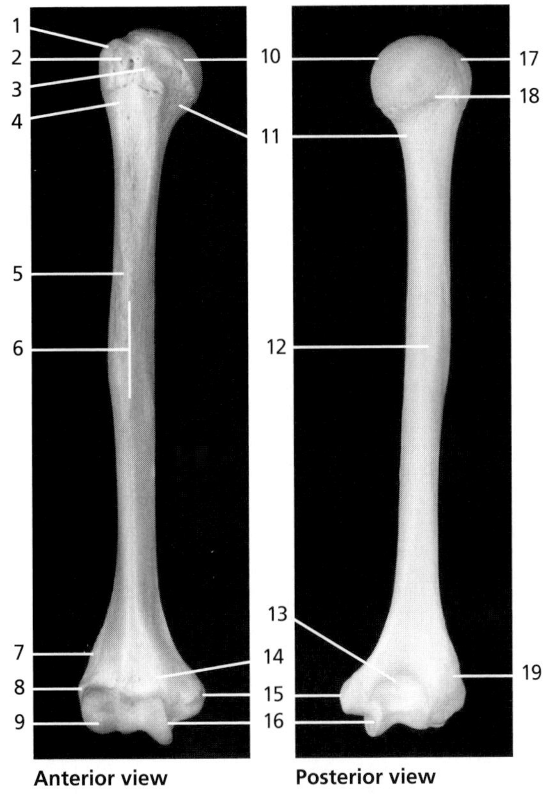

Figure 8.13.
The right humerus.

1. Greater tubercle
2. Intertubercular groove
3. Lesser tubercle
4. Nutrient foramen
5. Deltoid tuberosity
6. Anterior surface of humerus
7. Lateral supracondylar ridge
8. Lateral epicondyle
9. Capitulum
10. Head of humerus
11. Surgical neck
12. Posterior surface of humerus
13. Olecranon fossa
14. Coronoid fossa
15. Medial epicondyle
16. Trochlea
17. Greater tubercle
18. Anatomical neck
19. Lateral epicondyle

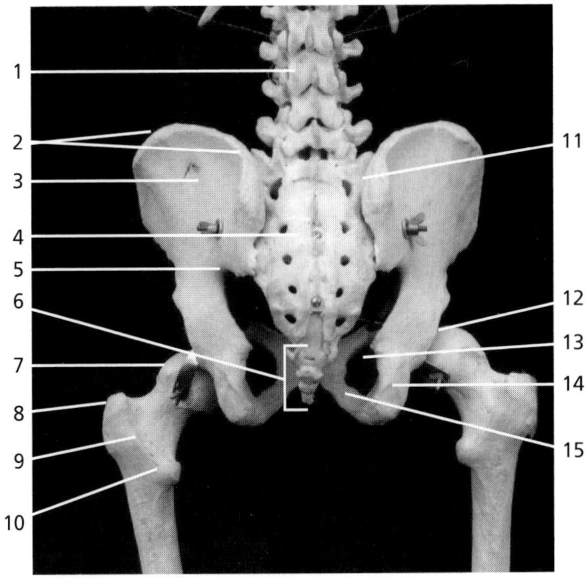

Anterior view   **Posterior view**

Figure 8.14. Articulated pelvic girdle showing the two coxal bones, the sacrum, and the two femora, posterior view.

1. Lumbar vertebra
2. Crest of the ilium
3. Ilium
4. Sacrum
5. Great sciatic notch
6. Coccyx
7. Head of femur
8. Greater trochanter
9. Intertrochanteric crest
10. Lesser trochanter
11. Sacroiliac joint
12. Acetabulum
13. Obturator foramen
14. Ischium
15. Pubis

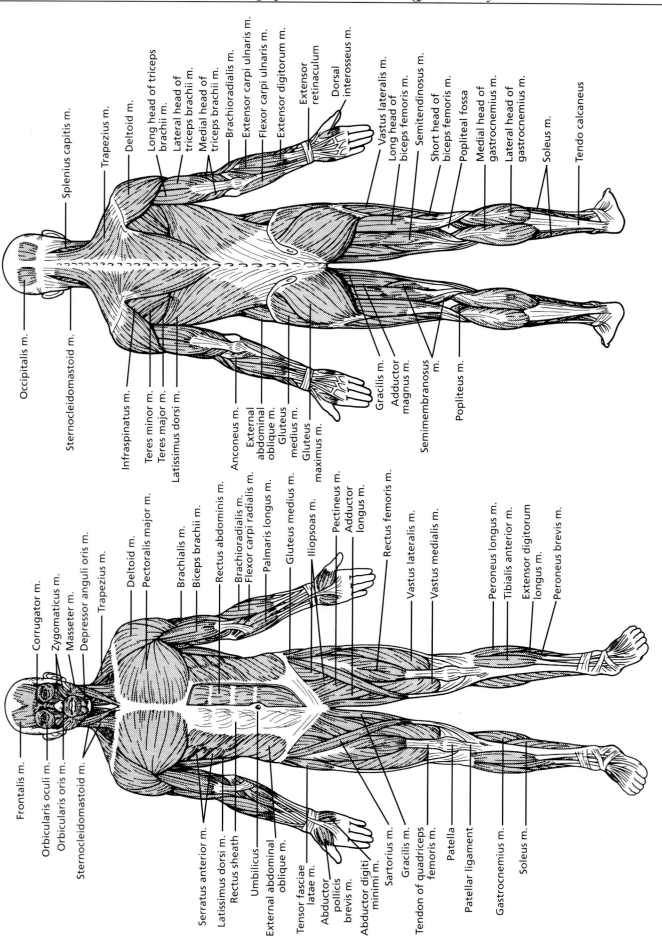

Figure 8.16. Human musculature, posterior view.

Figure 8.15. Human musculature, anterior view. (m. = muscle)

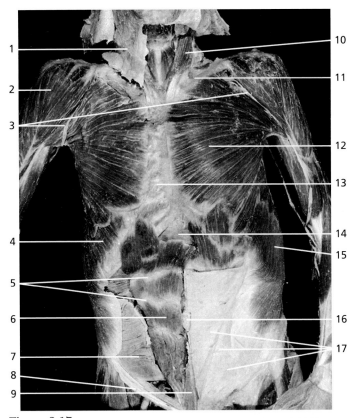

Figure 8.17.

Trunk, anterior view.

1. Platysma m.
2. Deltoid m.
3. Cephalic vein
4. External abdominal oblique m. (aponeurosis removed)
5. Tendinous inscriptions of rectus abdominis m.
6. Rectus abdominis m.
7. Internal abdominal oblique m.
8. Inguinal ligament
9. Pyramidalis m.
10. Sternocleidomastoid m.
11. Clavicle
12. Pectoralis major m.
13. Sternum
14. Xiphoid process
15. External abdominal oblique m.
16. Umbilicus
17. Aponeurosis of external abdominal oblique m.

Figure 8.18.

Trunk, posterolateral view.

1. Trapezius m.
2. Triangle of auscultation
3. Latissimus dorsi m.
4. Vertebral column (spinous processes)
5. Infraspinatus m.
6. Deltoid m.
7. Teres minor m.
8. Teres major m.
9. Serratus anterior m.
10. Rib
11. External abdominal oblique m.
12. Iliac crest
13. Gluteus medius m.
14. Gluteus maximus m.

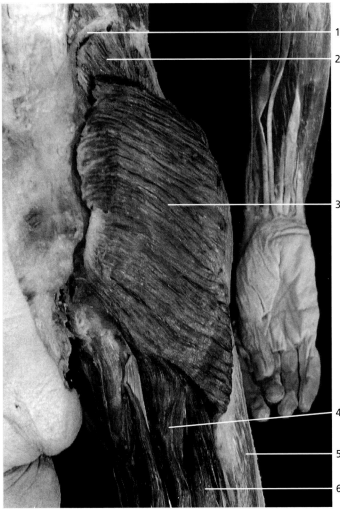

Figure 8.19

Superficial muscles of the gluteal region.

1. Iliac crest
2. Gluteus medius m.
3. Gluteus maximus m.
4. Semitendinosus m.
5. Vastus lateralis m.
6. Biceps femoris m. – long head

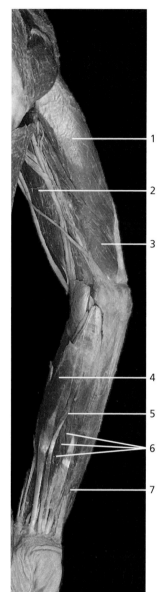

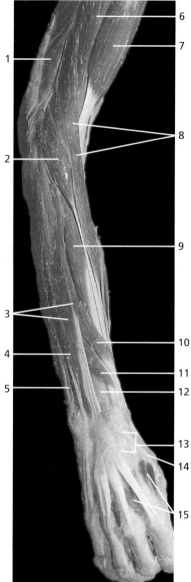

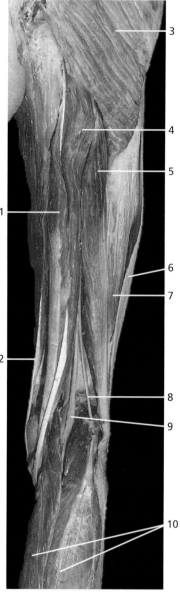

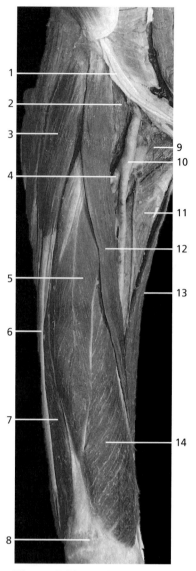

Figure 8.20. Medial brachium and superficial flexors of the right antebrachium.

1. Triceps brachii m. – long head
2. Biceps brachii m. – short head
3. Triceps brachii m. – medial head
4. Flexor carpi radialis m.
5. Palmaris longus m.
6. Superficial digital flexor m.
7. Flexor carpi ulnaris m.

Figure 8.21. Superficial muscles of the right forearm, posterior view.

1. Triceps brachii m. – medial head
2. Extensor carpi radialis longus m.
3. Extensor digitorum m.
4. Extensor digiti minimi m.
5. Extensor carpi ulnaris m.
6. Brachialis m.
7. Biceps brachii m. – long head
8. Brachioradialis m.
9. Extensor carpi radialis brevis m.
10. Abductor pollicis longus m.
11. Extensor pollicis brevis m.
12. Radius
13. Extensor retinaculum
14. Tendon of extensor pollicis longus m.
15. Dorsal interosseous m.

Figure 8.22. Right thigh, posterior view.

1. Semimembranosus m.
2. Gracilis m.
3. Gluteus maximus m.
4. Semitendinosus m.
5. Biceps femoris m. – long head
6. Iliotibial tract
7. Vastus lateralis m.
8. Common peroneal nerve
9. Tibial nerve
10. Gastrocnemius m. – lateral and medial heads

Figure 8.23. Right thigh, anterior view.

1. Inguinal ligament
2. Iliopsoas m.
3. Tensor fasciae latae m.
4. Deep femoral artery
5. Rectus femoris m.
6. Iliotibial tract
7. Vastus lateralis m.
8. Patella
9. Pectineus m.
10. Femoral artery
11. Adductor longus m.
12. Sartorius m.
13. Gracilis m.
14. Vastus medialis m.

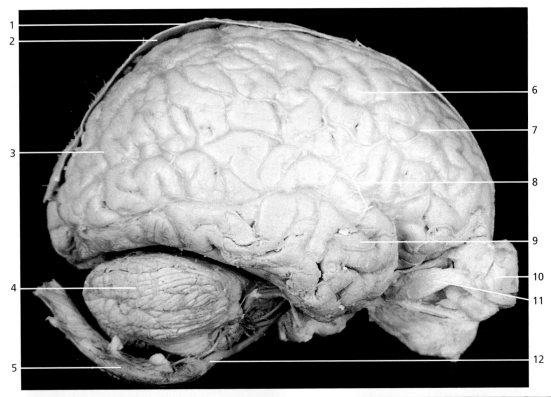

Figure 8.24. Lateral view of brain and upper spinal cord, with meninges partially removed. The eyes are attached to the brain via the optic nerves.

1. Dura mater
2. Subdural space
3. Occipital lobe of cerebrum
4. Cerebellum
5. Spinal cord
6. Gyrus
7. Sulcus
8. Arachnoid mater covering blood vessels
9. Temporal lobe of cerebrum
10. Superior palpebra muscle
11. Lateral rectus muscle
12. Vertebral artery

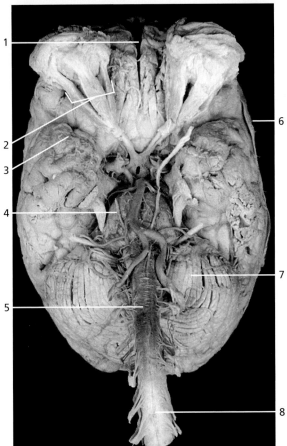

Figure 8.25. Inferior view of the brain.

1. Longitudinal cerebral fissure
2. Muscles of the eye
3. Temporal lobe
4. Pons
5. Medulla oblongata
6. Dura mater
7. Cerebellum
8. Spinal cord

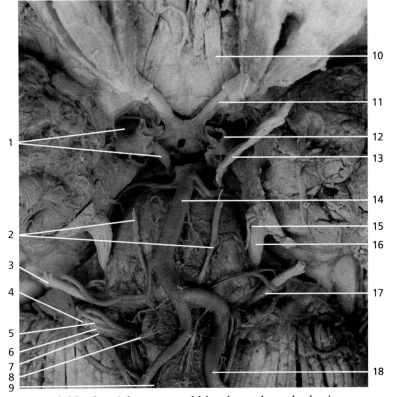

Figure 8.26. Cranial nerves and blood supply to the brain.

1. Cerebral arterial circle (circle of Willis)
2. Abducens nerves
3. Facial nerve
4. Vagus nerve
5. Glossopharyngeal nerve
6. Cochlear nerve
7. Vestibular nerve
8. Hypoglossal nerve
9. Accessory nerve
10. Olfactory tract
11. Optic nerve
12. Internal carotid artery
13. Oculomotor nerve
14. Basilar artery
15. Trochlear nerve
16. Trigeminal nerve
17. Facial nerve
18. Vertebral artery

Figure 8.27. Midsagittal view of the brain.

1. Truncus of corpus callosum
2. Cingulate gyrus
3. Septum pellucidum (partially removed)
4. Lateral ventricle
5. Genu of corpus callosum
6. Hypothalamus (in wall of third ventricle)
7. Optic chiasma
8. Infundibular stalk
9. Temporal lobe
10. Intermediate mass (interthalamic adhesion)
11. Medial aspect of thalamus
12. Splenium of corpus callosum
13. Superior and inferior colliculi (corpora quadrigemina)
14. Arbor vitae of cerebellum
15. Tegmentum of midbrain
16. Mammillary body
17. Cerebral aqueduct
18. Fourth ventricle
19. Pons
20. Medulla oblongata

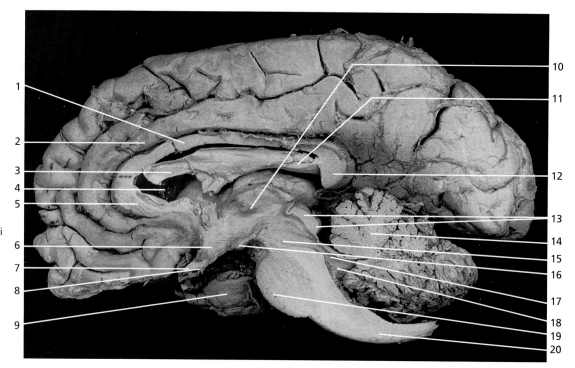

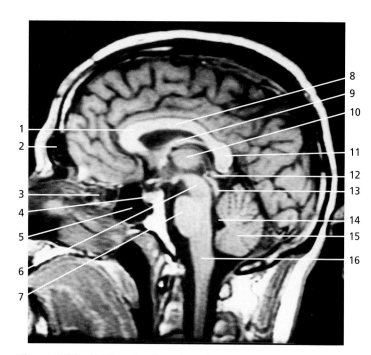

Figure 8.28. MRI sagittal section through the skull.

1. Genu of corpus callosum
2. Frontal sinus
3. Ethmoidal sinus
4. Pituitary gland
5. Sphenoidal sinus
6. Tegmentum (midbrain)
7. Pons
8. Truncus of corpus callosum
9. Fornix
10. Thalamus
11. Splenium of corpus callosum
12. Pineal gland
13. Superior and inferior colliculi
14. Fourth ventricle
15. Cerebellum
16. Medulla oblongata

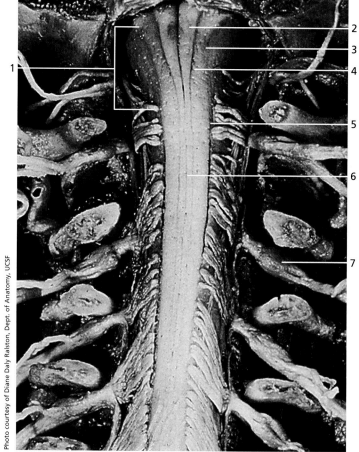

Photo courtesy of Diane Daly Ralston, Dept. of Anatomy, UCSF

Figure 8.29. Posterior view of cervical region of human spinal cord.

1. Medulla oblongata
2. Fasciculus gracilis
3. Fasciculus cuneatus
4. Posterior intermediate sulcus
5. Posterolateral sulcus
6. Posterior median sulcus
7. Posterior root ganglion

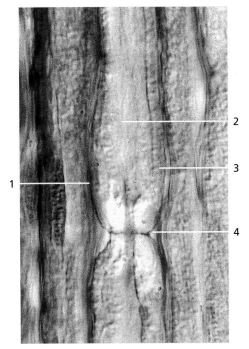

Figure 8.30. Histology of a myelinated nerve. (X282)

1. Neurolemmal sheath
2. Axon
3. Myelin layer
4. Neurofibril node (node of Ranvier)

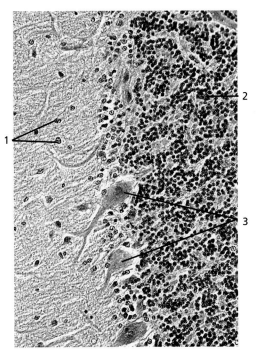

Figure 8.31. Histology of the cerebellum of the brain. (X50)

1. Outer stellate cells
2. Granule cells
3. Purkinje cells

Photo courtesy of Scott C. Miller

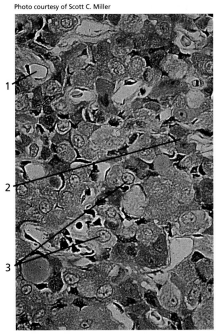

Figure 8.32. Histology of the human pituitary gland. (X200)

1. Capillary
2. Chromophils
3. Chromophobes

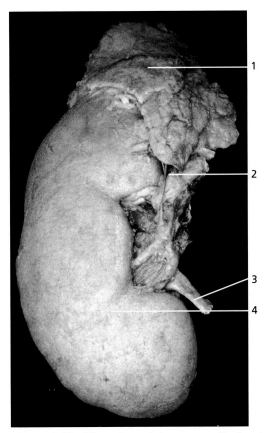

Figure 8.33. The adrenal (suprarenal) gland.

1. Adrenal gland   3. Ureter
2. Inferior suprarenal artery   4. Kidney

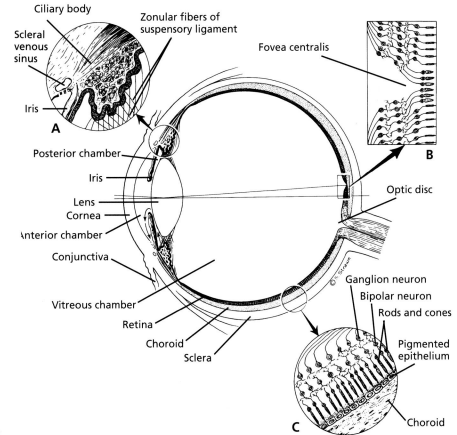

Figure 8.34. Structure of the eye. (A) cilary body; (B) fovea centralis; and (C) retina.

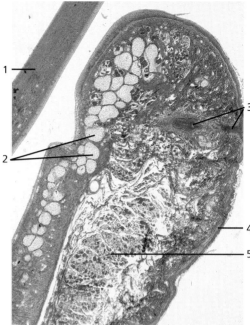

Figure 8.35. Eyelid and cornea. (X161)

1. Cornea
2. Tarsal glands
3. Eyelash within follicle
4. Epidermis
5. Skeletal muscle fibers

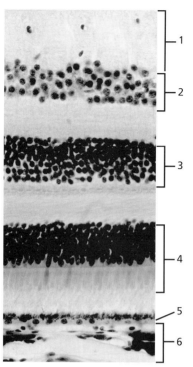

Figure 8.36. Histology of the anterior portion of the eye.

1. Iris
2. Anterior chamber
3. Cornea
4. Conjunctiva
5. Artifactual space
6. Lens
7. Posterior chamber
8. Ciliary processes
9. Smooth muscle fibers of ciliary body

Figure 8.37. A micrograph of the retina. (X250)

1. Fibers of the optic nerve
2. Ganglion neurons
3. Bipolar neurons
4. Photoreceptor neurons
5. Pigment epithelial layer
6. Choroid

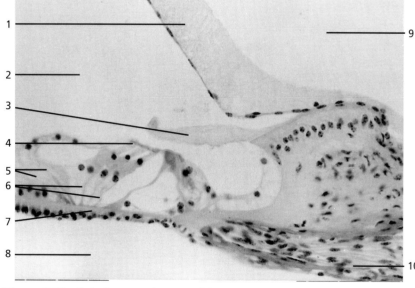

Figure 8.38. Spiral organ (organ of Corti) in the cochlea of the inner ear. (X50)

1. Vestibular membrane
2. Cochlear duct
3. Tectorial membrane
4. Dendritic endings of hair cells
5. Supporting cells
6. Nerve fibers
7. Basilar membrane
8. Scala tympani
9. Scala vestibuli
10. Cochlear nerve

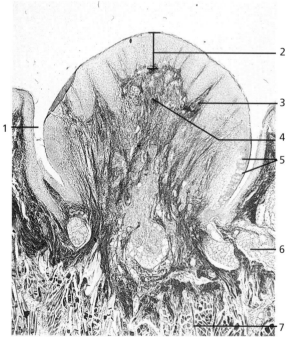

Figure 8.39. Vallate papilla of the tongue. (X32)

1. Circular furrow
2. Stratified squamous epithelium
3. Secondary papilla
4. Lamina propria
5. Taste buds
6. Serous acini of glands (von Ebner's)
7. Skeletal muscle fibers

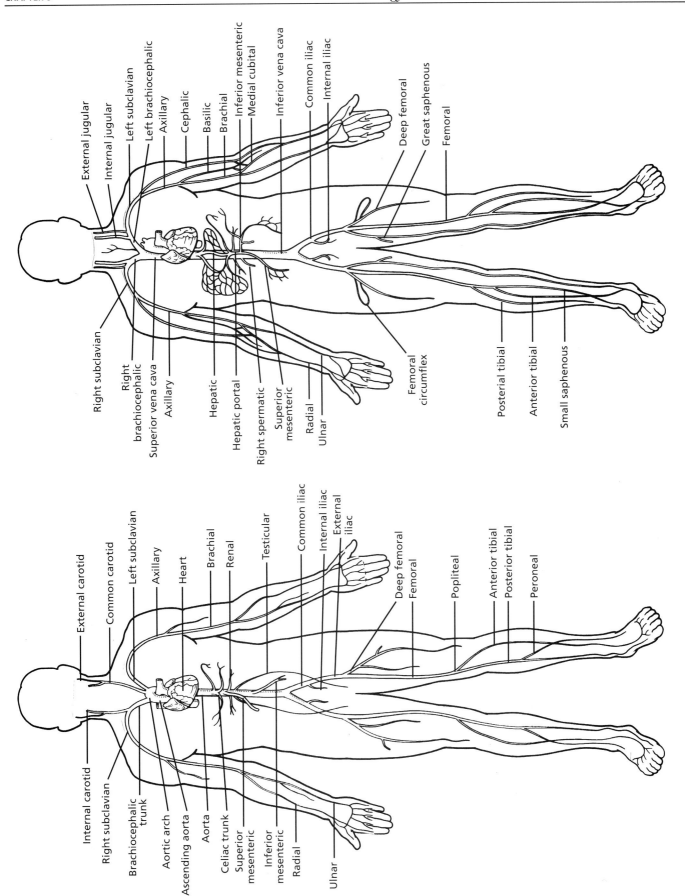

Figure 8.41. Principal veins of the body.

Figure 8.40. Principal arteries of the body.

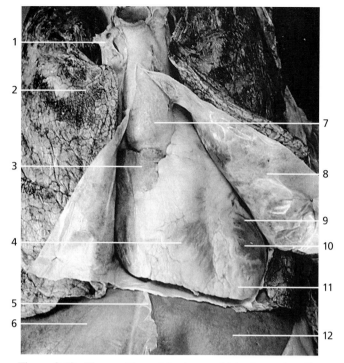

Figure 8.42. The position of the heart within the pericardium.

1. Superior vena cava
2. Right lung
3. Right atrium
4. Right ventricle
5. Falciform ligament
6. Right lobe of liver
7. Ascending portion of aorta
8. Pericardium (cut and reflected)
9. Coronary vein
10. Left ventricle
11. Apex of heart
12. Left lobe of liver

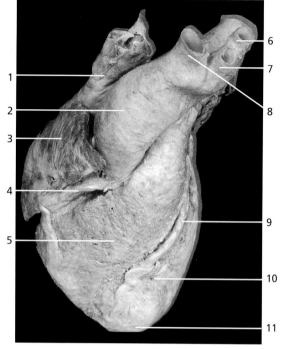

Figure 8.43. Anterior view of the heart and great vessels.

1. Superior vena cava
2. Ascending portion of aorta
3. Right atrium
4. Right coronary artery
5. Right ventricle
6. Left subclavian artery
7. Left common carotid artery
8. Brachiocephalic trunk
9. Anterior interventricular branch of the left coronary artery
10. Left ventricle
11. Apex of heart

Figure 8.44. Internal structure of the heart.

1. Left subclavian artery
2. Aortic arch
3. Pulmonary artery
4. Right atrium
5. Cusp of right atrioventricular valve
6. Interventricular septum
7. Trabeculae carneae
8. Apex
9. Left common carotid artery
10. Brachiocephalic trunk
11. Ascending aorta
12. Right auricle
13. Chordae tendineae
14. Papillary muscle
15. Left ventricle

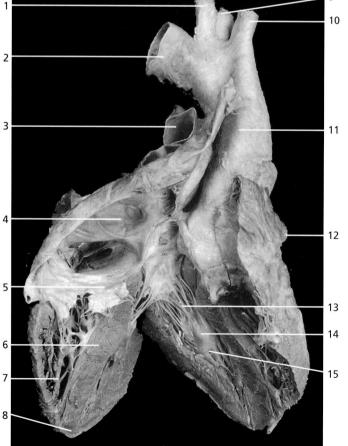

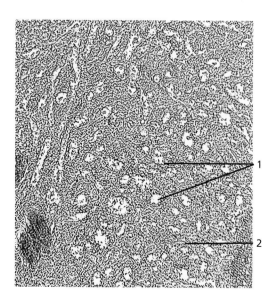

Figure 8.45. Histology of the spleen. (X20)

1. Vascular sinuses
2. Pulp

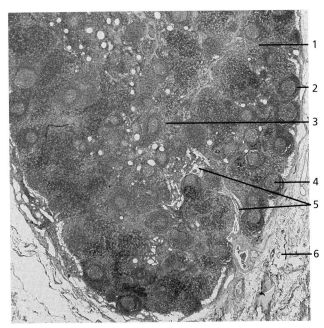

Figure 8.46. Histology of a lymph node. (X160)

1. Cortex of lymph node
2. Lymph nodule
3. Medulla of lymph node
4. Germinal center
5. Trabeculae
6. Pericapsular connective tissue

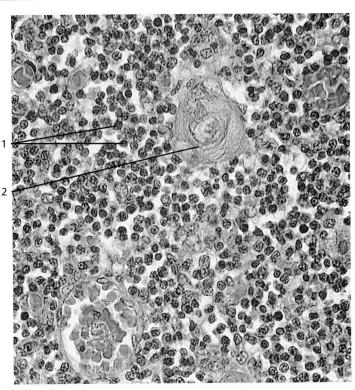

Figure 8.47. Histology of the thymus of the fetus.

1. Lymphocytes in the cortex of the thymus
2. Degenerating thymic corpuscle

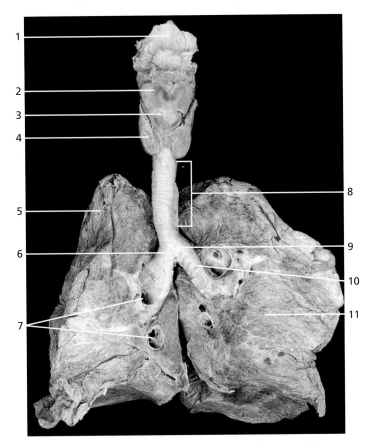

Figure 8.48. Anterior view of larynx, trachea, and lungs.

1. Epiglottis
2. Thyroid cartilage
3. Cricoid cartilage
4. Thyroid gland
5. Right lung
6. Right principal (primary) bronchus
7. Pulmonary vessels
8. Trachea
9. Carina
10. Left principal (primary) bronchus
11. Left lung

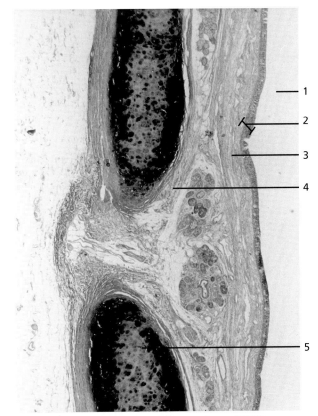

Figure 8.49. Histology of the trachea. (X20)

1. Lumen
2. Pseudostratified ciliated columnar epithelium
3. Lamina propria
4. Smooth muscle
5. Tracheal cartilage (hyaline cartilage)

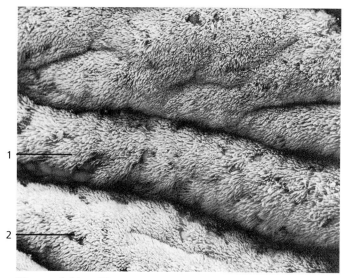

Figure 8.50. Electron micrograph of the lining of the trachea.

1. Cilia

2. Goblet cell

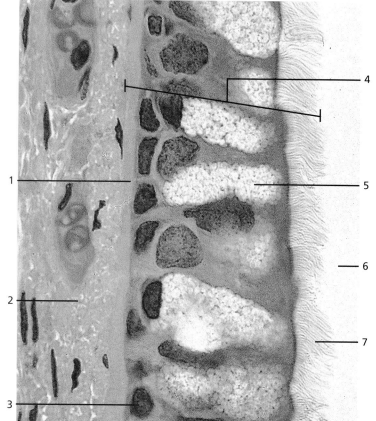

Figure 8.51. Pseudostratified ciliated columnar epithelium in a bronchus. (X450)

1. Basement membrane
2. Lamina propria
3. Nucleus
4. Pseudostratified squamous epithelium
5. Goblet cell
6. Lumen of bronchus
7. Cilia

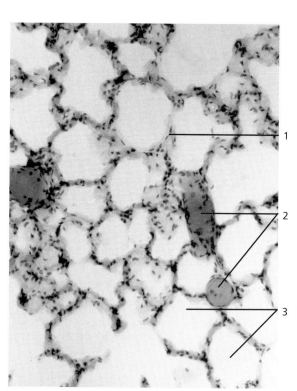

Figure 8.52. Alveoli of the lung. (X100)

1. Simple squamous epithelium
2. Blood vessels
3. Alveoli

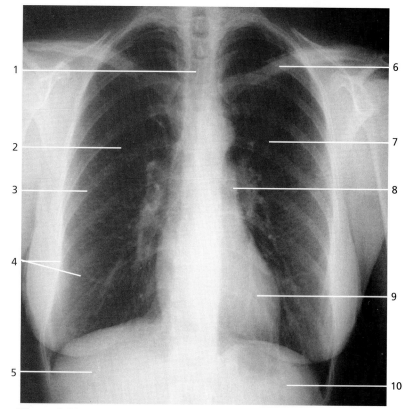

Figure 8.53. Radiograph of thorax, anterior view.

1. Thoracic vertebra
2. Right lung
3. Rib
4. Image of right breast
5. Diaphragm/liver
6. Clavicle
7. Left lung
8. Mediastinum
9. Heart
10. Diaphragm/stomach

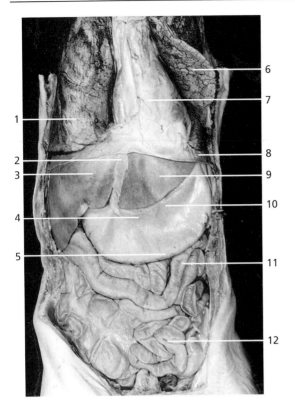

Figure 8.54. Anterior aspect of the trunk.

1. Right lung
2. Falciform ligament
3. Right lobe of liver
4. Body of stomach
5. Greater curvature of stomach
6. Left lung (reflected)
7. Heart surrounded by pericardium
8. Diaphragm
9. Left lobe of liver
10. Lesser curvature of stomach
11. Transverse colon
12. Small intestine

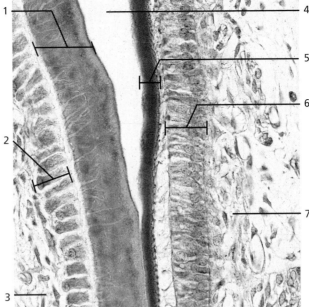

Figure 8.55. Developing tooth. (X112)

1. Dentin
2. Predentin (odontoblast)
3. Pulp
4. Artifactual space
5. Enamel
6. Ameloblasts
7. Stellate reticulum

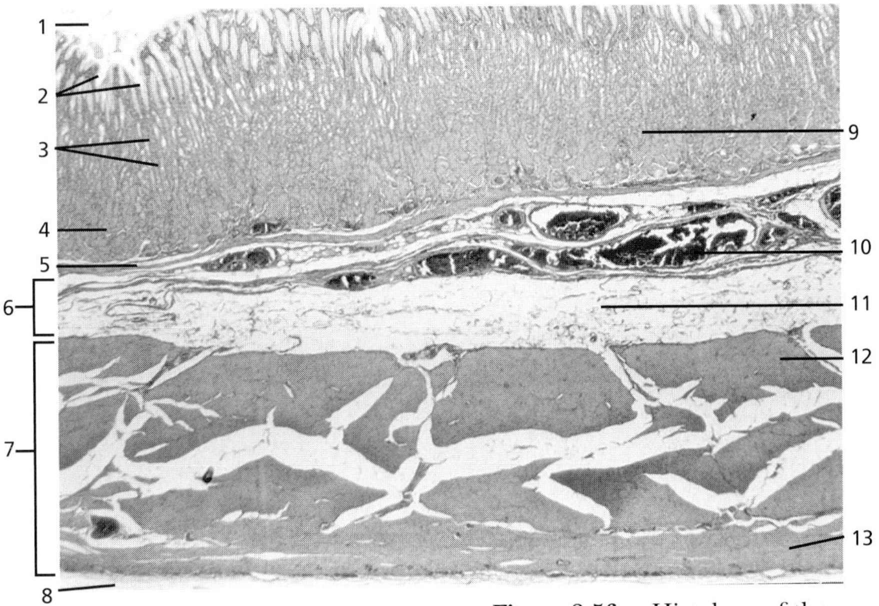

Figure 8.56. Histology of the fundic area of the stomach. (X15)

1. Surface epithelial cells
2. Gastric pits
3. Parietal cells
4. Chief cells
5. Muscularis mucosa
6. Submucosa
7. Muscularis externa
8. Serosa
9. Gastric glands
10. Oblique muscle layer
11. Loose connective tissue
12. Inner circular muscle layer
13. Outer longitudinal muscle layer

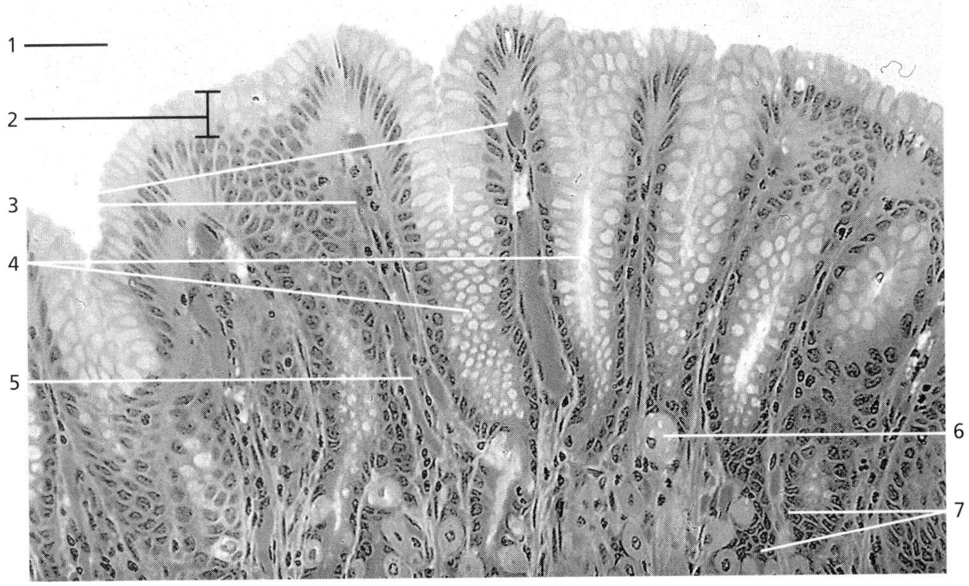

Figure 8.57. Histology of the cardiac region of the stomach. (X315)

1. Lumen of stomach
2. Surface epithelium
3. Mucosal ridges
4. Gastric pits
5. Lamina propria
6. Parietal cells
7. Chief (zymogenic) cells

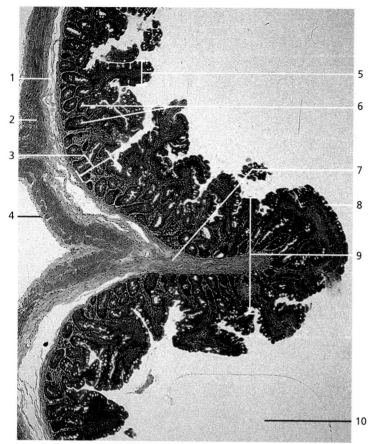

Figure 8.58. Histology of the jejunum of the small intestine. (X12)

1. Submucosa
2. Circular and longitudinal muscles
3. Mucosa
4. Serosa
5. Villus
6. Intestinal glands
7. Submucosa
8. Microvilli
9. Plica circulares
10. Lumen of small intestine

Figure 8.59. Histology of the ileum of the small intestine (ileum). (X100)

1. Lymphatic nodules (Peyer's patches)
2. Muscularis mucosa
3. Goblet cells
4. Simple columnar epithelial cells with striated border
5. Lamina propria
6. Arteriole
7. Venule
8. Lamina propria
9. Lacteal
10. Villus

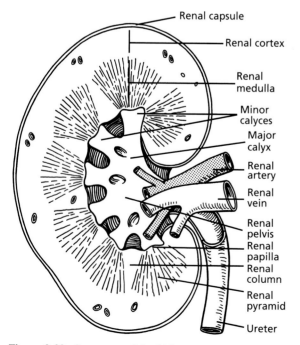

Figure 8.60. Structure of the kidney.

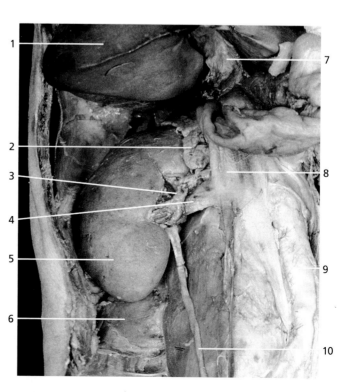

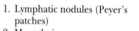

Figure 8.61. Kidney and ureter with overlaying viscera removed.

1. Liver
2. Suprarenal gland
3. Renal artery
4. Renal vein
5. Right kidney
6. Quadratus lumborum m.
7. Gallbladder
8. Inferior vena cava
9. Abdominal portion of aorta
10. Ureter

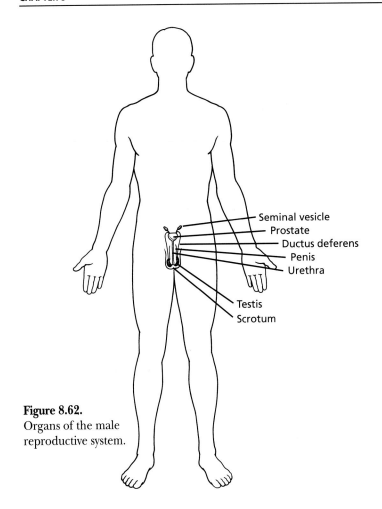

Figure 8.62.
Organs of the male
reproductive system.

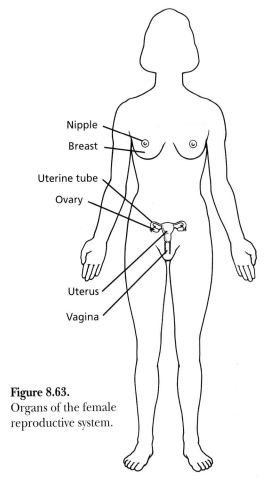

Nipple
Breast
Uterine tube
Ovary
Uterus
Vagina

Figure 8.63.
Organs of the female
reproductive system.

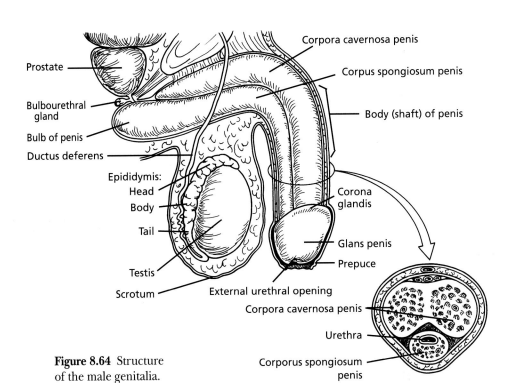

Prostate

Bulbourethral
gland

Bulb of penis

Ductus deferens

Epididymis:
Head
Body
Tail

Testis

Scrotum

Corpora cavernosa penis

Corpus spongiosum penis

Body (shaft) of penis

Corona
glandis

Glans penis

Prepuce

External urethral opening

Corpora cavernosa penis

Urethra

Corporus spongiosum
penis

Figure 8.64 Structure
of the male genitalia.

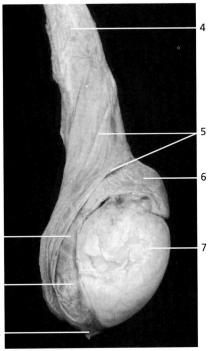

Figure 8.65. Testis and associated structures.

1. Body of epididymis
2. Tail of epididymis
3. Gubernaculum
4. Spermatic cord
5. Spermatic fascia
6. Head of epididymis
7. Testis

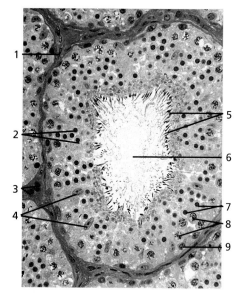

Figure 8.66. Histology of a seminiferous tubule the testis. (X256)

1. Fibroblast
2. Spermatids
3. Interstitial cells (of Leydig)
4. Sustentacular (nurse) cells
5. Maturing spermatozoa
6. Lumen of seminiferous tubule
7. Secondary spermatocytes
8. Primary spermatocytes
9. Spermatogonium

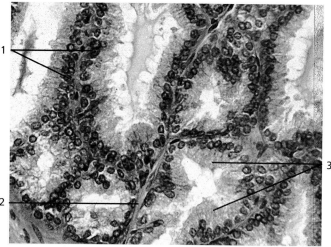

Figure 8.67. Histology of the prostate. (X125)

1. Glandular epithelium   2. Smooth muscle   3. Glandular acini

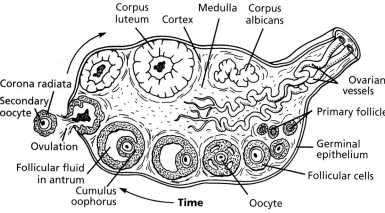

Figure 8.68. Structure of the ovary.

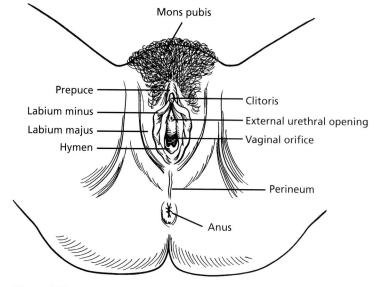

Figure 8.69. Uterus and vagina.

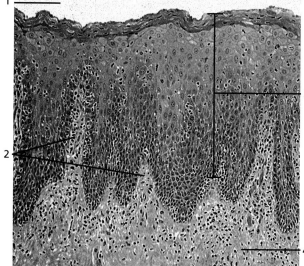

Figure 8.70. Histology of the vagina. (X80)

1. Lumen of vagina
2. Connective tissue papillae
3. Stratified squamous epithelium
4. Lamina propria

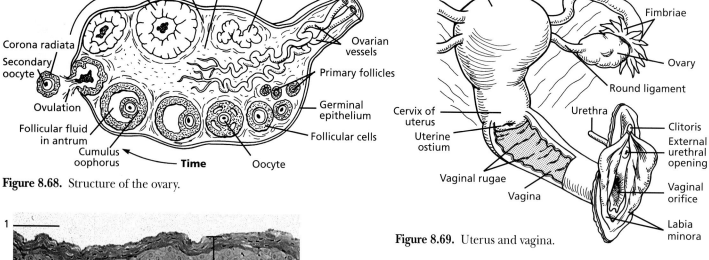

Figure 8.71. Female genitalia.

Vertebrate Dissections

An understanding of the structure of a vertebrate organism is requisite to learning about physiological mechanisms and about how the animal functions in its environment. The selective pressures that determine evolutionary changes frequently have an influence on anatomical structures. Studying dissected specimens, therefore, provides phylogenetic information about how groups of organisms are related.

Some biology laboratories have the resources to provide students with opportunities for doing selected vertebrate dissections. For these students, the photographs contained in this chapter will be a valuable source for identification of structures on your specimens as they are dissected and studied. If dissection specimens are not available, the excellent photographs of carefully dissected prepared specimens presented in this chapter will be an adequate substitute. Care has gone into the preparation of these specimens to depict and identify the principal body structures from representative specimens of each of the classes of vertebrates. Selected human cadaver dissections are shown in photographs contained in chapter 8. As the anatomy of vertebrate specimens is studied in this chapter, observe the photographs of human dissections in the previous chapter and note the similarities of body structure, particularly to those of another mammal.

CLASS AGNATHA

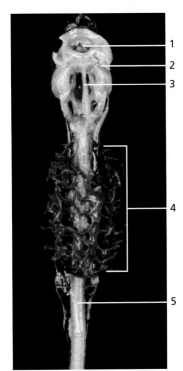

Figure 9.1. A cartilaginous skeleton of the marine lamprey, *Petromyzon marinus*, shown in ventral view.

1. Buccal cavity
2. Skull
3. Tongue support
4. Branchial basket
5. Notochord

CLASS CHONDRICHTHYES

Figure 9.2. External anatomy of a leopard shark, *Triakis semifasciata*, lateral view.

1. Lateral line
2. Spiracle
3. Eye
4. Gill slits
5. Pectoral fin
6. Dorsal fins
7. Caudal fin (heterocercal tail)
8. Pelvic fin

CLASS CHONDRICHTHYES

Figure 9.3. External anatomy of the dogfish shark, *Squalus acanthias*, lateral view.

1. Eye
2. Nostril
3. Mouth
4. Gill slits
5. Pectoral fin
6. Dorsal fins
7. Caudal fin (heterocercal tail
8. Pelvic fin

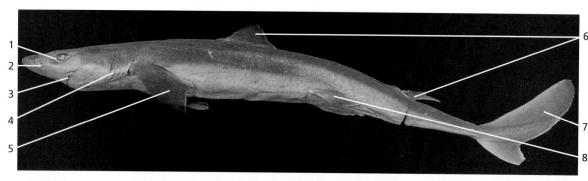

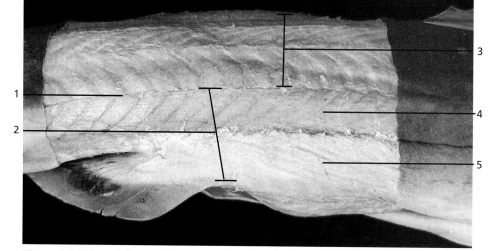

Figure 9.5. The axial musculature of the dogfish shark, lateral view.

1. Transverse septum
2. Hypaxial myotome portion
3. Epaxial myotome portion
4. Lateral bundle of myotomes
5. Ventral bundle of myotomes

Figure 9.4. A suspended cartilaginous skeleton of the male dogfish shark, *Squalus acanthias*, shown in ventral view.

1. Palatopterygoquadrate cartilage (upper jaw)
2. Hypobranchial cartilage
3. Caudal fin
4. Caudal vertebrae
5. Pelvic fin
6. Posterior dorsal fin
7. Rostrum
8. Chondrocranium
9. Meckel's cartilage (lower jaw)
10. Visceral arches
11. Pectoral girdle
12. Pectoral fin
13. Trunk vertebrae
14. Anterior dorsal fin
15. Pelvic girdle
16. Clasper

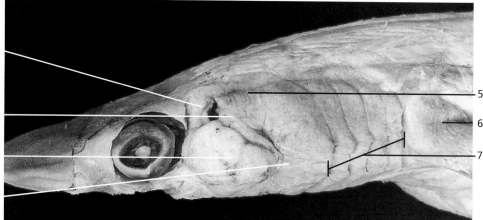

Figure 9.6. Jaw, gill, and pectoral fin musculature of the dogfish shark, lateral view.

1. Spiracular muscle
2. Facial nerve
3. Mandibular adductor
4. 2nd ventral constrictor
5. 2nd dorsal constrictor
6. Levator of pectoral fin
7. 3rd-to-6th ventral constrictors

CLASS CHONDRICHTHYES

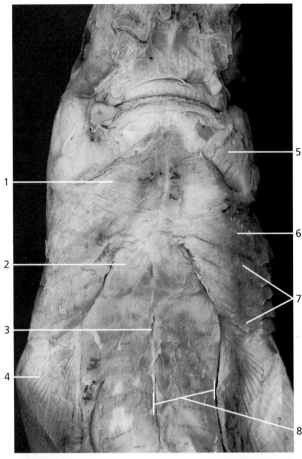

Figure 9.7. Hypobranchial musculature of the dogfish shark, ventral view.

1. 1st ventral constrictor
2. Common coracoarcual
3. Linea alba
4. Depressor of pectoral fin
5. Mandibular adductor
6. 2nd ventral constrictor
7. 3rd-to-6th ventral constrictors
8. Hypaxial muscle

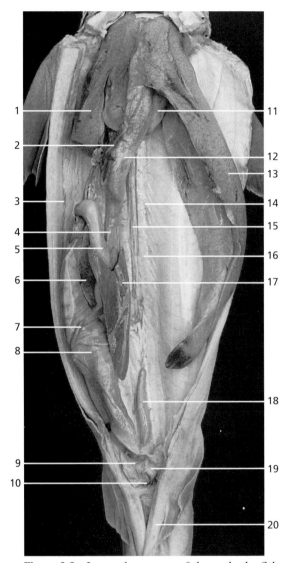

Figure 9.8. Internal anatomy of the male dogfish shark, ventral view.

1. Right lobe of liver (cut)
2. Hepatic portal vein
3. Lateral abdominal vein
4. Stomach (pyloric region)
5. Duodenum
6. Pancreas
7. Anterior intestinal vein
8. Ileum
9. Cloaca
10. Urogenital pore
11. Testis
12. Stomach (cardiac region)
13. Left lobe of liver
14. Mesonephric duct
15. Dorsal aorta
16. Kidney
17. Spleen
18. Rectal gland
19. Urogenital papilla
20. Clasper

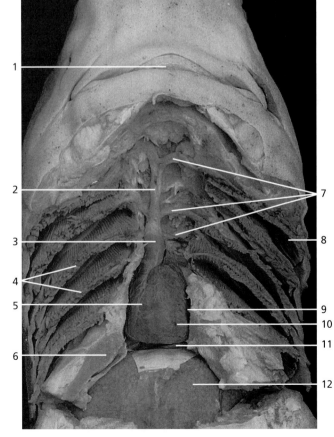

Figure 9.9. Heart and gills of the dogfish shark, ventral view.

1. Mouth
2. Ventral aorta
3. Conus ateriosus
4. Gills
5. Atrium
6. Pectoral girdle
7. Afferent branchial arteries
8. Gill cleft
9. Pericardial cavity
10. Ventricle
11. Sinus venosus
12. Liver

CLASS CHONDRICHTYES

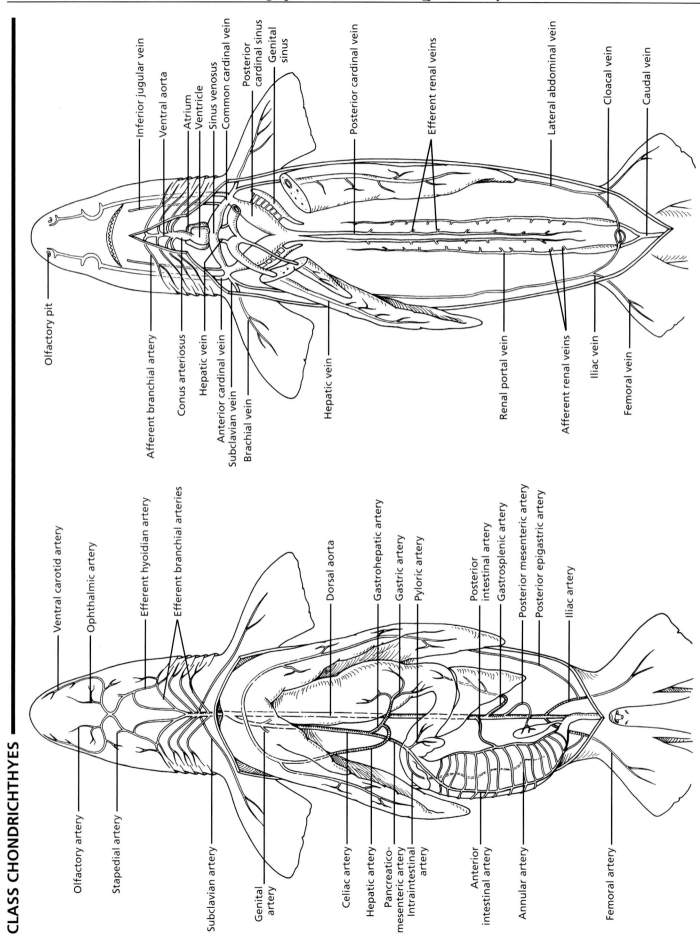

Inferior jugular vein
Ventral aorta
Atrium
Ventricle
Sinus venosus
Common cardinal vein
Posterior cardinal sinus
Genital sinus
Posterior cardinal vein
Efferent renal veins
Lateral abdominal vein
Cloacal vein
Caudal vein

Olfactory pit

Afferent branchial artery
Conus arteriosus
Hepatic vein
Anterior cardinal vein
Subclavian vein
Brachial vein
Hepatic vein
Renal portal vein
Afferent renal veins
Iliac vein
Femoral vein

Figure 9.11. Venous system of the shark

Ventral carotid artery
Ophthalmic artery
Efferent hyoidian artery
Efferent branchial arteries
Dorsal aorta
Gastrohepatic artery
Gastric artery
Pyloric artery
Posterior intestinal artery
Gastrosplenic artery
Posterior mesenteric artery
Posterior epigastric artery
Iliac artery

Olfactory artery
Stapedial artery
Subclavian artery
Genital artery
Celiac artery
Hepatic artery
Pancreatico-mesenteric artery
Intraintestinal artery
Anterior intestinal artery
Annular artery
Femoral artery

Figure 9.10. Arterial system of the shark.

CLASS CHONDRICHTHYES

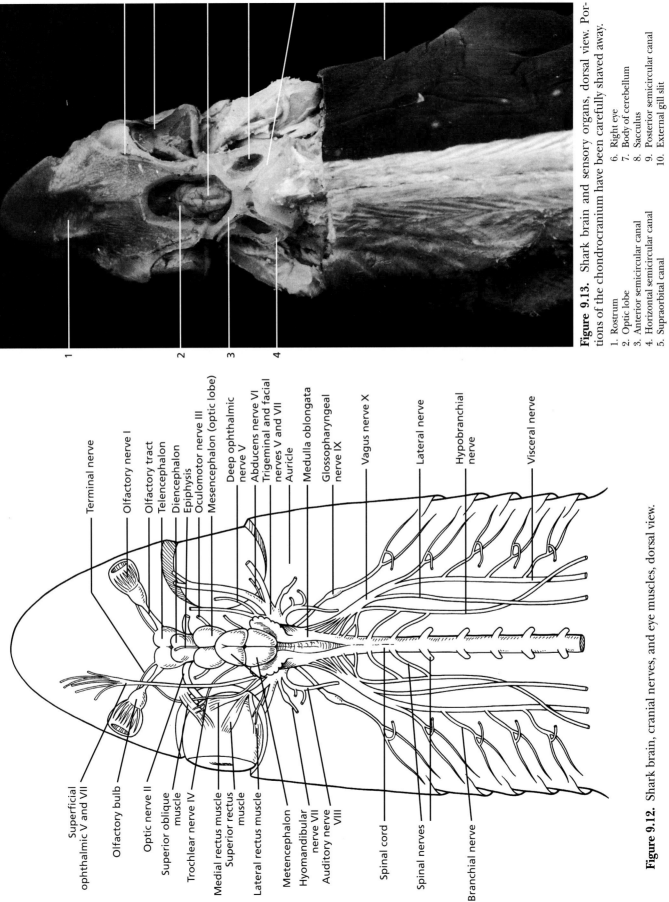

Figure 9.13. Shark brain and sensory organs, dorsal view. Portions of the chondrocranium have been carefully shaved away.

1. Rostrum
2. Optic lobe
3. Anterior semicircular canal
4. Horizontal semicircular canal
5. Supraorbital canal
6. Right eye
7. Body of cerebellum
8. Sacculus
9. Posterior semicircular canal
10. External gill slit

Figure 9.12. Shark brain, cranial nerves, and eye muscles, dorsal view.

Labels (Figure 9.12):

Terminal nerve
Olfactory nerve I
Olfactory tract
Telencephalon
Diencephalon
Epiphysis
Oculomotor nerve III
Mesencephalon (optic lobe)
Deep ophthalmic nerve V
Abducens nerve VI
Trigeminal and facial nerves V and VII
Auricle
Medulla oblongata
Glossopharyngeal nerve IX
Vagus nerve X
Lateral nerve
Hypobranchial nerve
Visceral nerve

Superficial ophthalmic V and VII
Olfactory bulb
Optic nerve II
Superior oblique muscle
Trochlear nerve IV
Medial rectus muscle
Superior rectus muscle
Lateral rectus muscle
Metencephalon
Hyomandibular nerve VII
Auditory nerve VIII
Spinal cord
Spinal nerves
Branchial nerve

CLASS OSTEICHTHYES

Figure 9.14. Skeleton of a perch.

1. Anterior dorsal fin
2. Posterior dorsal fin
3. Caudal fin
4. Orbit
5. Premaxilla
6. Dentary
7. Branchial skeleton
8. Opercular bones
9. Pectoral girdle
10. Pelvic girdle
11. Pectoral fin
12. Pelvic fin
13. Rib
14. Vertebral column
15. Anal fin
16. Haemal spine
17. Neural spine

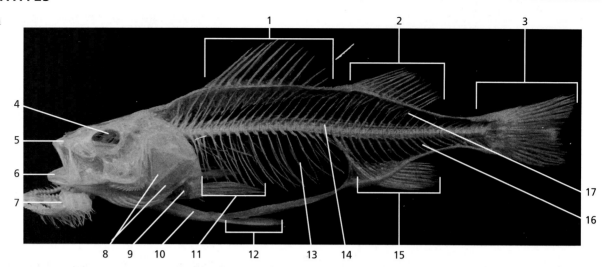

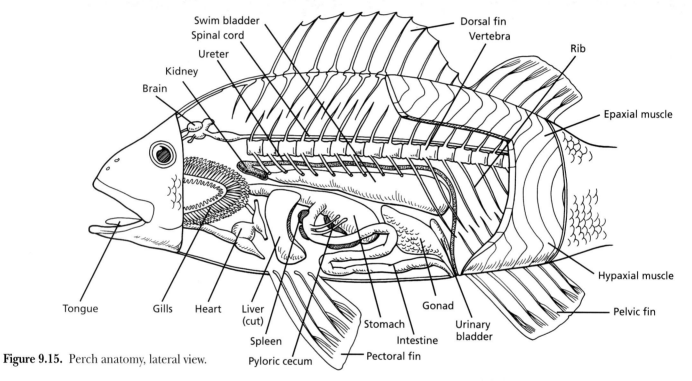

Figure 9.15. Perch anatomy, lateral view.

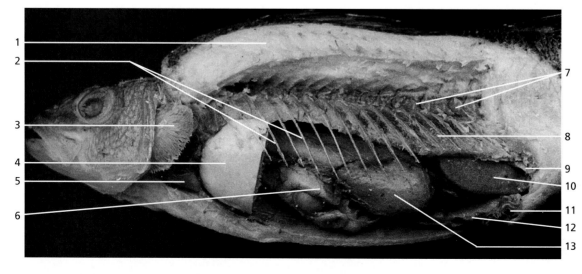

Figure 9.16.
Perch, visceral organs.

1. Epaxial muscles
2. Ribs
3. Gill
4. Liver (cut)
5. Heart
6. Pyloric cecum
7. Vertebrae
8. Swim bladder
9. Urinary bladder
10. Gonad
11. Anus
12. Large intestine
13. Stomach

CLASS AMPHIBIA ━━ **Order Anura**

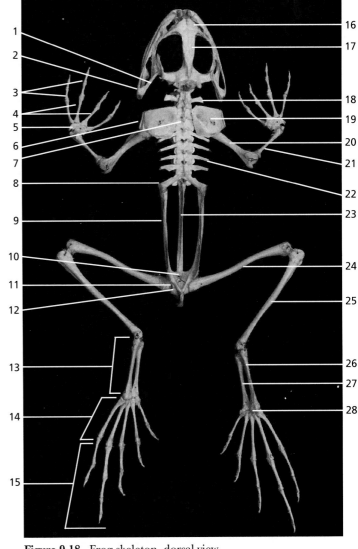

Figure 9.17. Surface anatomy and body regions of the leopard frog, *Rana pipiens*.

1. Ankle
2. Knee
3. Foot
4. Eyes
5. Nostril
6. Tympanic membrane
7. Brachium
8. Antebrachium
9. Digits

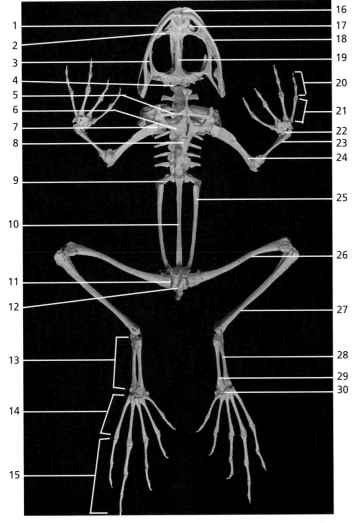

Figure 9.18. Frog skeleton, dorsal view.

1. Squamosal bone
2. Quadratojugal bone
3. Phalanges of digits
4. Metacarpal bones
5. Carpal bones
6. Scapula
7. Vertebra
8. Transverse process of ninth vertebra
9. Ilium
10. Pubis
11. Acetabulum
12. Ischium
13. Tarsal bones
14. Metatarsal bones
15. Phalanges of digits
16. Nasal bone
17. Frontoparietal bone
18. Transverse process
19. Suprascapula
20. Humerus
21. Radioulna
22. Transverse process
23. Urostyle
24. Femur
25. Tibiofibula
26. Fibulare (calcaneum)
27. Tibiale (astragalus)
28. Distal tarsal bones

Figure 9.19 Frog skeleton, ventral view.

1. Maxilla
2. Palatine
3. Pterygoid bone
4. Exoccipital bone
5. Clavicle
6. Coracoid
7. Glenoid fossa
8. Sternum
9. Transverse process of ninth vertebra
10. Urostyle
11. Acetabulum
12. Ischium
13. Tarsal bones
14. Metatarsal bones
15. Phalanges of digits
16. Premaxilla
17. Vomer
18. Dentary
19. Parasphenoid bone
20. Phalanges of digits
21. Metacarpal bones
22. Carpal bones
23. Radioulna
24. Humerus
25. Ilium
26. Femur
27. Tibiofibula
28. Fibulare (calcaneum)
29. Tibiale (astragalus)
30. Distal tarsal bones

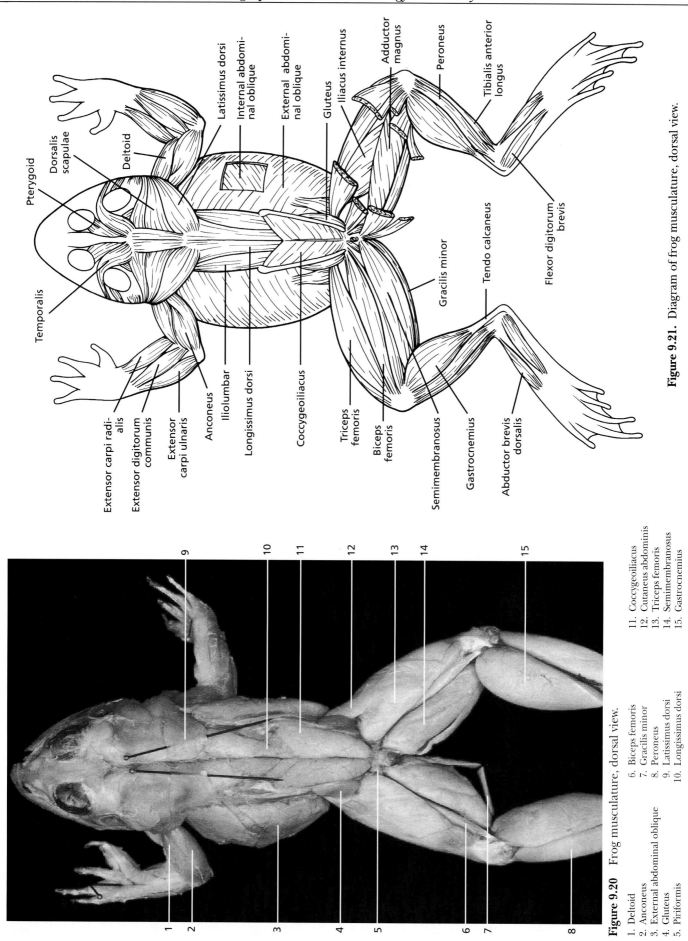

Figure 9.21. Diagram of frog musculature, dorsal view.

Figure 9.20 Frog musculature, dorsal view.

1. Deltoid
2. Anconeus
3. External abdominal oblique
4. Gluteus
5. Piriformis
6. Biceps femoris
7. Gracilis minor
8. Peroneus
9. Latissimus dorsi
10. Longissimus dorsi
11. Coccygeoiliacus
12. Cutaneus abdominis
13. Triceps femoris
14. Semimembranosus
15. Gastrocnemius

CLASS AMPHIBIA ━━━━━━━━━━━━━━━━━━━━━━━━━━━━━━━ Order Anura

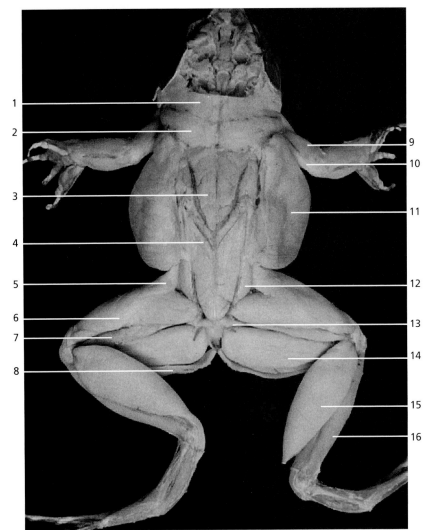

Figure 9.22. Frog musculature, dorsal view. (m. = muscle)

1. Dorsalis scapulae m.
2. Latissimus dorsi m.
3. Longissimus dorsi m.
4. Coccygeoiliacus m.
5. Cutaneus abdominis m.
6. Triceps femoris m.
7. Biceps femoris m.
8. Gracilis minor m.
9. Deltoid m.
10. Anconeus m.
11. External abdominal oblique m.
12. Gluteus m.
13. Piriformis m.
14. Semimembranosus m.
15. Gastrocnemius m.
16. Peroneus m.

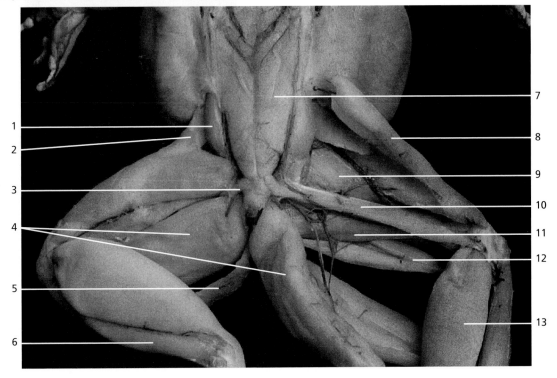

Figure 9.23. Frog musculature of the legs, dorsal view.

1. Gluteus m.
2. Cutaneus abdominis m.
3. Piriformis m.
4. Semimembranosus m.
5. Gracilis minor m.
6. Peroneus m.
7. Coccygeoiliacus m.
8. Triceps femoris m. (cut)
9. Iliacus internus m.
10. Biceps femoris m.
11. Adductor magnus m.
12. Semitendinosus m .
13. Gastrocnemius m.

Order Anura

CLASS AMPHIBIA

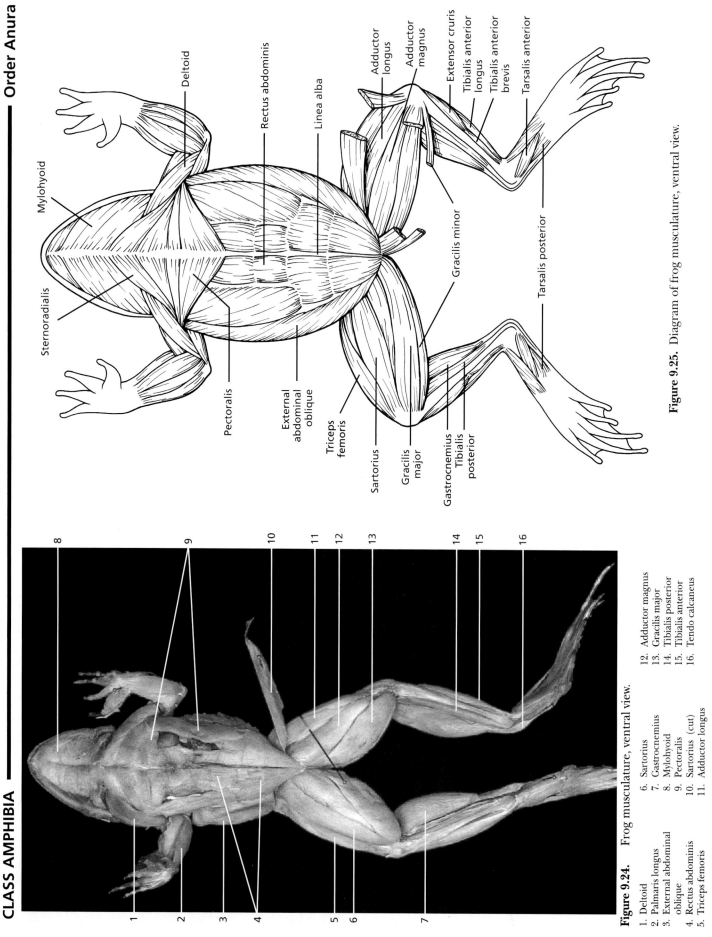

Figure 9.25. Diagram of frog musculature, ventral view.

Figure 9.24.   Frog musculature, ventral view.

1. Deltoid
2. Palmaris longus
3. External abdominal oblique
4. Rectus abdominis
5. Triceps femoris
6. Sartorius
7. Gastrocnemius
8. Mylohyoid
9. Pectoralis
10. Sartorius (cut)
11. Adductor longus
12. Adductor magnus
13. Gracilis major
14. Tibialis posterior
15. Tibialis anterior
16. Tendo calcaneus

CLASS AMPHIBIA ━━━━━━━━━━━━━━━━━━━━━━━━━━━━━━━━━━━━━ Order Anura

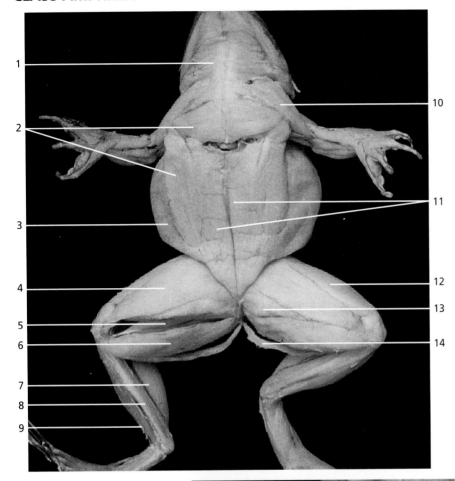

Figure 9.26. Frog musculature, ventral view.

1. Mylohyoid m.
2. Pectoralis m.
3. External abdominal oblique m.
4. Sartorius m.
5. Semitendinosus m.
6. Gracilis major m.
7. Gastrocnemius m.
8. Tibialis posterior m.
9. Tibialis anterior m.
10. Deltoid m.
11. Rectus abdominis mm.
12. Triceps femoris m.
13. Adductor magnus m.
14. Gracilis minor m.

Figure 9.27. Frog musculature of the legs, ventral view.

1. External abdominal oblique m.
2. Triceps femoris m.
3. Adductor longus m.
4. Adductor magnus m.
5. Semitendinosus m.
6. Semimembranosus m.
7. Gastrocnemius m.
8. Tibialis posterior m.
9. Rectus abdominis m.
10. Sartorius m.
11. Gracilis major m.
12. Gracilis minor m.
13. Extensor cruris m.
14. Tibialis anterior longus m.
15. Tibialis anterior brevis m.

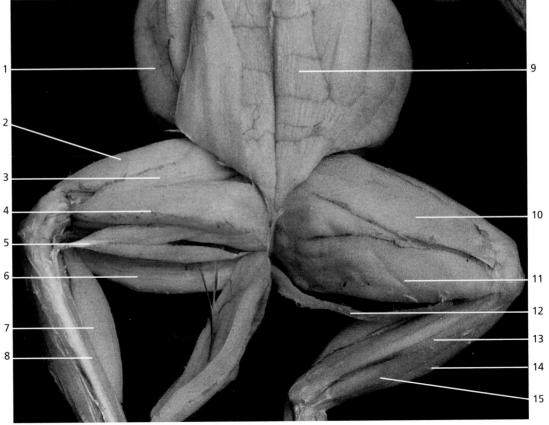

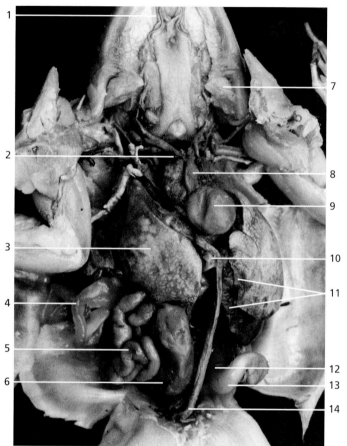

Figure 9.28. Frog viscera, ventral view.

1. External carotid artery
2. Right aortic arch
3. Right lobe of liver
4. Fat body
5. Small intestine
6. Large intestine
7. Vocal sac (male only)
8. Truncus arteriosus
9. Heart
10. Gallbladder
11. Left lobe of liver
12. Stomach
13. Duodenum
14. Ventral abdominal vein

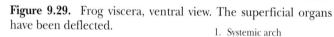

Figure 9.29. Frog viscera, ventral view. The superficial organs have been deflected.

1. Systemic arch
2. Pulmocutaneous arch
3. Right lung
4. Fat body
5. Testis
6. Right kidney
7. Common carotid artery
8. Heart
9. Liver
10. Gallbladder
11. Small intestine
12. Mesentery
13. Large Intestine

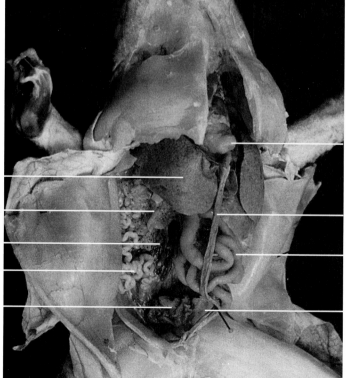

Figure 9.30. Frog, female reproductive organs, ventral view.

1. Liver
2. Ovary
3. Kidney
4. Oviduct
5. Uterus
6. Heart
7. Ventral abdominal vein
8. Small intestine
9. Urinary bladder (retracted)

CLASS AMPHIBIA — **Order Anura**

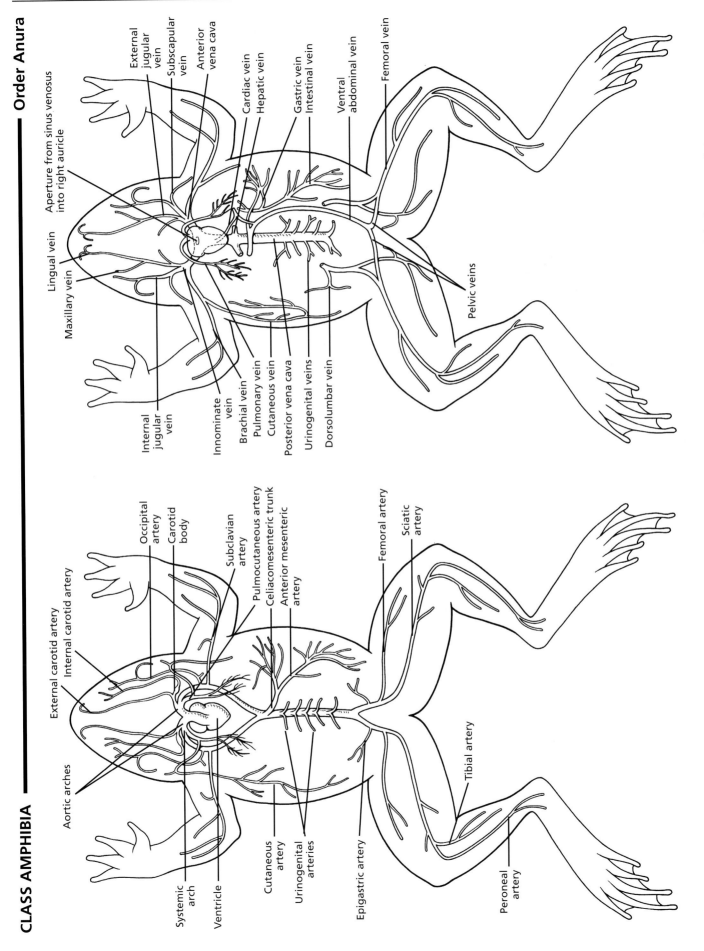

Figure 9.32. Veins of the frog, ventral view.

Figure 19.31. Arteries of the frog, ventral view.

CLASS REPTILIA ━━━━━━━━ **SUBCLASS ANAPSIDA** ━━━━━━━━ **Order Chelonia**

Figure 9.33. Turtle, dorsal view of the carapace.

1. Eye
2. Pentadactyl foot
3. Vertebral scales
4. Marginal scales (encircle the carapace)
5. Nostril
6. Head
7. Nuchal scale
8. Costal scales

Figure 9.34. Turtle, ventral view of the plastron.

1. Gular scales
2. Pectoral scales
3. Femoral scales
4. Tail
5. Humeral scales
6. Abdominal scales
7. Anal scales

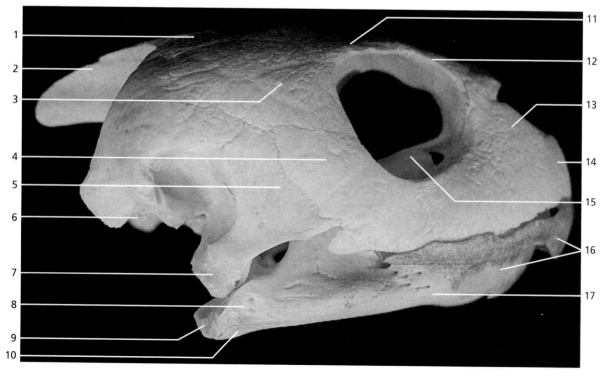

Figure 9.35. Turtle skull, lateral view.

1. Parietal bone
2. Supraoccipital bone
3. Postorbital bone
4. Jugal bone
5. Quadratojugal bone
6. Exoccipital bone
7. Quadrate bone
8. Supraangular bone
9. Articular bone
10. Angular bone
11. Frontal bone
12. Prefrontal bone
13. Maxilla
14. Premaxilla
15. Palatine bone
16. Beak
17. Dentary

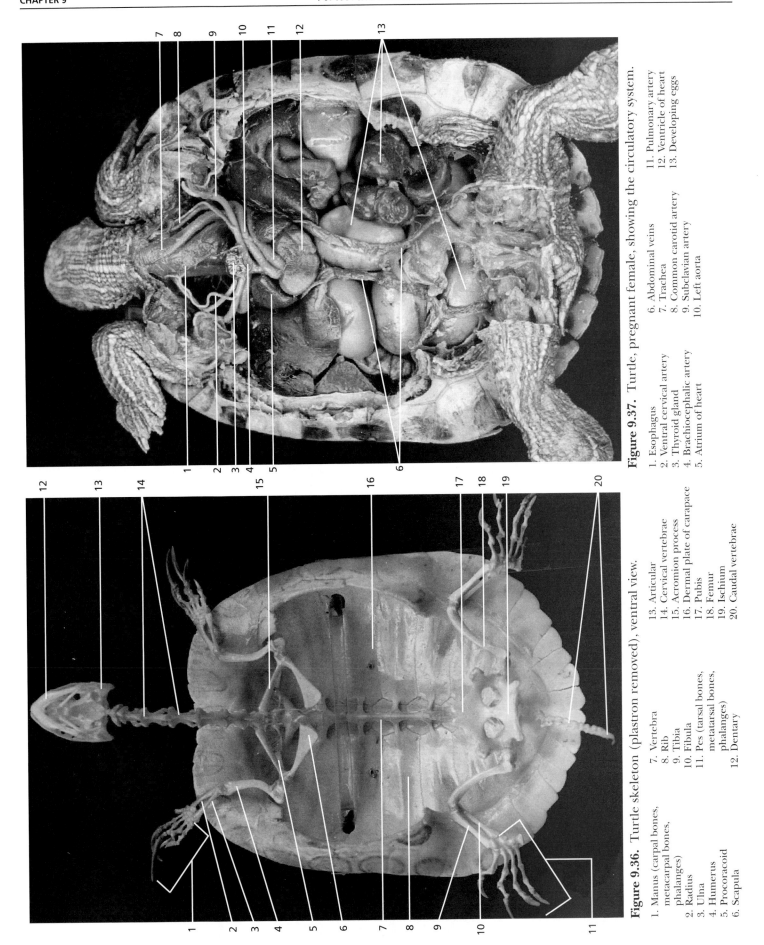

Figure 9.37. Turtle, pregnant female, showing the circulatory system.

1. Esophagus	6. Abdominal veins	11. Pulmonary artery
2. Ventral cervical artery	7. Trachea	12. Ventricle of heart
3. Thyroid gland	8. Common carotid artery	13. Developing eggs
4. Brachiocephalic artery	9. Subclavian artery	
5. Atrium of heart	10. Left aorta	

Figure 9.36. Turtle skeleton (plastron removed), ventral view.

1. Manus (carpal bones, metacarpal bones, phalanges)	7. Vertebra	13. Articular
	8. Rib	14. Cervical vertebrae
2. Radius	9. Tibia	15. Acromion process
3. Ulna	10. Fibula	16. Dermal plate of carapace
4. Humerus	11. Pes (tarsal bones, metatarsal bones, phalanges)	17. Pubis
5. Procoracoid		18. Femur
6. Scapula	12. Dentary	19. Ischium
		20. Caudal vertebrae

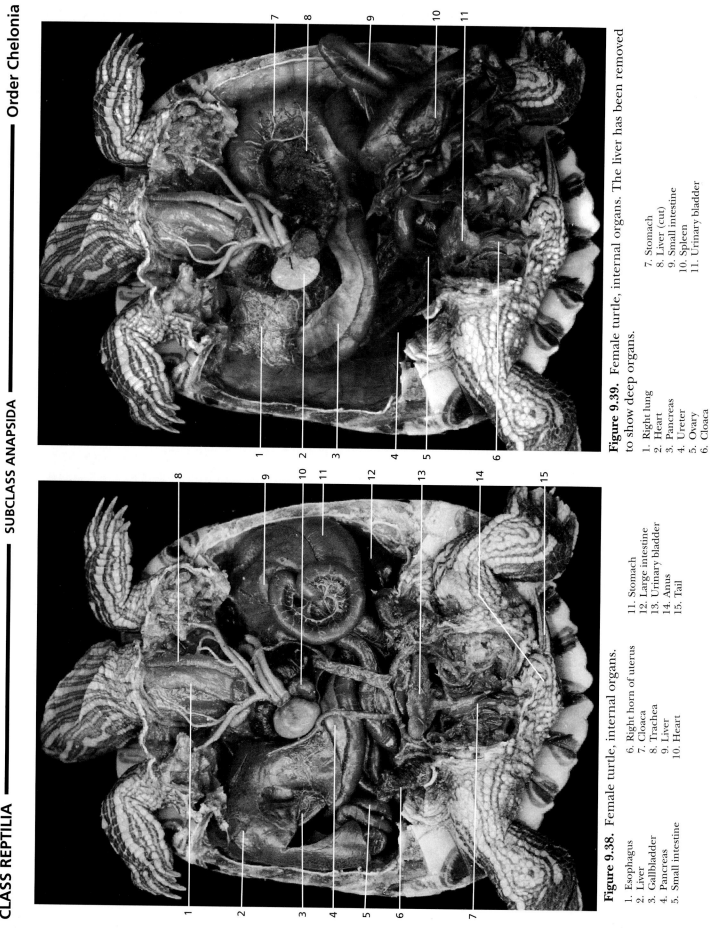

Figure 9.38. Female turtle, internal organs.

1. Esophagus
2. Liver
3. Gallbladder
4. Pancreas
5. Small intestine
6. Right horn of uterus
7. Cloaca
8. Trachea
9. Liver
10. Heart
11. Stomach
12. Large intestine
13. Urinary bladder
14. Anus
15. Tail

Figure 9.39. Female turtle, internal organs. The liver has been removed to show deep organs.

1. Right lung
2. Heart
3. Pancreas
4. Ureter
5. Ovary
6. Cloaca
7. Stomach
8. Liver (cut)
9. Small intestine
10. Spleen
11. Urinary bladder

CLASS REPTILIA ══════════ SUBCLASS LEPIDOSAURIA ══════════ Order Squamata

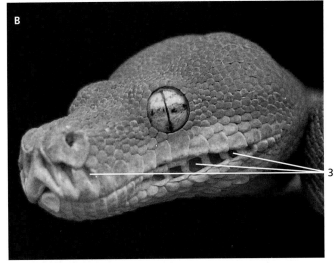

Figure 9.40. A comparison of the external anatomy of the head of a lizard and a snake. A lizard (A) such as the savannah monitor (*Varanus exanthematicus*) has eyelids and external ears. A snake (B) such as the green tree python (*Chondropython viridis*) lacks eyelids and external ears. Note the heat pits that are characteristic of pythons and pit vipers. The presence of heat pit receptors is a specialization for predation on warm-blooded vertebrates.

1. Eyelids   2. External ear   3. Heat pits

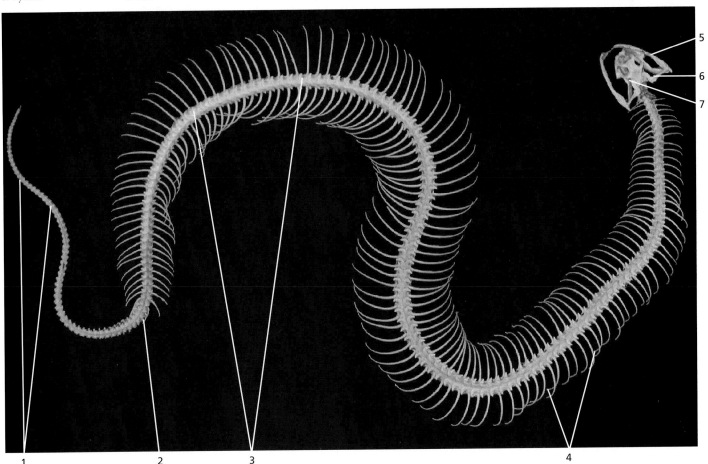

Figure 9.41. Skeleton of a snake (python), dorsal view.

1. Caudal vertebrae   5. Dentary
2. Vestigial pelvic girdle   6. Quadrate bone
3. Trunk vertebrae   7. Supratemporal bone
4. Ribs

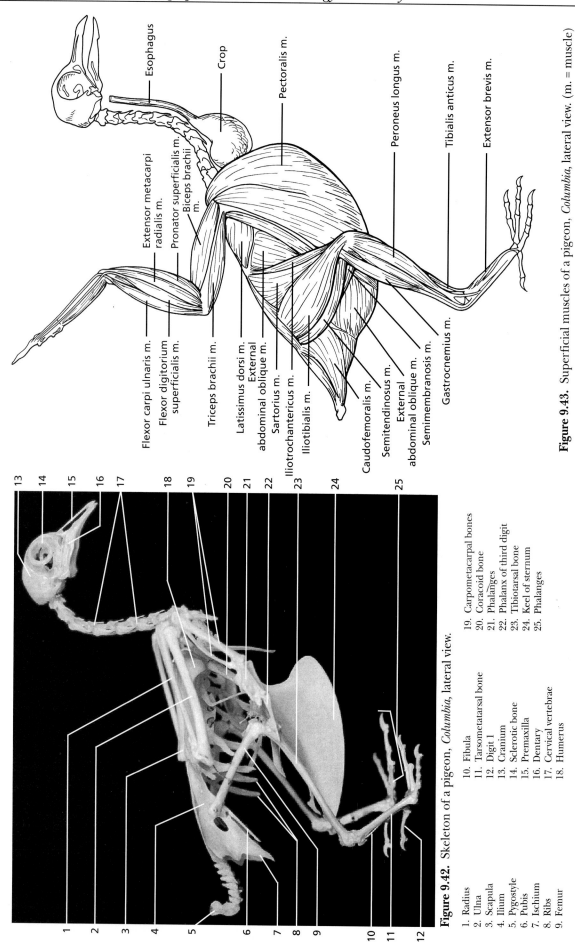

Esophagus

Crop

Pectoralis m.

Peroneus longus m.

Tibialis anticus m.

Extensor brevis m.

Extensor metacarpi radialis m.

Pronator superficialis m.

Biceps brachii m.

Flexor carpi ulnaris m.

Flexor digitorium superficialis m.

Triceps brachii m.

Latissimus dorsi m.

External abdominal oblique m.

Sartorius m.

Iliotrochantericus m.

Iliotibialis m.

Caudofemoralis m.

Semitendinosus m.

External abdominal oblique m.

Semimembranosis m.

Gastrocnemius m.

Figure 9.43. Superficial muscles of a pigeon, *Columbia*, lateral view. (m. = muscle)

Figure 9.42. Skeleton of a pigeon, *Columbia*, lateral view.

1. Radius
2. Ulna
3. Scapula
4. Ilium
5. Pygostyle
6. Pubis
7. Ischium
8. Ribs
9. Femur
10. Fibula
11. Tarsometatarsal bone
12. Digit 1
13. Cranium
14. Sclerotic bone
15. Premaxilla
16. Dentary
17. Cervical vertebrae
18. Humerus
19. Carpometacarpal bones
20. Coracoid bone
21. Phalanges
22. Phalanx of third digit
23. Tibiotarsal bone
24. Keel of sternum
25. Phalanges

CLASS AVES

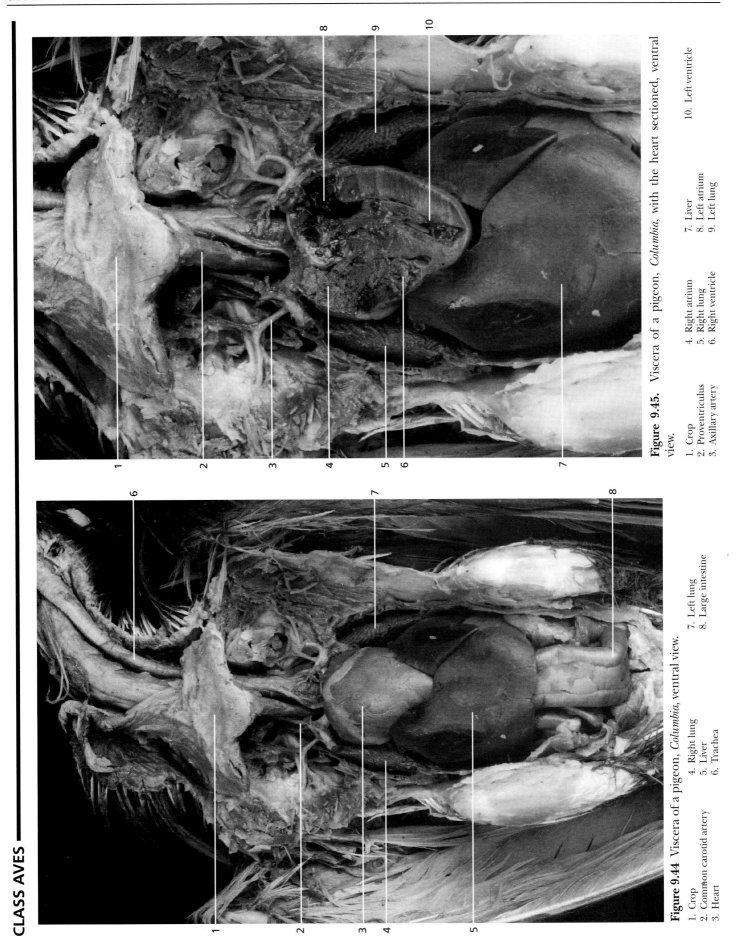

Figure 9.44 Viscera of a pigeon, *Columbia*, ventral view.

1. Crop
2. Common carotid artery
3. Heart
4. Right lung
5. Liver
6. Trachea
7. Left lung
8. Large intestine

Figure 9.45. Viscera of a pigeon, *Columbia*, with the heart sectioned, ventral view.

1. Crop
2. Proventriculus
3. Axillary artery
4. Right atrium
5. Right lung
6. Right ventricle
7. Liver
8. Left atrium
9. Left lung
10. Left ventricle

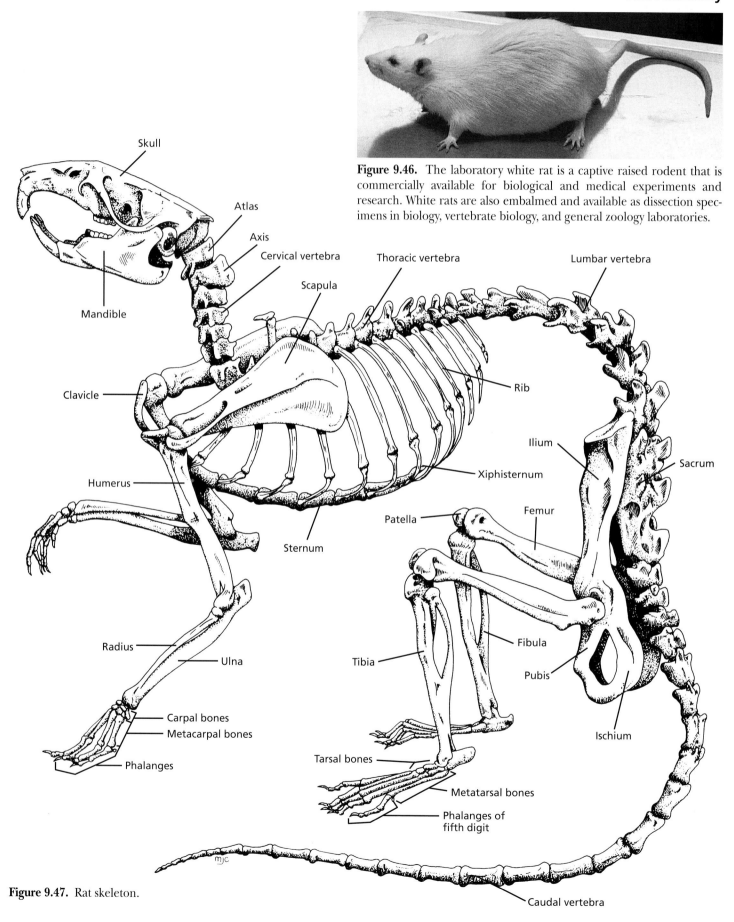

Figure 9.46. The laboratory white rat is a captive raised rodent that is commercially available for biological and medical experiments and research. White rats are also embalmed and available as dissection specimens in biology, vertebrate biology, and general zoology laboratories.

Figure 9.47. Rat skeleton.

RAT ANATOMY

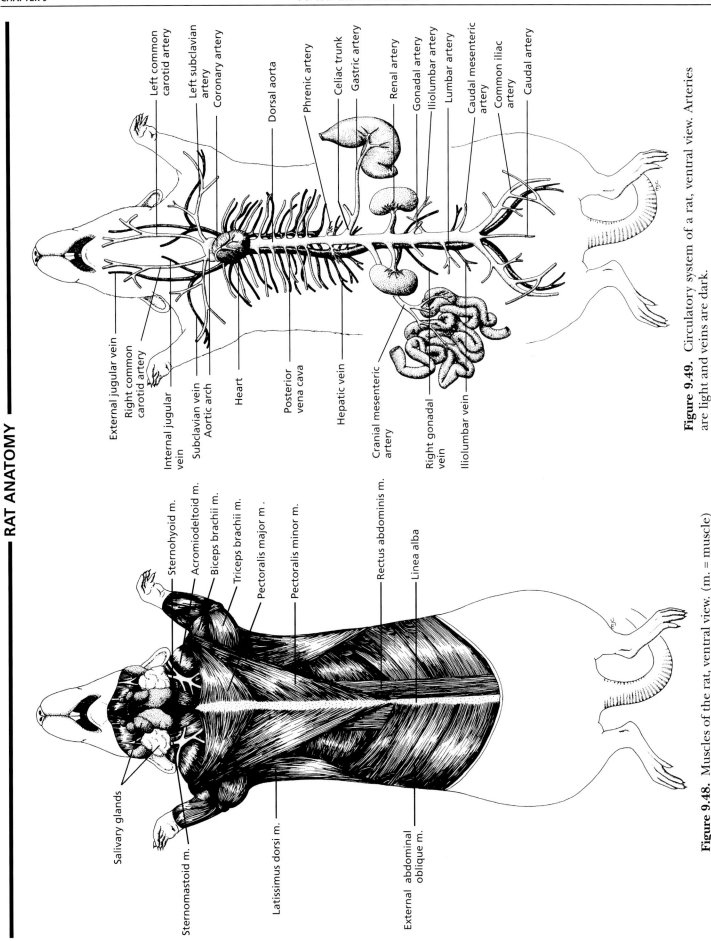

Left common carotid artery
Left subclavian artery
Coronary artery
Dorsal aorta
Phrenic artery
Celiac trunk
Gastric artery
Renal artery
Gonadal artery
Iliolumbar artery
Lumbar artery
Caudal mesenteric artery
Common iliac artery
Caudal artery

External jugular vein
Right common carotid artery
Internal jugular vein
Subclavian vein
Aortic arch
Heart
Posterior vena cava
Hepatic vein
Cranial mesenteric artery
Right gonadal vein
Iliolumbar vein

Figure 9.49. Circulatory system of a rat, ventral view. Arteries are light and veins are dark.

Sternohyoid m.
Acromiodeltoid m.
Biceps brachii m.
Triceps brachii m.
Pectoralis major m.
Pectoralis minor m.
Rectus abdominis m.
Linea alba

Salivary glands
Sternomastoid m.
Latissimus dorsi m.
External abdominal oblique m.

Figure 9.48. Muscles of the rat, ventral view. (m. = muscle)

RAT ANATOMY

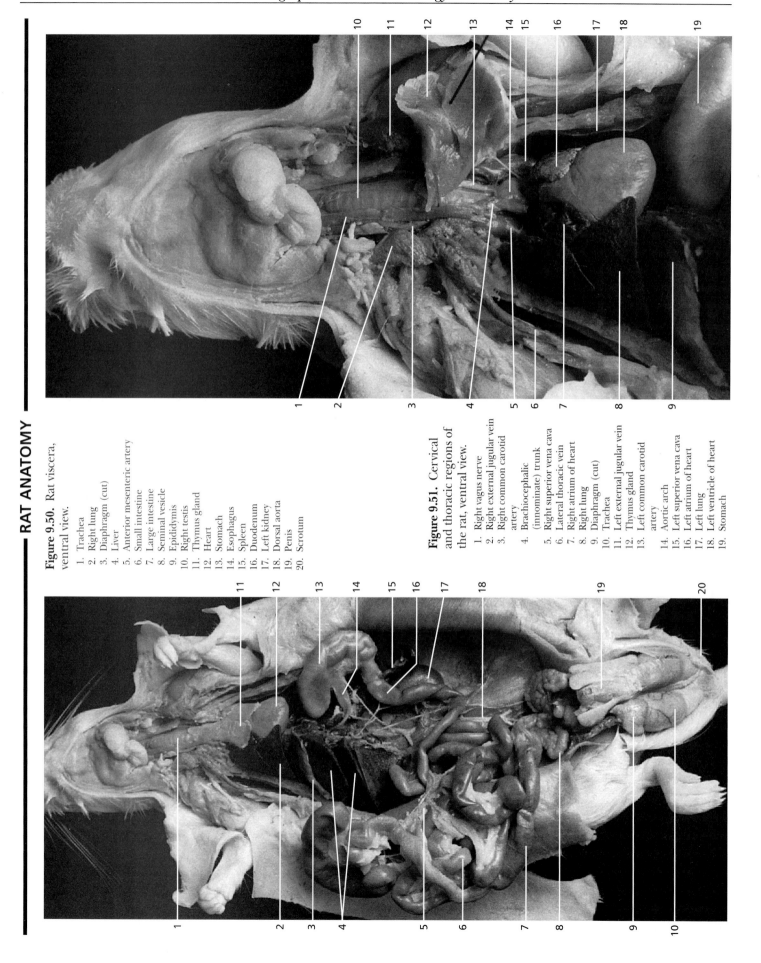

Figure 9.50. Rat viscera, ventral view.

1. Trachea
2. Right lung
3. Diaphragm (cut)
4. Liver
5. Anterior mesenteric artery
6. Small intestine
7. Large intestine
8. Seminal vesicle
9. Epididymis
10. Right testis
11. Thymus gland
12. Heart
13. Stomach
14. Esophagus
15. Spleen
16. Duodenum
17. Left kidney
18. Dorsal aorta
19. Penis
20. Scrotum

Figure 9.51. Cervical and thoracic regions of the rat, ventral view.

1. Right vagus nerve
2. Right external jugular vein
3. Right common carotid artery
4. Brachiocephalic (innominate) trunk
5. Right superior vena cava
6. Lateral thoracic vein
7. Right atrium of heart
8. Right lung
9. Diaphragm (cut)
10. Trachea
11. Left external jugular vein
12. Thymus gland
13. Left common carotid artery
14. Aortic arch
15. Left superior vena cava
16. Left atrium of heart
17. Left lung
18. Left ventricle of heart
19. Stomach

RAT ANATOMY

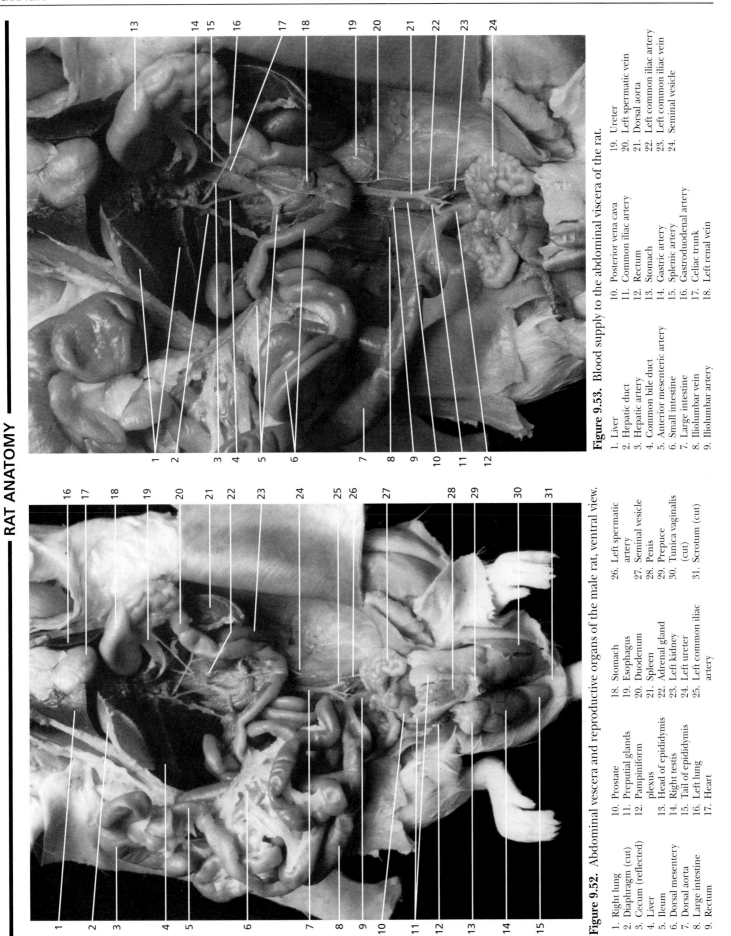

Figure 9.53. Blood supply to the abdominal viscera of the rat.

1. Liver
2. Hepatic duct
3. Hepatic artery
4. Common bile duct
5. Anterior mesenteric artery
6. Small intestine
7. Large intestine
8. Iliolumbar vein
9. Iliolumbar artery
10. Posterior vena cava
11. Common iliac artery
12. Rectum
13. Stomach
14. Gastric artery
15. Splenic artery
16. Gastroduodenal artery
17. Celiac trunk
18. Left renal vein
19. Ureter
20. Left spermatic vein
21. Dorsal aorta
22. Left common iliac artery
23. Left common iliac vein
24. Seminal vesicle

Figure 9.52. Abdominal vescera and reproductive organs of the male rat, ventral view.

1. Right lung
2. Diaphragm (cut)
3. Cecum (reflected)
4. Liver
5. Ileum
6. Dorsal mesentery
7. Dorsal aorta
8. Large intestine
9. Rectum
10. Prostate
11. Preputial glands
12. Pampiniform plexus
13. Head of epididymis
14. Right testis
15. Tail of epididymis
16. Left lung
17. Heart
18. Stomach
19. Esophagus
20. Duodenum
21. Spleen
22. Adrenal gland
23. Left kidney
24. Left ureter
25. Left common iliac artery
26. Left spermatic artery
27. Seminal vesicle
28. Penis
29. Prepuce
30. Tunica vaginalis (cut)
31. Scrotum (cut)

RAT ANATOMY

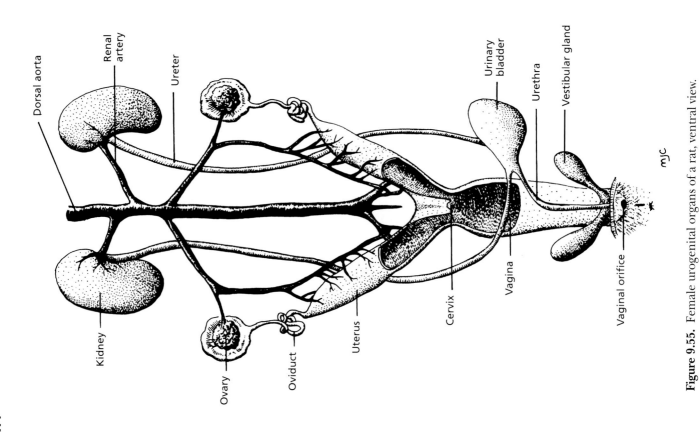

Figure 9.55. Female urogenital organs of a rat, ventral view.

Dorsal aorta

Renal artery

Ureter

Urinary bladder

Urethra

Vestibular gland

Kidney

Ovary

Oviduct

Uterus

Cervix

Vagina

Vaginal orifice

mjc

Figure 9.54. Male urogenital organs of a rat, ventral view.

Adrenal gland

Coagulating gland

Seminal vesicle

Urinary bladder

Ductus (vas) deferens

Prostate

Bulbourethral gland

Preputial gland

Urogenital orifice

Kidney

Ureter

Ampullary gland

Urethra

Penis

Testis

Epididymis

mjc

CAT ANATOMY

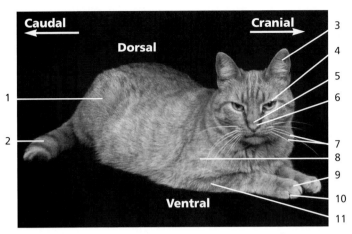

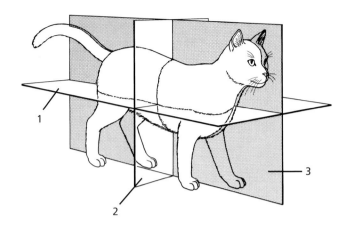

Figure 9.56. Directional terminology and superficial structures in a cat (quadrupedal vertebrate).

1. Thigh	7. Vibrissae
2. Tail	8. Brachium
3. Auricle (pinna)	9. Manus (front foot)
4. Superior palpebra (superior eyelid)	10. Claw
5. Bridge of nose	11. Antebrachium
6. Naris (nostril)	

Figure 9.57. Planes of reference in a cat (quadrupedal vertebrate).

1. Coronal plane (frontal plane)
2. Transverse plane (cross sectional plane)
3. Midsagittal plane (median plane)

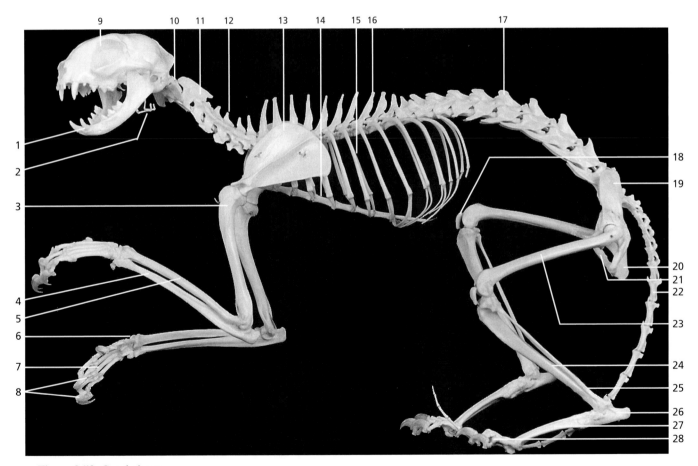

Figure 9.58. Cat skeleton.

1. Mandible	7. Metacarpal bones	13. Scapula	19. Ilium	25. Fibula
2. Hyoid bone	8. Phalanges	14. Sternum	20. Ischium	26. Tarsal bones
3. Humerus	9. Skull	15. Rib	21. Pubis	27. Metatarsal bones
4. Ulna	10. Atlas	16. Thoracic vertebra	22. Caudal vertebra	28. Phalanges
5. Radius	11. Axis	17. Lumbar vertebra	23. Femur	
6. Carpal bones	12. Cervical vertebra	18. Patella	24. Tibia	

CAT ANATOMY

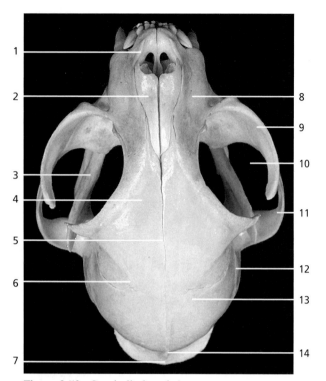

Figure 9.59. Cat skull, dorsal view.

1. Premaxilla
2. Nasal bone
3. Mandible
4. Frontal bone
5. Sagittal suture
6. Coronal suture
7. Nuchal crest
8. Maxilla
9. Zygomatic (malar) bone
10. Orbit
11. Zygomatic arch
12. Temporal bone
13. Parietal bone
14. Interparietal bone

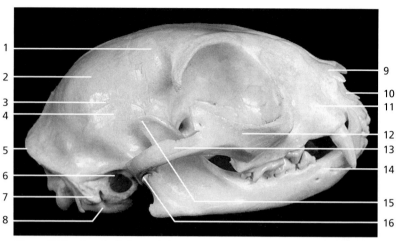

Figure 9.60. Cat skull, lateral view.

1. Frontal bone
2. Parietal bone
3. Squamosal suture
4. Temporal bone
5. Nuchal crest
6. External acoustic meatus
7. Mastoid process
8. Tympanic bulla
9. Nasal bone
10. Premaxilla
11. Maxilla
12. Zygomatic (malar) bone
13. Zygomatic arch
14. Mandible
15. Coronoid process of mandible
16. Condyloid process of mandible

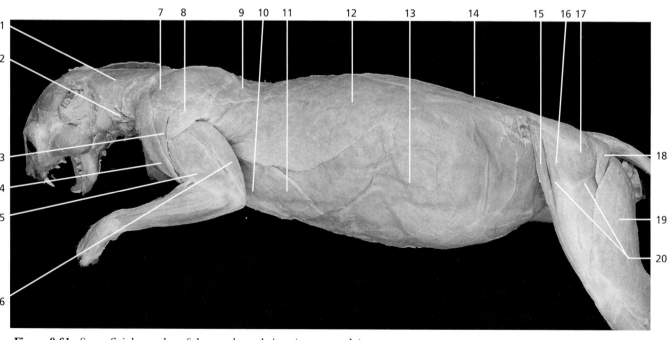

Figure 9.61. Superficial muscles of the cat, lateral view. (m. = muscle)

1. Clavotrapezius m.
2. Sternomastoid m.
3. Acromiodeltoid m.
4. Clavobrachialis m.
5. Triceps brachii m., lateral head
6. Triceps brachii m., long head
7. Acromiotrapezius m.
8. Spinodeltoid m.
9. Spinotrapezius m.
10. Pectoralis minor m.
11. Xiphihumeralis m.
12. Latissimus dorsi m.
13. External abdominal oblique m.
14. Lumbodorsal fascia
15. Sartorius m.
16. Gluteus medius m.
17. Gluteus maximus m.
18. Caudofemoralis m.
19. Biceps femoris m.
20. Tensor fasciae latae

CAT ANATOMY

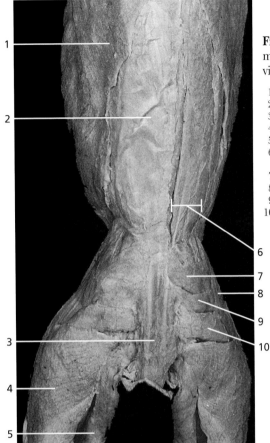

Figure 9.62. Superficial muscles of the cat, dorsal view.

1. Latissimus dorsi m.
2. Lumbodorsal fascia
3. Caudal muscles
4. Biceps femoris m.
5. Semitendinosus m.
6. Extensor dorsi communis m. (sacrospinalis)
7. Gluteus medius m.
8. Tensor fasciae latae m.
9. Gluteus maximus m.
10. Caudofemoralis m.

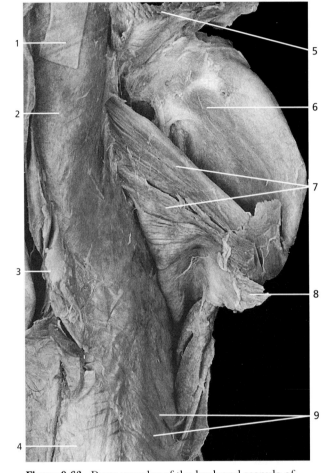

Figure 9.63. Deep muscles of the back and scapula of the cat.

1. Rhomboideus capitis m.
2. Splenius m.
3. Rhomboideus m. (cut)
4. Spinalis dorsi m.
5. Trapezius m. (cut)
6. Subscapularis m.
7. Serratus posterior m.
8. Rhomboideus m. (cut)
9. Serratus anterior m.

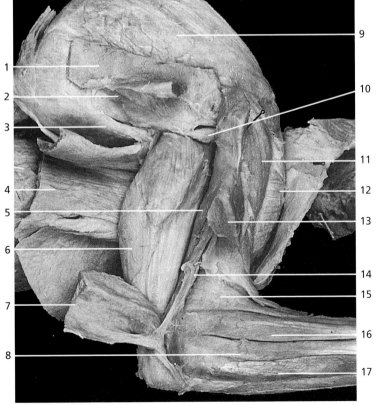

Figure 9.64. Deep muscles of the cat shoulder and arm, lateral view.

1. Acromiotrapezius m.
2. Infraspinatus m.
3. Teres major m.
4. Latissimus dorsi m.
5. Triceps brachii m., medial head
6. Triceps brachii m., long head
7. Triceps brachii m., lateral head (cut)
8. Extensor digitorum lateralis m.
9. Supraspinatus m.
10. Teres minor m.
11. Brachialis m.
12. Pectoralis major m.
13. Triceps brachii m., lateral head (cut)
14. Radial nerve
15. Extensor carpi radialis longus m.
16. Extensor digitorum communis m.
17. Extensor carpi ulnaris m.

CAT ANATOMY

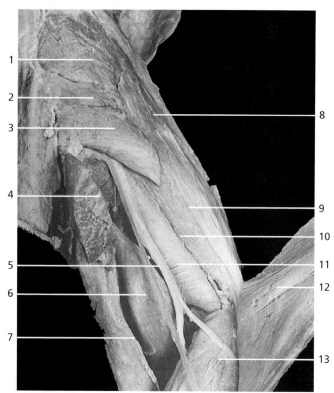

Figure 9.65. Deep muscles of the cat thigh, lateral view.

1. Gluteus medius m.
2. Gluteus maximus m.
3. Caudofemoralis m.
4. Biceps femoris m.,
   (cut)
5. Sciatic nerve
6. Semimembranosus m.
7. Semitendinosus m.
8. Tensor fascae latae m.
9. Vastus lateralis m.
10. Adductor femoris m.
11. Sciatic nerve
12. Biceps femoris m.,
    (cut)
13. Gastrocnemius m.

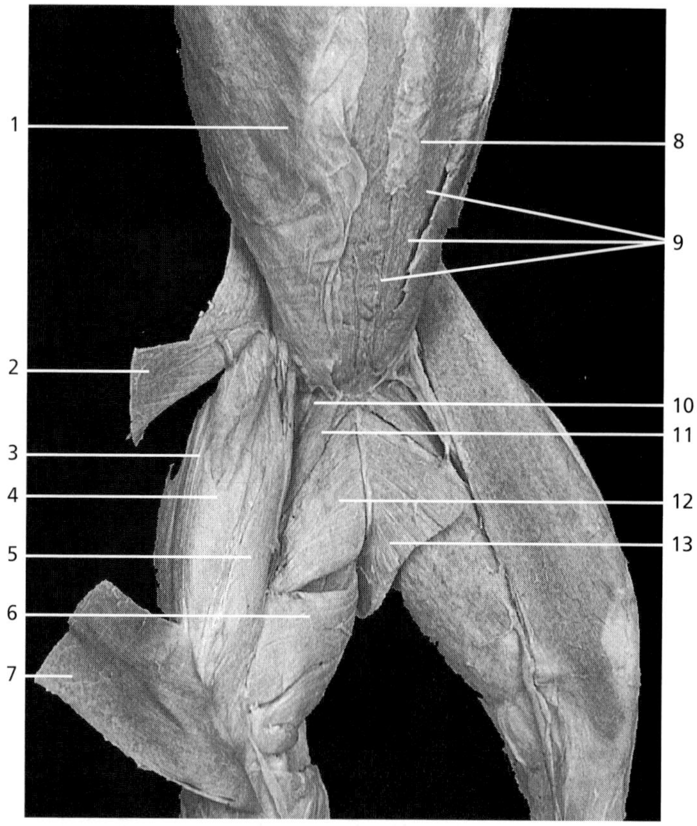

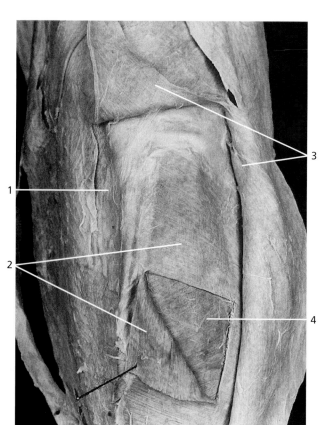

Figure 9.66. Deep muscles of the cat thigh, medial view.

1. External abdominal oblique m.
2. Sartorius m., (cut)
3. Vastus lateralis m.
4. Rectus femoris m.
5. Vastus medialis m.
6. Semimembranosus m.
7. Sartorius m., (cut)
8. Linea alba
9. Rectus abdominus m.
10. Pectineus m.
11. Adductor longus m.
12. Adductor femoris m.
13. Gracilis m., (cut)

Figure 9.67. Abdominal muscles of the cat, ventral view.

1. Rectus abdominus m.
2. Internal abdominal oblique m.
3. External abdominal oblique m.
4. Transverse abdominus m.

CAT ANATOMY

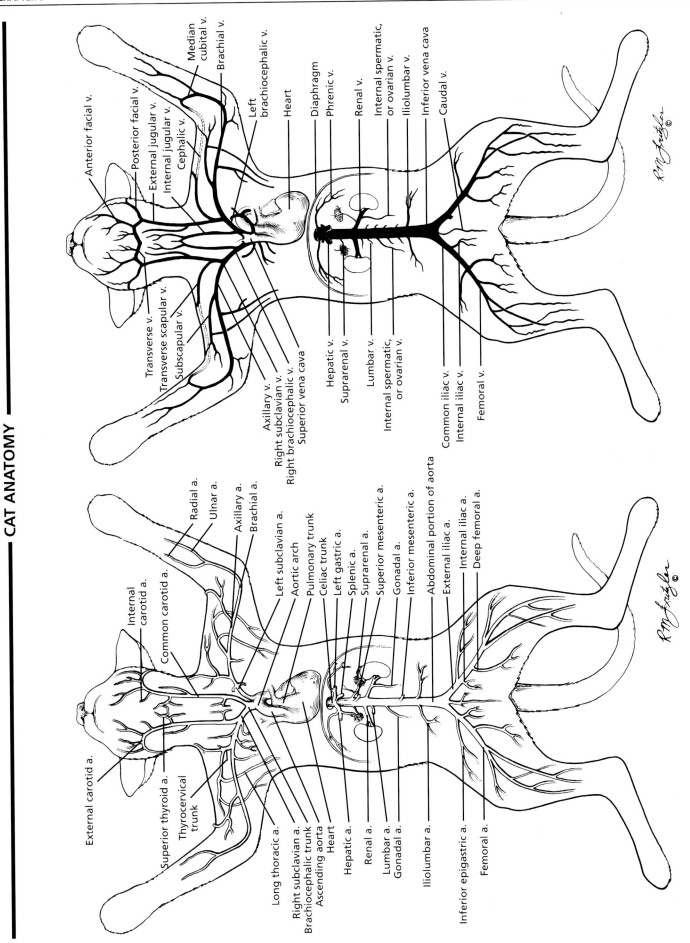

Figure 9.69. Principal veins of the cat, ventral view. (v. = vein)

Figure 9.68. Principal arteries of the cat, ventral view. (a. = artery)

CAT ANATOMY

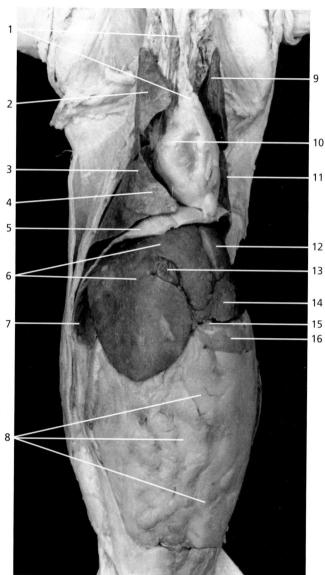

Figure 9.70. Cat dissection, ventral view.

1. Thymus
2. Right anterior lobe of lung
3. Right middle lobe of lung
4. Right posterior lobe of lung
5. Diaphragm
6. Right median lobe of liver
7. Right lateral lobe of liver
8. Greater omentum
9. Left anterior lobe of lung
10. Heart
11. Left middle lobe of lung
12. Left median lobe of liver
13. Gallbladder
14. Left lateral lobe of liver
15. Lesser omentum
16. Stomach

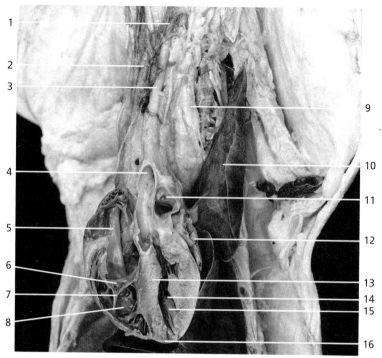

Figure 9.71. Heart and associated structures of the cat, ventral view.

1. Brachiocephalic vein
2. Superior vena cava
3. Brachiocephalic artery
4. Aortic arch
5. Right atrium
6. Cusp of right atrioventricular (tricuspid) valve
7. Chordae tendinae
8. Right ventricle
9. Esophagus
10. Left lung
11. Pulmonary trunk
12. Left auricle
13. Interventricular septum
14. Left ventricle
15. Papillary muscle
16. Apex of heart

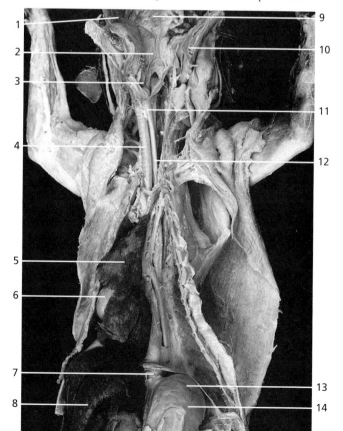

Figure 9.72. Upper GI tract of the cat, ventral view.

1. Tongue
2. Soft palate
3. Epiglottis
4. Trachea
5. Right lung
6. Heart
7. Lower esophageal sphincter
8. Liver
9. Hard palate
10. Mandible (split)
11. Larynx
12. Esophagus
13. Cardia of stomach
14. Stomach

CAT ANATOMY

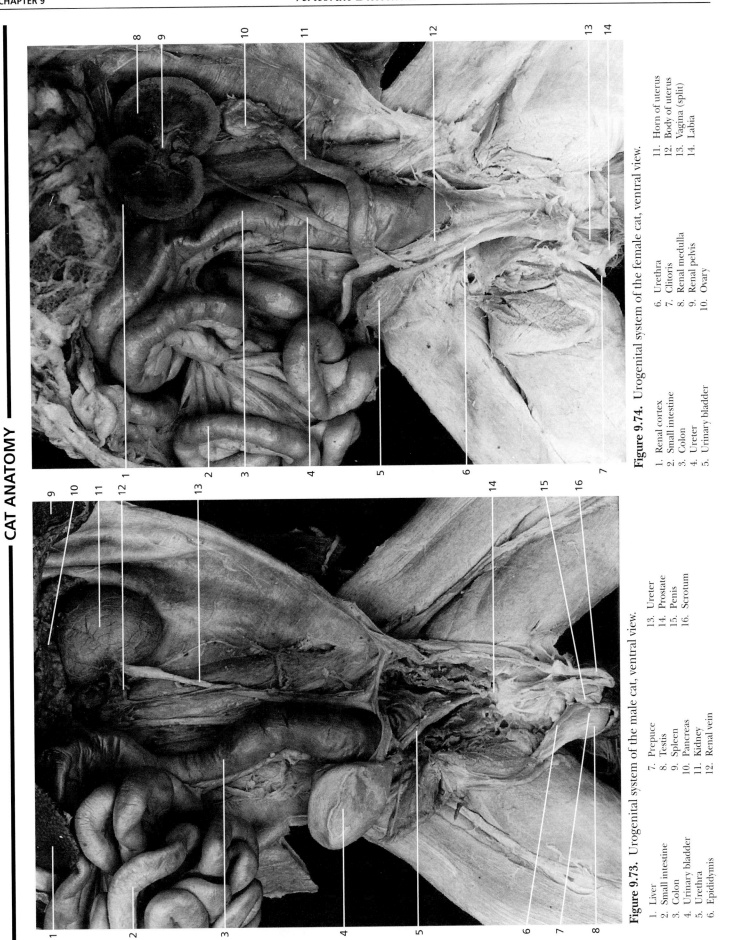

Figure 9.73. Urogenital system of the male cat, ventral view.

1. Liver
2. Small intestine
3. Colon
4. Urinary bladder
5. Urethra
6. Epididymis
7. Prepuce
8. Testis
9. Spleen
10. Pancreas
11. Kidney
12. Renal vein
13. Ureter
14. Prostate
15. Penis
16. Scrotum

Figure 9.74. Urogenital system of the female cat, ventral view.

1. Renal cortex
2. Small intestine
3. Colon
4. Ureter
5. Urinary bladder
6. Urethra
7. Clitoris
8. Renal medulla
9. Renal pelvis
10. Ovary
11. Horn of uterus
12. Body of uterus
13. Vagina (split)
14. Labia

FETAL PIG ANATOMY

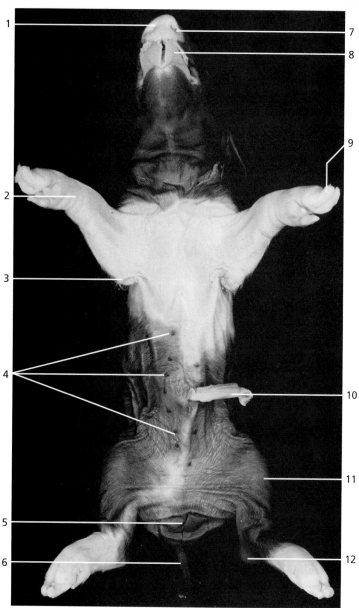

Figure 9.75. Surface anatomy of fetal pig, ventral view.

1. Nose (snout)
2. Wrist
3. Elbow
4. Teats
5. Scrotum
6. Tail
7. Naris (nostril)
8. Tongue
9. Digit
10. Umbilical cord
11. Knee
12. Ankle

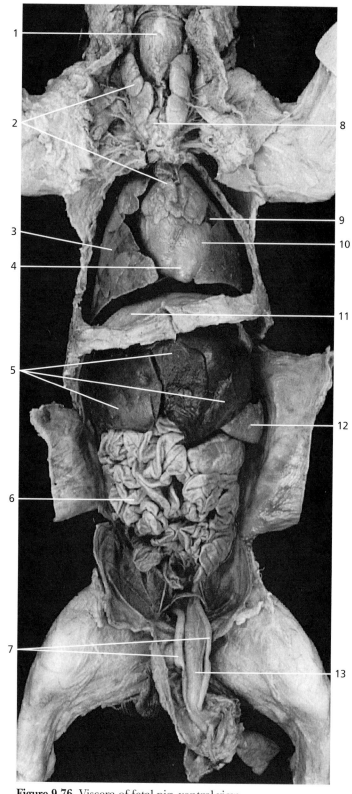

Figure 9.76. Viscera of fetal pig, ventral view.

1. Larynx
2. Thymus
3. Right lung
4. Ventricle of the heart
5. Liver
6. Small intestine
7. Umbilical arteries
8. Thyroid gland
9. Atrium of the heart
10. Left lung
11. Diaphragm
12. Spleen
13. Urinary bladder

FETAL PIG ANATOMY

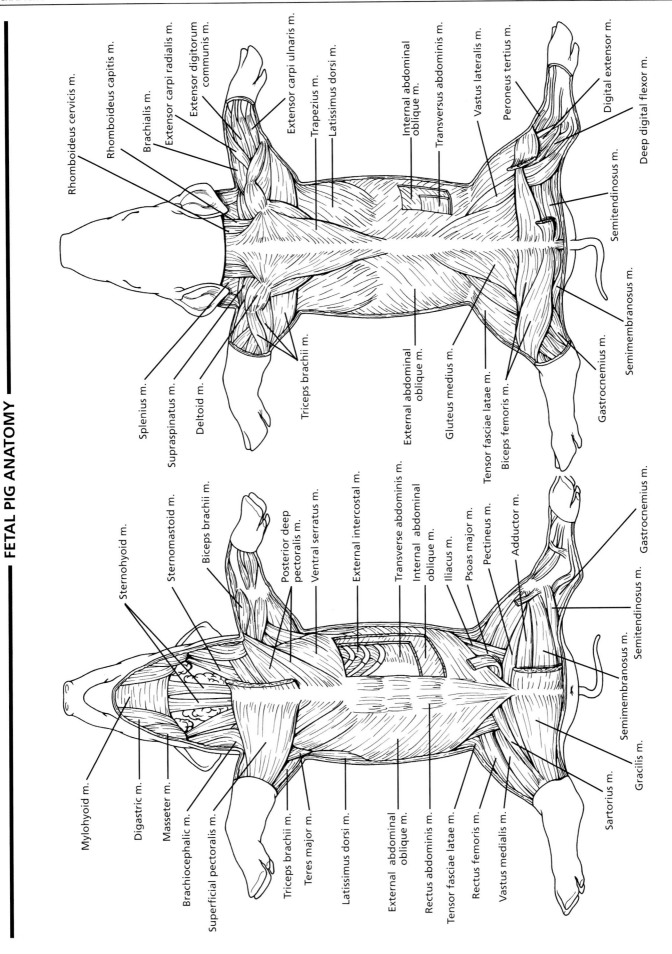

Rhomboideus cervicis m.

Rhomboideus capitis m.

Brachialis m.

Extensor carpi radialis m.

Extensor carpi ulnaris m.

Extensor digitorum communis m.

Trapezius m.

Latissimus dorsi m.

Internal abdominal oblique m.

Transversus abdominis m.

Vastus lateralis m.

Peroneus tertius m.

Digital extensor m.

Deep digital flexor m.

Semitendinosus m.

Semimembranosus m.

Gastrocnemius m.

Biceps femoris m.

Tensor fasciae latae m.

Gluteus medius m.

External abdominal oblique m.

Triceps brachii m.

Deltoid m.

Supraspinatus m.

Splenius m.

Figure 9.78. Muscles of the fetal pig, dorsal view.

Mylohyoid m.

Digastric m.

Masseter m.

Brachiocephalic m.

Superficial pectoralis m.

Triceps brachii m.

Teres major m.

Latissimus dorsi m.

External abdominal oblique m.

Rectus abdominis m.

Tensor fasciae latae m.

Rectus femoris m.

Vastus medialis m.

Sternohyoid m.

Sternomastoid m.

Biceps brachii m.

Posterior deep pectoralis m.

Ventral serratus m.

External intercostal m.

Transverse abdominis m.

Internal abdominal oblique m.

Iliacus m.

Psoas major m.

Pectineus m.

Adductor m.

Sartorius m.

Gracilis m.

Semimembranosus m.

Semitendinosus m.

Gastrocnemius m.

Figure 9.77. Muscles of the fetal pig, ventral view. (m. = muscle)

FETAL PIG ANATOMY

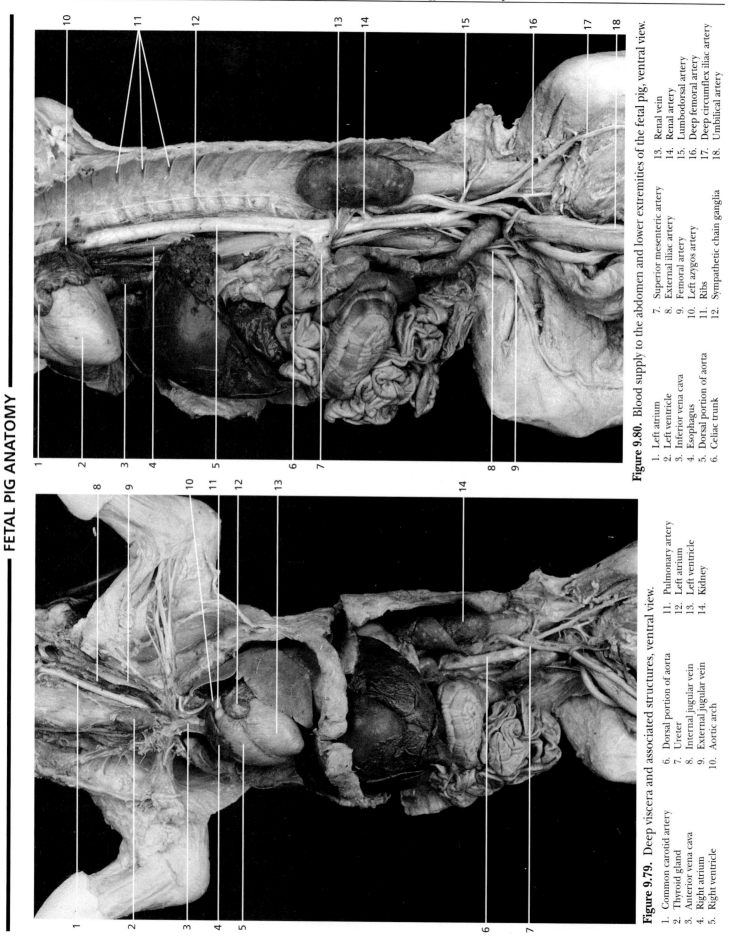

Figure 9.79. Deep viscera and associated structures, ventral view.

1. Common carotid artery	6. Dorsal portion of aorta	11. Pulmonary artery
2. Thyroid gland	7. Ureter	12. Left atrium
3. Anterior vena cava	8. Internal jugular vein	13. Left ventricle
4. Right atrium	9. External jugular vein	14. Kidney
5. Right ventricle	10. Aortic arch	

Figure 9.80. Blood supply to the abdomen and lower extremities of the fetal pig, ventral view.

1. Left atrium	7. Superior mesenteric artery	13. Renal vein
2. Left ventricle	8. External iliac artery	14. Renal artery
3. Inferior vena cava	9. Femoral artery	15. Lumbodorsal artery
4. Esophagus	10. Left azygos artery	16. Deep femoral artery
5. Dorsal portion of aorta	11. Ribs	17. Deep circumflex iliac artery
6. Celiac trunk	12. Sympathetic chain ganglia	18. Umbilical artery

FETAL PIG ANATOMY

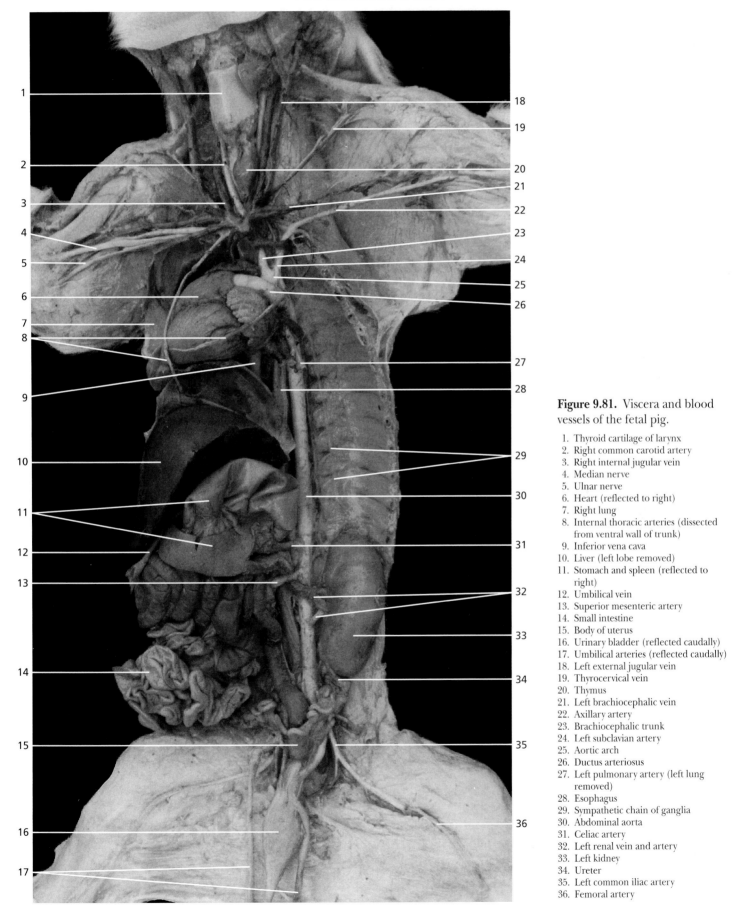

Figure 9.81. Viscera and blood vessels of the fetal pig.

1. Thyroid cartilage of larynx
2. Right common carotid artery
3. Right internal jugular vein
4. Median nerve
5. Ulnar nerve
6. Heart (reflected to right)
7. Right lung
8. Internal thoracic arteries (dissected from ventral wall of trunk)
9. Inferior vena cava
10. Liver (left lobe removed)
11. Stomach and spleen (reflected to right)
12. Umbilical vein
13. Superior mesenteric artery
14. Small intestine
15. Body of uterus
16. Urinary bladder (reflected caudally)
17. Umbilical arteries (reflected caudally)
18. Left external jugular vein
19. Thyrocervical vein
20. Thymus
21. Left brachiocephalic vein
22. Axillary artery
23. Brachiocephalic trunk
24. Left subclavian artery
25. Aortic arch
26. Ductus arteriosus
27. Left pulmonary artery (left lung removed)
28. Esophagus
29. Sympathetic chain of ganglia
30. Abdominal aorta
31. Celiac artery
32. Left renal vein and artery
33. Left kidney
34. Ureter
35. Left common iliac artery
36. Femoral artery

FETAL PIG ANATOMY

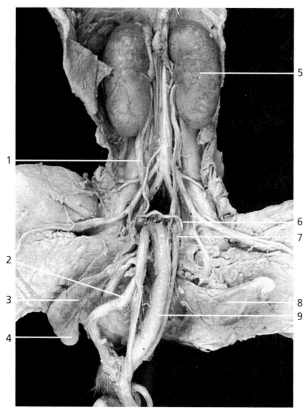

Figure 9.82. Urogenital system of the male fetal pig, ventral view.

1. Ureter
2. Penis
3. Testis
4. Epididymis
5. Kidney
6. Ductus (vas) deferens
7. Umbilical artery
8. Spermatic cord
9. Urinary bladder

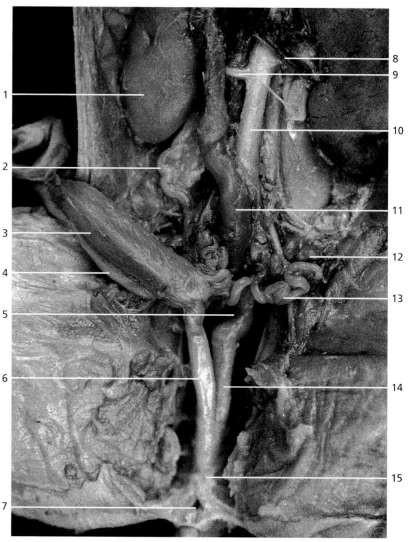

Figure 9.83. Urogenital system of the female fetal pig, ventral view. The left ureter and left umbilical artery have been cut.

1. Right kidney
2. Ureter
3. Urinary bladder
4. Umbilical artery
5. Uterus
6. Urethra
7. Urogenital orifice
8. Renal vein
9. Renal artery
10. Abdominal descending aorta
11. Rectum
12. Ovary
13. Horn of uterus
14. Vagina
15. Urogenital sinus

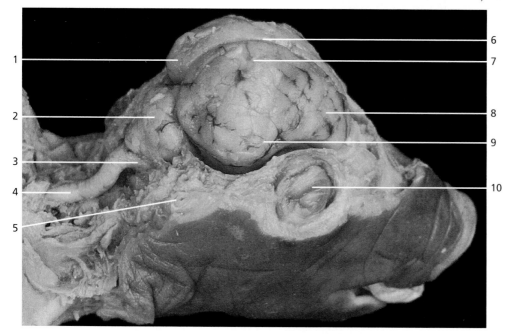

Figure 9.84. General structures of the fetal pig brain. Because the cerebrum is less defined in pigs, the regions are not known as lobes as they are in humans.

1. Occipital region of cerebrum
2. Cerebellum
3. Medulla oblongata
4. Spinal cord
5. External auditory meatus
6. Longitudinal fissure
7. Parietal region of cerebrum
8. Frontal region of cerebrum
9. Temporal region of cerebrum
10. Eye

SHEEP

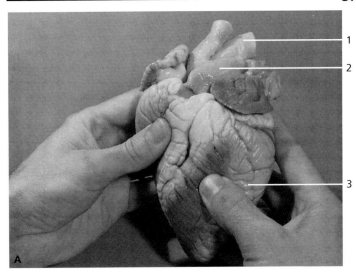

Position the heart so the ventral surface faces you. Notice the thicker ventricular walls, especially the left ventricle.

1. Aortic arch
2. Pulmonary trunk
3. Left ventricle

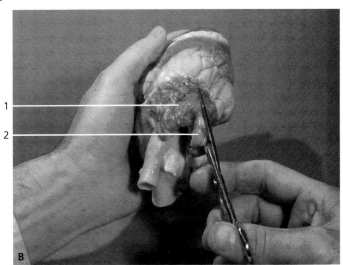

Insert scissors into the superior vena cava. The cut should expose the interior of the right atrium. Notice the tricuspid valve.

1. Right atrium
2. Superior vena cava

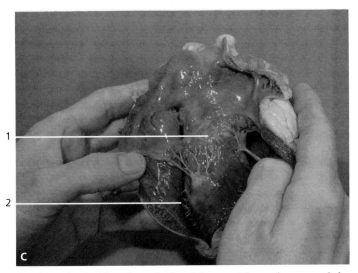

Continue the incision through the right ventricle to the apex of the heart. Observe the structure of the valve.

1. Tricuspid valve
2. Right ventricle

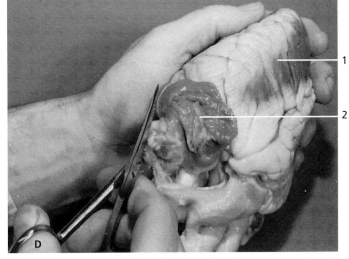

Begin the next incision in the left atrium. This time, continue through both the atrium and the ventricle.

1. Left ventricle
2. Left atrium

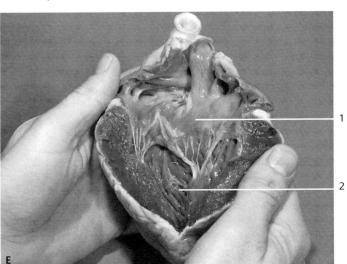

Figure 9.85. Sheep heart dissection.

Expose the left ventricle and atrium. Notice the difference between the right and left ventricle, especially the thicker muscular wall.

1. Bicuspid valve
2. Left ventricle

Glossary of Prefixes and Suffixes

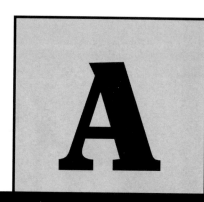

Element	Definition and Example	Element	Definition and Example	Element	Definition and Example
a-	absent, deficient or without: atrophy	cata-	lower, under, against: catabolism	ede-	swelling: edema
ab-	off, away from: abduct	-coel	swelling, and enlarged space or cavity: blastocoele	-emia	pertaining to a condition of the blood: lipemia
abdomin-	abdomen			end-	within: endoderm
-able	capable of: viable	cephal-	head: cephalis	entero-	intestine: enteritis
ac-	toward, to: actin	cerebro-	brain: cerebrospinal fluid	epi-	upon, in addition: epidermis
acou-	hear, acoustic	chol-	bile: cholic	erythro-	red: erythrocyte
ad-	denoting to, toward: adduct	chondr-	cartilage: chondrocyte	ex-	out of: excise
af-	movement toward a central point: afferent artery	chrom-	color: chromocyte	exo-	outside: exocrine
		-cid	destroy: germicide	extra-	outside of, beyond, in addition: extracellular
alba-	pale or white: linea alba	circum-	around: circumduct		
-alg	pain: neuralgia	-cis	cut, kill: excision		
ambi-	both: ambidextrous	co-	together: copulation	fasci-	band: fascia
angi-	pertaining to vessel: angiology	coel-	hollow cavity: coelom	febr-	fever: febrile
ante-	before: antebrachium	con-	with, together: congenital	-ferent	bear, carry: efferent arteriole
anti-	against: anticoagulant	contra-	against, opposite: contraception	fiss-	split: fissure
aqua-	water: aqueous			for-	opening: foramen
archi-	to be first: archeteron	corn-	denoting hardness: cornified	-form	shape: fusiform
arthri-	joint: arthritis	corp-	body: corpus		
-asis	condition or state of: homeostasis	crypt-	hidden: cryptorchism	gastro-	stomach: gastrointestinal
		cyan-	blue color: cyanosis	-gen	an agent that produces or originates: pathogen
aud-	pertaining to ear: auditory	cysti-	sac or bladder: cystoscope		
auto-	self: autolysis	cyto-	cell: cytology	-genic	produced from, producing: carcinogenic
bi-	two: bipedal	de-	down, from: descent	gloss-	tongue: glossopharyngeal
bio-	life: biology	derm-	skin: dermatology	glyco-	sugar: glycosuria
blast-	generative or germ bud: osteoblast	di-	two: diarthrotic	-gram	a record, recording: myogram
		dipl-	double: diploid	gran-	grain, particle: agranulocyte
brachi-	arm: brachialis	dis-	apart, away from: disarticulate	-graph	instrument for recording: electrocardiograph
brachy-	short: brachydont	duct-	lead, conduct: ductus deferens		
brady-	slow: bradycardia	dur-	hard: dura mater	grav-	heavy: gravid
bucc-	cheek: buccal cavity	-dynia	pain	gyn-	female sex: gynecology
		dys-	bad, difficult, painful: dysentery		
cac-	bad, ill: cachexia			hema(o)-	blood: hematology
calci-	stone: calculus	e-	out, from: eccrine	haplo-	simple or single: haploid
capit-	head: capitis	ecto-	outside, outer, external: ectoderm	hemi-	half: hemiplegia
carcin-	cancer: carcinogenic			hepat-	liver: hepatic portal
cardi-	heart: cardiac			hetero-	other, different: heterosexual
caud-	tail: cauda equina	-ectomy	surgical removal: tonsillectomy	histo-	webb, tissue: histology

Element	Definition and Example
holo-	whole, entire: holocrine
homo-	same, alike: homologous
hydro-	water: hydrocoel
hyper-	beyond, above, excessive: hypertension
hypo-	under, below: hypoglycemia
-ia	state or condition: hypoglycemia
-iatrics	medical specialties: pediatrics
idio-	self, separate, distinct: idiopathic
ilio-	ilium: iliosacral
infra-	beneath: infraspinatus
inter-	among, between
intra-	inside, within: intracellular
-ion	process: acromion
iso-	equal, like: isotonic
-ism	condition or state: rheumatism
-itis	inflammation: meningitis
labi-	lip: labium majus
lacri-	tears: nasolacrimal
later-	side: lateral
leuc-	white: leucocyte
lip-	fat: lipid
-logy	science of: morphology
-lysis	solution, dissolve: hemolysis
macro-	large, great: macrophage
mal-	bad, abnormal, disorder: malignant
medi-	middle: medial
mega-	great, large: megakaryocyte
meso-	middle or moderate: mesoderm
meta-	after, beyond: metatarsal
micro-	small: microtome
mito-	thread: mitosis
mono-	alone, one, single: monocyte
mons-	mountain: mons pubis
morph-	form, shape: morphology
multi-	many, much: multinuclear
myo-	muscle: myology
narc-	numbness, stupor: narcotic
neo-	new, young: neonatal
necro-	corpse, dead: necrosis
nephro-	kidney: nephritis
neuro-	nerve: neurolemma
noto-	back: notochord
ob-	against, toward, in front of: obturator

Element	Definition and Example
oc-	against: occlusion
-oid	resembling, likeness: sigmoid
oligo-	few, small: oligodendrocyte
-oma	tumor: lymphoma
oo-	egg: oocyte
or-	mouth: oral
orchi-	testicles: cryptorchidism
-ory	pertaining to: sensory
osteo-	bone: osteoblast
-ose	full of: adipose
oto	ear: otolith
ovo-	egg: ovum
para-	give birth to, bear: parturition
para-	near, beyond, beside: paranasal
path-	disease, that which undergoes sickness: pathology
-pathy	abnormality, disease: neuropathy
ped-	children: pediatrician
pen-	need, lack: penicillin
-penia	deficiency: thrombocytopenia
per-	through: percutaneous
peri-	near, around: pericardium
phag-	to eat: phagocyte
-phil	have an affinity for: neutrophil
phlebo-	vein: phlebitis
-phobe	abnormal fear, dread: hydrophobia
-plasty	reconstruction of: rhinoplasty
platy-	flat, side: platysma
-plegia	stroke, paralysis: paraplegia
-pnea	to breathe: apnea
pneumato-	breathing: pneumonia
pod-	foot: podiatry
-poieis	formation of: hematopoiesis
poly-	many, much: polyploid
post-	after, behind: post natal
pre-	before in time or place: prenatal
prim-	first: primitive
pro-	before in time or place: prosect
proct-	anus: proctology
pseudo-	false: pseudostratified
psycho-	mental: psychology
pyo-	pus: pyoculture
quad-	fourfold: quadriceps femoris
re-	back, again: repolarization
rect-	straight: rectus abdominis
reno-	kidney: renal
rete-	network: retina

Element	Definition and Example
retro-	backward: retroperitoneal
rhin-	nose: rhinitis
-rrhage	excessive flow: hemorrhage
-rrhea	flow, or discharge: diarrhea
sanguin-	blood: sanguiferous
sarc-	flesh: sarcoplasm
-scope	instrument for examination of a part: stethoscope
-sect	cut: dissect
semi-	half: semilunar
serrate-	saw-edged: serratus anterior
-sis	state or condition: dialysis
steno-	narrow: stenohaline
-stomy	surgical opening: tracheotomy
sub-	under, beneath, below: subcutaneous
super-	above, beyond, upper: superficial
supra-	above, over: suprarenal
syn (sym)	together, joined, with: synapse
tachy-	swift, rapid: tachometer
tele-	far: telencephalon
tens-	stretch: tensor fascia lata
tetra-	four: tetrad
therm-	heat: thermogram
thorac-	chest: thoracic cavity
thrombo-	lump, clot: thrombocyte
-tomy	cut: appendectomy
tox-	poison: toxic
tract-	draw, drag: traction
trans-	across, over: transfuse
tri-	three: trigone
trich-	hair: trichology
-trophy	a state relating to nutrition: hypertrophy
-tropic	turning toward, changing: gonadotropic
ultra-	beyond, excess: ultrasonic
uni-	one: unicellular
uro-	urine, urinary organs or tract: uroscope
-uria	urine: polyuria
vas-	vessel: vasoconstriction
vermi-	worm: vermiform
viscer-	organ: visceral
vit-	life: vitamin
zoo-	animal: zoology
zygo-	union, join: zygote

Glossary of Terms

abdomen (ab-do´men): the portion of the trunk of the mammalian body located between the diaphragm and the pelvis, that contains the abdominal cavity and its visceral organs; one of the three principal body regions (head, thorax, and abdomen) of many animals.

abduction (ab-duk´shun): a movement away from the axis or midline of the body; opposite of adduction, a movement of a digit away from the axis of a limb.

abiotic (ab´e-ot´ik): without living organisms; non-living portions of the environment.

abscission: the shedding of leaves, flowers, fruits, or other plant parts, usually following the formation of an abscission zone.

absorption: movement of a substance into a cell or an organism, or through a surface within an organism.

acapnia (ah-kap´ne-ah): a decrease in normal amount of CO_2 in the blood.

accommodation (ah-kom-o-da´shun): a change in the shape of the lens of the eye so that vision is more acute; the focusing for various distances.

acetone (as´e-tone): an organic compound that may be present in the urine of diabetics; also called ketone body.

acetylcholine (as´e-til-ko´len): a neurotransmitter chemical secreted at the terminal ends of many neurons, responsible for postsynaptic transmission; also called ACh.

acetylcholinesterase (as´e-til-ko´lin-es´ ter-as): an enzyme that breaks down acetylcholine; also called AChE.

Achilles tendon (ah-kil´ez): see *tendo calcaneus*.

acid (as´id): a substance that releases hydrogen ions (H^+) in a solution.

acidosis (as-i-do´sis): a disorder of body chemistry in which the alkaline substances of the blood are reduced in amount below normal.

acoelomate (a-sel´o-mate): without a coelomic cavity; as in flatworms.

acoustic (ah-koos´tik): referring to sound or the sense of hearing.

actin (ak´tin): a protein in muscle fibers that together with myosin is responsible for contraction.

action potential (ak´shun po-ten´shal): the change in ionic charge propagated along the membrane of a neuron; the nerve impulse.

active transport: movement of a substance into or out of a cell from a lesser to a greater concentration, requiring a carrier molecule and expenditure of energy.

adaptation (ad´ap-ta´shun): structural, physiological, or behavioral traits of an organism that promote its survival and contribute to its ability to reproduce under specific environmental conditions.

adduction (ah-duk´shun): a movement toward the axis or midline of the body; opposite of abduction, a movement of a digit toward the axis of a limb.

adenohypophysis (ad´e-no-hi-pof´i-sis): anterior pituitary gland.

adenoid (ad´e-noid): paired lymphoid structures in the nasopharynx; also called pharyngeal tonsils.

adenosine triphosphate (ATP) (ah-den´o-sen tri-fos´fate): a chemical compound that provides energy for cellular use.

adhesion: the attraction between unlike substances.

adipose (ad´e-pose): fat, or fat-containing, such as adipose tissue.

adrenal glands (ah-dre´nal): endocrine glands; one superior to each kidney; also called *suprarenal glands*.

adventitious root (ad´ven-tish´us root): supportive root developing from the stem of a plant.

aerobic (a-er-o´bik): requiring free O_2 for growth and metabolism as in the case of certain bacteria called *aerobes*.

agglutination (ah-gloo´ti-na´shun): clumping of cells; particular reference to red blood cells in an antigen-antibody reaction.

aggregate fruit: ripened ovaries from a single flower with several separate carpels.

aggression (ah-gresh´un): provoking, domineering behavior.

alga (*pl.* **algae**): any of a diverse group of aquatic photosynthesizing organisms that are either unicellular or are multicellular; algae comprise the phytoplankton and seaweeds of the Earth.

alkaline: a substance having a pH greater than 7.0; basic.

allantois (ah-lan´to-is): an extraembryonic membranous sac that forms blood cells and gives rise to the fetal umbilical arteries and vein. It also contributes to the formation of the urinary bladder.

allele (ah-lel´): an alternative form of a gene occurring at a given chromosome site, or locus.

all-or-none response: functioning completely when exposed to a stimulus of threshold strength; applies to action potentials through neurons and muscle fiber contraction.

alpha helix (al´fah he´liks): right-handed spiral typical in proteins and DNA.

alternation of generations: two-phased life cycle characteristic of many plants in which there are sporophyte and gametophyte generations.

altruism (al´troo-ism): behavior benefiting other organisms without regard to its possible advantage or detrimental effect on the performer.

alveolus (al-ve´o-lus): an individual air capsule within the lung. Alveoli are the basic functional units of respiration. Also, the socket that secures a tooth.

amino acid (ah-me´no as´id): a unit of protein that contains an amino group (NH_2) and an acid group (COOH).

amnion (am´ne-on): a membrane that surrounds the fetus to contain the amniotic fluid.

amniote (am´ne-oat): an animal that has an amnion during embryonic development; reptiles, birds, and mammals.

amoeba (ah-me´bah): protozoans that move by means of pseudopodia.

amphiarthrosis (am´fe-ar-thro´sis): a slightly moveable joint in a functional classification of joints.

anaerobic respiration (an-a´er-ob´ik res´pi-ra´shun): metabolizing and growing in the absence of oxygen.

analogous (ah-nal´o-gus): similar in function regardless of developmental origin; generally in reference to similar adaptations.

anatomical position (an´ah-tom´e-kal): the position in human anatomy in which there is an erect body stance with the eyes directed forward, the arms at the sides, and the palms of the hands facing forward.

anatomy (ah-nat´o-me): the branch of science concerned with the structure of the body and the relationship of its organs.

angiosperm: flowering plant, having double fertilization resulting in development of specialized seeds within fruits.

annual: a flowering plant that completes its entire life cycle in a single year or growing season.

annual rings: yearly growth demarcations in woody plants formed by buildup of secondary xylem.

annulus: a ringlike segment, such as body rings on leeches.

antebrachium (an´te-bra´ke-um): the forearm.

antenna: a sensory appendage on many species of invertebrate animals.

anterior (ventral) (an-te´re-or): toward the front; the opposite of *posterior (dorsal)*.

anther (an´ther): the portion of a plant stamen in which pollen is produced.

antheridium (an´ther-id´e-um): male reproductive organ in certain nonseed plants and algae where motile sperm are produced.

anticodon (an´ti-ko´don): three ("a triplet") nucleotide sequence in transfer RNA that pairs with a complementary codon (triplet) in messenger RNA.

antigen (an´ti-jen): a foreign material, usually a protein, that triggers the immune system to produce antibodies.

anus (a´nus): the terminal end of the GI tract, opening of the anal canal.

aorta (a-or´tah): the major systemic vessel of the arterial portion of the circulatory system, emerging from the left ventricle.

apical meristem: embryonic plant tissue in the tip of a root, bud, or shoot where continual cell divisions cause growth in length.

apocrine gland (ap´o-krin): a type of sweat gland that functions in evaporative cooling.

apopyle: opening of the radial canal into the spongocoel of sponges.

appeasement (ah-pez´ment): submissive behavior, usually soliciting an end of aggression.

appendix (ah-pen´diks): a short pouch that attaches to the cecum.

aqueous humor (a´kwe-us hu´mor): the watery fluid that fills the anterior and posterior chambers of the eye.

arachnoid mater (ah-rak´noid): the weblike middle covering (meninx) of the central nervous system.

arbor vitae (ar´bor vi´tah): the branching arrangement of white matter within the cerebellum.

archaebacteria (ar´ke-bak-te´re-ah): organisms within the kingdom Monera that represent an early group of simple life forms.

archegonium (ar´ke-go´ne-um): female reproductive organ in certain nonseed plants and algae where eggs are produced.

archenteron (ar-ken´ter-on): the principal cavity of an embryo during the gastrula stage. Lined with endoderm, the archenteron develops into the digestive tract.

areola (ah-re´o-lah): the pigmented ring around the nipple.

artery (ar´ter-e): a blood vessel that carries blood away from the heart.

articular cartilage (ar-tik´u-lar ker´ti-lij): a hyaline cartilaginous covering over the articulating surface of bones of synovial joints.

ascending colon (ko´lon): the portion of the large intestine between the cecum and the hepatic (right colic) flexure.

asexual: lacking distinct sexual organs and lacking the ability to produce gametes.

aster (as´ter): minute rays of microtubules at the ends of the spindle apparatus in animal cells during cell division.

asymmetry (a-sim´e-tre): not symmetrical.

atom (at´om): the smallest unit of an element that can exist and still have the properties of the element; collectively, atoms form molecules in a compound.

atomic number (ah-tom´ik num´ber): the weight of the atom of a particular element.

atomic weight (ah-tom´ik wait): the number of protons together with the number of neutrons within the nucleus of an atom.

ATP (adenosine triphospate): a compound of adenine, ribose, and three phosphates; it is the energy carrier for most cellular processes.

atrium (a´tre-um): either of two superior chambers of the heart that receive venous blood.

atrophy (at´ro-fe): a wasting away or decrease in size of a cell or organ.

auditory tube (aw´di-to´re): a narrow canal that connects the middle ear chamber to the pharynx; also called the *eustachian canal.*

autonomic (aw-to-nom´ik): self-governing; pertaining to the division of the nervous system which controls involuntary activities.

autosome (aw´to-som): a chromosome other than a sex chromosome.

autotroph (aw´to-trof): an organism capable of synthesizing its own organic molecules (food) from inorganic molecules.

axilla (ak-sil´ah): the depressed hollow under the arm; the armpit.

axillary bud (ak´si-lar´e bud): a group of meristematic cells at the junction of a leaf and stem which develops branches or flowers; also called *lateral bud.*

axon (ak´son): The elongated process of a neuron (nerve cell) that transmits an impulse away from the cell body.

bacillus (*pl.* **bacilli**): a rod-shaped bacterium.

bacteria (bak-te´re ah): prokaryotes within the kingdom Monera, lacking the organelles of eukaryotic cells.

bark: outer tissue layers of a tree consisting of cork, cork cambium, cortex, and phloem.

basal: at or near the base or point of attachment, as of a plant shoot.

base: a substance that contributes or liberates hydroxide ions in a solution.

basement membrane: a thin sheet of extracellular substance to which the basal surfaces of membranous epithelial cells are attached.

basidia (bah-sid´e-ah): club-shaped reproductive structures of club fungi that produce basidiospores during sexual reproduction.

basophil (ba´so-fil): a granular leukocyte that readily stains with basophilic dye.

belly: the thickest circumference of a skeletal muscle.

benign: (be-nine´): nonmalignant; a confined tumor.

berry: a simple fleshy fruit.

biennial: a plant that lives through two growing seasons; generally, these plants have only vegetative growth during the first season, and flower and set seed during the second.

bilateral symmetry (bi-lat´er-al sim´e-tre): the morphologic condition of having similar right and left halves.

binary fission (bi´na-re fish´un): a process of sexual reproduction that does not involve a mitotic spindle.

binomial nomenclature (bi-no´me-al sis´tem): assignment of two names to an organism, the first of which is the genus and the second the species; together, constituting the scientific name.

biome (bi´om): a major climax community characterized by a particular group of plants and animals.

biosphere (bi´o-sfer): the portion of the earth's atmosphere and surface where living organisms exist.

biotic (bi-ot´ik): pertaining to aspects of life, especially to characteristics of ecosystems.

bisexual flower: a flower that contains both male and female sexual structures.

blade: the broad expanded portion of a leaf.

blastocoel (blas´to-sel): the cavity of a blastocyst.

blastula (blas´tu-lah): an early stage of prenatal development between the morula and embryonic stages.

blood: the fluid connective tissue that circulates through the cardiovascular system to transport substances throughout the body.

bolus (bo´lus): a moistened mass of food that is swallowed from the oral cavity into the pharynx.

bone: an organ composed of solid, rigid connective tissue, forming a component of the skeletal system.

Bowman's capsule (bo´manz kap´sul): see *glomerular capsule.*

brain: the enlarged superior portion of the central nervous system, located in the cranial cavity of the skull.

brain stem: the portion of the brain consisting of the medulla oblongata, pons, and midbrain.

bronchial tree (brong´ke-al): the bronchi and their branching bronchioles.

bronchiole (brong´ke-ol): a small division of a bronchus within the lung.

bronchus (bron´kus): a branch of the trachea that leads to a lung.

budding: a type of asexual reproduction in which outgrowths from the parent plant pinch off to live independently or else remain attached forming colonies.

buccal cavity (buk´al): the mouth, or oral cavity.

buffer (buf´er): a compound or substance that prevents large changes in the pH of a solution.

bulb: a thickened underground stem often enclosed by enlarged, fleshy leaves containing stored food.

bursa (ber´sah): a saclike structure filled with synovial fluid, which occurs around joints.

buttock (but´ok): the rump or fleshy mass on the posterior aspect of the lower trunk, formed primarily by the gluteal muscles.

calorie (kal´o-re): the heat required to raise one kilogram of water one degree centigrade.

calyx (ka´liks): a cup-shaped portion of the renal pelvis that encircles renal papillae; the collective term for the sepals of a flower.

cambium: the layer of meristematic tissue in roots and stems of many vascular plants that continues to produce tissue.

cancellous bone (kan´se-lus): spongy bone; bone tissue with a latticelike structure.

capillary (kap´i-lar´e): a microscopic blood vessel that connects an arteriole and a venule; the functional unit of the circulatory system.

carapace: protective covering over the dorsal part of the body of certain crustaceans and turtles.

carcinogenic (kar-si-no-jen´ik): stimulating or causing the growth of a malignant tumor, or cancer.

carnivore (kar´ni-vor): any animal that feeds upon another; specifically, flesh-eating mammal.

carpus (kar´pus): the proximal portion of the hand that contains the eight carpal bones.

carrying capacity: the maximum number of organisms of a species that can be maintained indefinitely in an ecosystem.

cartilage (kar´ti´lij): a type of connective tissue with a solid elastic matrix.

catalyst: a chemical, such as an enzyme, that accelerates the rate of a reaction of a chemical process but is not used up in the process.

caudal (kaw´dal): referring to a position more toward the tail.

cecum (se´kum): the pouchlike portion of the large intestine to which the ileum of the small intestine is attached.

cell: the structural and functional unit of an organism; the smallest structure capable of performing all the functions necessary for life.

cell wall: a rigid protective structure of a plant cell surrounding the plasma membrane; often composed of cellulose fibers embedded in a polysaccharide/protein matrix.

cellular respiration: the reactions of glycolysis, Krebs cycle, and electron transport system that provide cellular energy and accompanying reactions to produce ATP.

cellulose: a polysaccharide produced as fibers that forms a major part of the rigid cell wall around a plant cell.

central nervous system (CNS): the brain and the spinal cord.

centromere (sen´tro-mer): a portion of the chromosome to which a spindle fiber attaches during mitosis or meiosis.

centrosome (sen´tro-som): a dense body near the nucleus of a cell that contains a pair of centrioles.

cephalothorax (sef´ah-lo-tho´raks): fusion of the head and thoracic regions characteristic of certain arthropods.

cercaria: larva of trematodes (flukes).

cerebellum (ser´e-bel´um): the portion of the brain concerned with the coordination of movements and equilibrium.

cerebrospinal fluid (ser´e-bro-spi´nal): a fluid that buoys and cushions the central nervous system.

cerebrum (ser´e-brum): the largest portion of the brain, composed of the right and left hemispheres.

cervical (ser´vi-kal): pertaining to the neck or a necklike portion of an organ.

chelipeds: front pair of pincerlike legs in most decapod crustaceans, adapted for seizing and crushing.

chitin (ki´tin): strong, flexible polysaccharide forming the exoskeleton of arthropods.

chlorophyll (klo´ro-fil): green pigment in photosynthesizing organisms that absorbs energy from the sun's rays.

chloroplast (klo´ro-plast): a membrane-enclosed organelle which contains chlorophyll and is the site of photosynthesis.

choanae (ko-a´na): the two posterior openings from the nasal cavity into the nasopharynx.

cholesterol (ko-les´ter-ol): a lipid used in the synthesis of steroid hormones.

chondrocyte (kon´dro-site): a cartilage cell.

chorion (ko´re-on): an extraembryonic membrane that participates in the formation of the placenta.

choroid (ko´roid): the vascular, pigmented middle layer of the wall of the eye.

chromatin (kro´mah-tin): threadlike network of DNA and proteins within the nucleus.

chromosome (kro´mo-som): structure in the nucleus that contains the genes for genetic expression.

chyme (kime): the mass of partially digested food that passes from the stomach into the duodenum of the small intestine.

cilia (sil´e-ah): microscopic, hairlike processes that move in a wavelike manner on the exposed surfaces of certain epithelial cells.

ciliary body (sil´e-er´e): a portion of the choroid layer of the eye that secretes aqueous humor and contains the ciliary muscle.

ciliates (sil´e-ats): protozoans that move by means of cilia.

circadian rhythm (ser´kah-de´an rith´um): a daily physiological or behavioral event, occurring on an approximate 24 hour cycle.

circumduction (ser´kum-duk´shun): a conelike movement of a body part, such that the distal end moves in a circle while the proximal portion remains relatively stable.

clitoris (kli´to-ris): a small, erectile structure in the vulva of the female.

cochlea (kok´le-ah): the spiral portion of the inner ear that contains the spiral organ (organ of Corti).

climax community (kli´maks ko-mu´ni-te): the final, stable stage in succession.

clone (klon): asexually produced organisms having a consistent genetic constitution.

cnidarian (ni-dah´re-an): small aquatic organisms having radial symmetry and stinging cells with nematocysts.

cocoon: protective, or resting, stage of development in certain invertebrate animals.

codon (ko´don): a "triplet" of three nucleotides in RNA that directs the placement of an amino acid into a polypeptide chain.

coelom (se´lom): body cavity of higher animals, containing visceral organs.

collar cells (kol´ler selz): flagella-supporting cells in the inner layer of the wall of sponges.

colon (ko´lon): the first portion of the large intestine.

colony: an aggregation of organisms living together in close proximity.

common bile duct: a tube that is formed by the union of the hepatic duct and cystic duct, transports bile to the duodenum.

community: an ecological unit composed of all the populations of organisms living and interacting in a given area.

compact bone: tightly packed bone that is superficial to spongy bone; also called *dense bone.*

competition: interaction between individuals of the same or different species for a mutually necessary resource.

complete flower: a flower that has four whorls of floral components including sepals, petals, stamens, and carpels.

compound eye: arthropod eye consisting of multiple lenses.

compound leaf: a leaf blade divided into distinct leaflets.

condyle (kon´dile): a rounded process at the end of a long bone that forms an articulation.

conidia (ko-nid´e-ah): spores produced by fungi during asexual reproduction.

conifer (kon´i-fer): a cone-bearing seed plant, such as pine, fir, and spruce.

conjugation (kon´ju-ga´shun): sexual union in which the nuclear material of one cell enters another.

connective tissue: one of the four basic tissue types within the body. It is a binding and supportive tissue with abundant matrix.

consumer (kon-su´mer): an organism that derives nutrients by feeding upon another.

control: a sample in an experiment that undergoes all the steps in the experiment except the one being investigated.

convergent evolution: the evolution of similar structures in different groups of organisms exposed to similar environments.

coral (kor´al): a cnidarian that has a calcium carbonate skeleton whose remains contribute to form reefs.

cork: the protective outer layer of bark of trees, composed of dead cells that may be sloughed off.

cornea (kor´ne-ah): the transparent convex, anterior portion of the outer layer of the eye.

cortex (kor´teks): the outer layer of an organ such as the convoluted cerebrum, adrenal gland, or kidney.

costal cartilage (kos´tal): the cartilage that connects the ribs to the sternum.

cranial (kra´ne-al): pertaining to the cranium.

cranial nerve: one of twelve pairs of nerves that arise from the inferior surface of the brain.

cranium (kra´ne-um): the bones of the skull that enclose the brain and support the organs of sight, hearing, and balance.

crossing over: the exchange of corresponding chromatid segments of genetic material of homologous chromosomes during synapsis in meiosis I.

cuticle (ku´te-kl): waxlike covering on the epidermis of nonwoody plants to prevent water loss.

cyanobacteria (si´ah-no-bak´te´re-ah): photosynthetic prokaryotes that have chlorophyll and release oxygen.

cytokinesis (si´to-ki-ne´sis): division of the cellular cytoplasm.

cytology (si-tol´o-je): the science dealing with the study of cells.

cytoplasm (si´to-plazm´): the protoplasm of a cell located outside of the nucleus.

cytoskeleton (si´to-skel´e-ton): protein filaments throughout the cytoplasm of certain cells that help maintain the cell shape and provide movement.

deciduous (de-sid´you-us): plants that seasonally shed their leaves.

dendrite (den´drite): a nerve cell process that transmits impulses toward a neuron cell body.

denitrifying bacteria (de-ni´tri-fi-ing bak-te´re ah): single-cellular organisms of the kingdom Monera that convert nitrate to atmospheric nitrogen.

dentin (den´tine): the principal substance of a tooth, covered by enamel over the crown and by cementum on the root.

dermis (der´mis): the second, or deep, layer of skin beneath the epidermis.

descending colon: the segment of the large intestine that descends on the left side from the level of the spleen to the level of the left iliac crest.

detritus (de-tri´tus): non-living organic matter that is important in the nutrient cycle in soil formation.

diaphragm (di´ah-fram): a flat dome of muscle and connective tissue that separates the thoracic and abdominal cavities.

diaphysis (di-af´i-sis): the shaft of a long bone.

diastole (di-as´to-le): the portion of the cardiac cycle during which the ventricular heart chamber wall is relaxed.

diarthrosis (di´ar-thro´sis): a freely movable joint in a functional classification of joints.

diatoms (di´ah-tomz): aquatic unicellular algae characterized by a cell wall composed of two silica impregnated valves.

dicot (di´kot): a kind of angiosperm characterized by the presence of two cotyledons in the seed; also called *dicotyledon.*

diffusion: movement of molecules from an area of greater concentration to an area of lesser concentration.

dihybrid cross (di-hi´brid): a breeding experiment in which parental varieties differing in two traits are mated.

dimorphism (dimorphism): two distinct forms within a species, with regards to size, color, organ structure, and so on.

diphyodent: two sets of teeth, deciduous and permanent.

diploid (dip´loid): having two copies of each different chromosome, pairs of homologous chromosomes (2N).

distal (dis´tal): away from the midline or origin; the opposite of *proximal.*

division: a major taxonomic grouping of plants that includes classes sharing certain features with close biological relationships.

dominant (dom´i-nant): a hereditary characteristic that expresses itself even when the genotype is heterozygous.

dormancy: a period of suspended activity and growth.

dorsal (dor´sal): pertaining to the back or posterior portion of a body part; the opposite of *ventral.*

double helix (he´liks): a double spiral used to describe the three-dimensional shape of DNA.

ductus deferens (duk´tus def´er-enz): a tube that carries spermatozoa from the epididymis to the ejaculatory duct: also called the *vas deferens* or *seminal duct.*

duodenum (du´o-de-num): the first portion of the small intestine.

dura mater (du´rah ma´ter): the outermost meninx covering the central nervous system.

eccrine gland (ek´rin): a sweat gland that functions in body cooling.

ecology (e-kol´o-je): the study of the relationship of organisms and the physical environment and their interactions.

ecosystem (ek´o-sis´tem): a biological community and its associated abiotic environment.

ectoderm (ek´to-derm): the outermost of the three primary embryonic germ layers.

edema (e-de´mah): an excessive retention of fluid in the body tissues.

effector (ef-fek´tor): an organ such as a gland or muscle that responds to motor stimulation.

efferent (ef´er-ent): conveying away from the center of an organ or structure.

ejaculation (e-jak´u-la´shun): the discharge of semen from the male urethra during climax.

electrocardiogram (e-lek´tro-kar´de-o-gram´): a recording of the electrical activity that accompanies the cardiac cycle; also called *ECG* or *EKG*.

electroencephalogram (e-lek´tro-en-sef´ah-lo-gram): a recording of the brain wave pattern; also called *EEG*.

electromyogram (e-lek´tro-mi´o-gram): a recording of the activity of a muscle during contraction: also called *EMG*.

electrolyte (e-lek´tro-lite): a solution that conducts electricity by means of charged ions.

electron (e-lek´tron): the unit of negative electricity.

element (el´e-ment): a structure comprised of only one type of atom (i.e., carbon, hydrogen, oxygen).

embryo: a plant or an animal at an early stage of development.

emulsification (e-mul´si-fi´ka´shun): the process of dispersing one liquid in another.

enamel (e-nam´el): the outer, dense substance covering the crown of a tooth.

endocardium (en´do-kar´de-um): the fibrous lining of the heart chambers and valves.

endochondral bone (en´do-kon´dral): bones that form as hyaline cartilage models first and then are ossified.

endocrine gland (en´do-krin): a hormone-producing gland that secretes directly into the blood or body fluids.

endoderm (en´do-derm): the innermost of the three primary germ layers of an embryo.

endodermis (en´do-der´mis): a plant tissue composed of a single layer of cells that surrounds and regulates the passage of materials into the vascular cylinder of roots.

endometrium (en´do-me´tre-um): the inner lining of the uterus.

endoskeleton (en´do-skel´e-ton): hardened, supportive internal tissue of echinoderms and vertebrates.

endosperm: a plant tissue of angiosperm or gymnosperm seeds that stores nutrients; the endosperm of angiosperms is 3n in chromosome number.

endothelium (en´do-the´le-um): the layer of epithelial tissue that forms the thin inner lining of blood vessels and heart chambers.

enzyme (en´zim): a protein catalyst that activates a specific reaction.

eosinophil (e´o-sin´o-fil): a type of white blood cell that becomes stained by acidic eosin dye; constitutes about 2%–4% of the white blood cells.

epicardium (ep´i-kar´de-um): the thin, outer layer of the heart: also called the *visceral pericardium*.

epicotyl (ep´i-kot´il): plant embryo portion that contributes to stem development.

epidermis (ep´i-der´mis): the outermost layer of the skin, composed of stratified squamous epithelium.

epididymis (ep´i-did´i-mis): a coiled tube located along the posterior border of the testis; stores spermatozoa and discharges them during ejaculation.

epidural space (ep´i-du´ral): a space between the spinal dura mater and the bone of the vertebral canal.

epiglottis (ep´i-glot´is): a leaflike structure positioned on top of the larynx that covers the glottis during swallowing.

epinephrine (ep´i-nef´rin): a hormone secreted from the adrenal medulla resulting in actions similar to those from sympathetic nervous system stimulation; also called *adrenaline*.

epiphyseal plate (ep´i-fize-al): a cartilaginous layer located between the epiphysis and diaphysis of a long bone. Functions in longitudinal bone growth.

epiphysis (e-pif´i-sis): the end segment of a long bone, distinct in early life but later becoming part of the larger bone.

epiphyte (ep´i-fit): nonparasitic plant, such as orchid and Spanish moss, that grows on the surface of other plants.

epithelial tissue (ep´i-the´le-al): one of the four basic tissue types; the type of tissue that covers or lines all exposed body surfaces.

erection (e-rek´shun): a response within an organ, such as the penis, when it becomes turgid and erect as opposed to being flaccid.

erythrocyte (e-rith´ro-site): a red blood cell.

esophagus (e-sof´ah-gus): a tubular organ of the GI tract that leads from the pharynx to the stomach.

estrogen (es´tro-jen): female sex hormone secreted from the ovarian (Graafian) follicle.

estuary (es´tu-a-re): a zone of mixing between fresh water and seawater.

eukaryotic (u´kar-e-ot´ik): possessing the membranous organelles characteristic of complex cells.

eustachian canal (u-sta´ke-an): see *auditory tube*.

evolution: genetic and phenotypic changes occurring in populations of organisms through time, generally resulting in increased adaptation for continued survival.

excretion (eks-kre´shun): discharging waste material.

exocrine gland (ek´so-krin): a gland that secretes its product to an epithelial surface, directly or through ducts.

exoskeleton: an outer, hardened supporting structure secreted by ectoderm or epidermis.

expiration (ek´spi-ra´shun): the process of expelling air from the lungs through breathing out; also called *exhalation*.

extension (ek-sten´shun): a movement that increases the angle between two bones of a joint.

external ear: the outer portion of the ear, consisting of the auricle (pinna), and the external auditory canal, and tympanum.

extracellular (esk-trah-sel´u-lar): outside a cell or cells.

extraembryonic membranes (eks´trah-em´bre-on´ik): membranes that are not a part of the embryo but are essential for the health and development of the organism.

extrinsic (eks-trin´sik): pertaining to an outside or external origin.

facet (fas´et): a small, smooth surface of a bone where articulation occurs.

facilitated transport (fah-sil´i-tat-ed): transfer of a particle into or out of a cell along a concentration gradient by a process requiring a carrier.

fallopian tube (fal-lo´pe-an): see *uterine tube*.

fascia (fash´e-ah): a tough sheet of fibrous connective tissue binding the skin to underlying muscles or supporting and separating muscle.

fasciculus (fah-sik´u-lus): a bundle of muscle or nerve fibers.

feces (fe´seez): waste material expelled from the GI tract during defecation, composed of food residue, bacteria, and secretions; also called *stool*.

fertilization: the fusion of two haploid gamete nuclei to form a diploid zygote nucleus.

fetus (fe´tus): the unborn offspring during the last stage of prenatal development.

fibrous root (fi´brus): an intertwining mass of many roots of about equal size.

filament: a long chain of cells.

filter feeder: an animal that obtains food by straining it from the water.

filtration (fil-tra´shun): the passage of a liquid through a filter or a membrane.

fimbriae (fim´bre-e): fringelike extensions from the borders of the open end of the uterine tube.

fissure (fish´ure) a groove or narrow cleft that separates two parts of an organ.

flagella (flah-jel´ah): long slender locomotor processes characteristic of flagellate protozoans, certain bacteria, and sperm.

flexion (flek´shun): a movement that decreases the angle between two bones of a joint; opposite of extension.

flora (floor´uh): a general term for plant life.

flower (flow´er): the blossom of an angiosperm that contains the reproductive organs.

fluke (flook): a parasitic flatworm within the class Trematoda.

follicle (fol´i-kl): the portion of the ovary that produces the egg and the female sex hormone, estrogen; the depression that supports and develops a feather or hair.

fontanel (fon´tah-nel): a membranous-covered region on the skull of a fetus or baby where ossification has not yet occurred: also called a *soft spot*.

food web: the food links among populations in a community.

foot: the terminal portion of the lower extremity, consisting of the tarsus, metatarsus, and digits.

foramen (fo-ra´men): an opening in an anatomical structure for the passage of a blood vessel or a nerve.

foramen ovale (o-val´e): the opening through the interatrial septum of the fetal heart.

fossa (fos´ah): a depressed area, usually on a bone.

fossil (fos´l): any preserved ancient remains or impressions of an organism.

fourth ventricle (ven´tri-k´l): a cavity within the brain containing cerebrospinal fluid.

fovea centralis (fo´ve-ah sen´tra´lis): a depression on the macula lutea of the eye where only cones are located, which is the area of keenest vision.

frond (frond): the leaf of a fern containing many leaflets.

fruit: a mature ovary enclosing a seed or seeds.

gallbladder: a pouchlike organ, attached to the inferior side of the liver, which stores and concentrates bile.

gamete (gam´ete): a haploid sex cell, sperm or egg.

gametophyte: the haploid, gamete-producing generation in the life cycle of a plant.

gamma globulins (gam´mah glob´u-lins): protein substances that act as antibodies often found in immune serums.

ganglion (gang´gle-on): an aggregation of nerve cell bodies outside the central nervous system.

gastrointestinal tract (gas´tro-in-tes´tin-al): the tubular portion of the digestive system that includes the stomach and the small and large intestines; also called the *GI tract*.

gene (jeen): part of the DNA molecule located in a definite position on a certain chromosome and coding for a specific product.

gene pool: the total of all the genes of the individuals in a population.

genetic drift: evolution by chance process.

genetics (je-net´iks): the study of heredity.

genotype (je´no-tip): the genetic makeup of an organism.

genus (je´nus): the taxonomic category above species and below family.

geotropism: plant growth oriented with respect to gravity; stems grow upward, roots grow downward.

germ cells: gametes or the cells that give rise to gametes or other cells.

germination: the process by which a spore or seed ends dormancy and initiates normal metabolism, development, and growth.

gill: a gas exchange organ characteristic of fishes and other aquatic or semiaquatic animals.

gingiva (jin´ji-vah): the fleshy covering over the mandible and maxilla through which the teeth protrude within the mouth; also called the *gum*.

girdling: removal of a strip of bark from around a tree down to the wood layer.

gland: an organ that produces a specific substance or secretion.

glans penis (glanz pe´nis): the enlarged, distal end of the penis.

glomerular capsule: (glo-mer´u-lar): the double-walled proximal portion of a renal tubule that encloses the glomerulus of a *nephron;* also called *Bowman's capsule.*

glomerulus (glo-mer´u´lus): a coiled tuft of capillaries that is surrounded by the glomerular capsule and filters urine from the blood.

glottis (glot´is): a slitlike opening into the larynx, positioned between the vocal folds.

glycogen (gli´ko-jen): the principal storage carbohydrate in animals. It is stored primarily in the liver and is made available as glucose when needed by the body cells.

goblet cell: a unicellular gland within columnar epithelia that secretes mucus.

gonad (go´nad): a reproductive organ, testis or ovary, that produces gametes and sex hormones.

granum: a "stack" of membrane flattened disks within the chloroplast that contain chlorophyll.

gray matter: the portion of the central nervous system that is composed of nonmyelinated nervous tissue.

grazer (gra´zer): an animal that feeds on low growing vegetation, such as grasses.

greater omentum (o-men´tum): a double-layered paritoneal membrane that originates on the greater curvature of the stomach and extends over the abdominal viscera.

growth ring: a growth layer of secondary xylem (wood) or secondary phloem in gymnosperms or angiosperms.

guard cell: an epidermal cell to the side of a leaf stoma that helps to control the stoma size.

gut: pertaining to the intestine; generally a developmental term.

gymnosperm: a vascular seed-producing plant.

gyrus (ji´rus): a convoluted elevation or ridge.

habitat (hab´i-tat): the ecological abode of a particular organism.

hair: an epidermal structure consisting of keratinized dead cells that have been pushed up from a dividing basal layer.

hair cells: specialized receptor nerve endings for responding to sensations, such as in the spiral organ of the inner ear.

hair follicle (fol´li-k´l): a tubular depression in the skin in which a hair develops.

hand: the terminal portion of the upper extremity, consisting of the carpus, metacarpus, and digits.

haploid (hap´loid): having one copy of each different chromosome.

hard palate (pal´at): the bony partition between the oral and nasal cavities, formed by the maxillae and palatine bones.

haustra (haws´trh): sacculations or pouches of the colon.

haversian system (ha-ver´shan): see *osteon.*

heart: a muscular, pumping organ positioned in the thoracic cavity.

hematocrit (he-mat´o-krit): the volume percentage of red blood cells in whole blood.

hemoglobin (he´mo-glo´bin): the pigment of blood cells that transports O_2 and CO_2.

hemopoiesis (hem´ah-poi-e´sis): production of red blood cells.

hepatic portal circulation (por´tal): the return of venous blood from the digestive organs and spleen through a capillary network within the liver before draining into the heart.

herbaceous: a nonwoody plant.

herbaceous stem (her-ba´shus): stem of a non-woody plant.

herbivore: an organism that feeds exclusively on plants.

heredity (he-red´i-te): the transmission of certain characteristics, or traits, from parents to offspring, via the genes.

heterodont: having teeth differentiated into incisors, canines, premolars, and molars for specific functions.

heterotroph (het´er-o-trof): an organism that utilizes preformed food.

heterozygous (het´er-o-zi´gus): having two different alleles (i.e., *Bb*) for a given trait.

hiatus (hi-a´tus): an opening or fissure.

hilum (hi´lum): a concave or depressed area where vessels or nerves enter or exit an organ.

histology (his-tol´o-je): microscopic anatomy of the structure and function of tissues.

holdfast: basal extension of a multicellular alga that attaches it to a solid object.

homeostasis (ho-me-o-sta´sis): a consistency and uniformity of the internal body environment which maintains normal body function.

homologous (ho-mol´o-gus): similar in developmental origin and sharing a common ancestry.

homothallic: species in which individuals produce both male and female reproductive structures and are self fertile.

hormone (hor´mone): a chemical substance that is produced in an endocrine gland and secreted into the bloodstream to cause an effect in a specific target organ.

host: an organism on or in which another organism lives.

hyaline cartilage (hi´ah-line): the most common kind of cartilage in the body, occurring at the articular ends of bones, in the trachea, and within the nose, and the precursor to most of the bones of the skeleton.

hybrid (hi´brid): an offspring from the crossing of genetically different strains or species.

hymen (hi´men): a developmental remnant (vestige) of membranous tissue that partially covers the vaginal opening.

hyperextension (hi´per-ek-sten´shun): extension beyond the normal anatomical position of 180˚.

hypocotyl (hi´po-kot´il): portion of plant embryo that contributes to stem development.

hypothalamus (hi´po-thal´ah-mus): a structure within the brain below the thalamus, which functions as an autonomic center and regulates the pituitary gland.

hypotonic solution (hi´po-ton´ik): a fluid environment that has a greater concentration of water and a lesser concentration of solute than the cell.

ileocecal valve (il´e-o-se´kal): a specialization of the mucosa at the junction of the small and large intestines that forms a one-way passage and prevents the backflow of food materials.

ileum (il´e-um): the terminal portion of the small intestine between the jejunum and cecum.

imprinting: a type of learned behavior occurring during a limited critical period.

indigenous: organisms that are native to a particular region; not introduced.

inferior vena cava (ve´nah ka´vah): a systemic vein that collects blood from the body regions inferior to the level of the heart and returns it to the right atrium.

inguinal (ing´gwi-nal): pertaining to the groin region.

inguinal canal: the passage in the abdominal wall through which a testis descends into the scrotum.

insertion: the more movable attachment of a muscle, usually more distal in location.

inspiration (in´spi-ra´shun): the act of breathing air into the alveoli of the lungs; also called *inhalation.*

instar: stage of insect or other arthropod development between molts.

integument (in-teg´u-ment): pertaining to the skin.

internal ear: the innermost portion or chamber of the ear, containing the cochlea and the vestibular organs.

internode: region between stem nodes.

interstitial (in-ter-stish´al): pertaining to spaces or structures between the functioning active tissue of any organ.

intracellular (in-trah-sel´u-lar): within the cell itself.

intervertebral disc (in´ter-ver´te-bral): a pad of fibrocartilage between the bodies of adjacent vertebrae.

intestinal gland (in-tes´ti-nal): a simple tubular digestive gland that opens onto the surface of the intestinal mucosa and secretes digestive enzymes; also called *crypt of Lieberkuhn.*

intrinsic (in-trin´sik): situated or pertaining to internal origin.

invertebrate (in-ver´te-brat): an animal that lacks a vertebral column.

iris (i´ris): the pigmented vascular tunic portion of the eye that surounds the pupil and regulates its diameter.

islets of Langerhans (i´lets of lahng´er-hanz): see *pancreatic islets.*

isotope (i´so-tope): a chemical element that has the same atomic number as another but a different atomic weight.

isthmus (is´mus): a narrow neck or portion of tissue connecting two structures.

jejunum (je-joo´num): the middle portion of the small intestine, located between the duodenum and the ileum.

joint capsule (kap´sule): the fibrous tissue that encloses the joint cavity of a synovial joint.

jugular (jug´u-lar): pertaining to the veins of the neck which drain the areas supplied by the carotid arteries.

karyotype (kar´e-o-tip): the arrangement of chromosomes that is characteristic of the species or of a certain individual.

keratin (ker´ah-tin): an insoluble protein present in the epidermis and in epidermal derivatives such as hair and nails.

kidney (kid´ne): one of the paired organs of the urinary system that contains nephrons and filters wastes from the blood in the formation of urine.

kingdom: a taxonomic category grouping related divisions (plants) or phyla (animals).

labia majora (la´be-ah ma-jor´ah): a portion of the external genitalia of a female, consisting of two longitudinal folds of skin extending downward and backward from the mons pubis.

labia minora (ma-nor´ah): two small folds of skin, devoid of hair and sweat glands, lying between the labia majora of the external genitalia of a female.

lacrimal gland (lak´ri-mal): a tear-secreting gland, located on the superior lateral portion of the eyeball underneath the upper eyelid.

lactation (lak-ta´shun): the production and secretion of milk by the mammary glands.

lacteal (lak´te-al): a small lymphatic duct within a villus of the small intestine.

lacuna (lah-ku´nah): a hollow chamber that houses an osteocyte in mature bone tissue or a chondrocyte in cartilage tissue.

lamella (lah-mel´ah): a concentric ring of matrix surrounding the central canal in an osteon of mature bone tissue.

large intestine: the last major portion of the GI tract, consisting of the cecum, colon, rectum, and anal canal.

larva: an immature, developmental stage that is quite different from the adult.

larynx (lar´inks): the structure located between the pharynx and trachea that houses the vocal folds (cords); commonly called the *voice box.*

lateral root: a secondary root that arises by branching from an older root.

leaf veins: plant structures that contain the vascular tissues in a leaf.

legume: a member of the pea, or bean, family

lens (lenz): a transparent refractive structure of the eye, derived from ectoderm and positioned posterior to the pupil and iris.

lenticel (len´ti-sel): spongy area in the bark of a stem or root that permits interchange of gases between internal tissues and the atmosphere.

leukocyte (lu´ko-site): a white blood cell; also spelled *leucocyte.*

lichen (li´ken): algae or bacteria and fungi coexisting in a mutualistic relationship.

ligament (lig´ah-ment): a fibrous band or cord of connective tissue that binds bone to bone to strengthen and provide support to the joint; also may support viscera.

limbic system (lim´bik): a portion of the brain concerned with emotions and autonomic activity.

linea alba (lin´e-ah al´bah): a fibrous band extending down the anterior medial portion of the abdominal wall.

locus (lo´kus): the specific location or site of a gene within the chromosome.

lumbar (lum´bar): pertaining to the region of the loins.

lumen (lu´men): the space within a tubular structure through which a substance passes.

lung: one of the two major organs of respiration within the thoracic cavity.

lymph (limf): a clear fluid that flows through lymphatic vessels.

lymph node: a small, ovoid mass located along the course of lymph vessels.

lymphocyte (lim´fo-site): a type of white blood cell characterized by a granular cytoplasm.

macula lutea (mak´u-lah lu´te-ah): a depression in the retina that contains the fovea centralis, the area of keenest vision.

malignant (mah-lig´nant): a disorder that becomes worse and eventually causes death, as in cancer.

malnutrition (mal-nu-trish´un): any abnormal assimilation of food; receiving insufficient nutrients.

mammary gland (mam´er-e): in mammals, the gland of the female breast responsible for lactation and nourishment of the young.

mantle (man´t´l): fleshy fold of the body wall of a mollusk, typically involved in shell formation.

marine: pertaining to the sea or ocean.

marrow (mar´o): the soft vascular tissue that occupies the inner cavity of certain bones and produces blood cells.

matrix (ma´triks): the intercellular substance of a tissue.

meatus (me-a´tus): an opening or passageway into a structure.

mediastinum (me´de-as-ti´num): the partition in the center of the thorax between the two pleural cavities.

medulla (me-dul´ah): the center portion of an organ.

medulla oblongata (ob´long-ga´tah): a portion of the brain stem between the pons and the spinal cord.

medullary cavity (med´u-lar´e): the hollow center of the diaphysis of a long bone, occupied by marrow.

megaspore: a plant spore that will germinate to become a female gametophyte.

meiosis (mi-o´sis): cell division by which haploid cells are formed from a diploid cell.

melanocyte (mel´ah-no-site): a pigment-producing cell in the deepest epidermal layer of the skin.

membranous bone (mem´brah-nus): bone that forms from membranous connective tissue rather than from cartilage.

menarche (me-nar´ke): the first menstrual discharge.

meninges (me-nin´jez): a group of three fibrous membranes that cover the central nervous system.

meniscus (men-is´kus): wedge-shaped cartilage in certain synovial joints.

menopause (men´o-pawz): the cessation of menstrual periods in the human female.

menses (men´sez): the monthly flow of blood from the human female genital tract.

menstrual cycle (men´stru-al): the rhythmic female reproductive cycle, characterized by changes in hormone levels and physical changes in the uterine lining.

menstruation (men´stru-a´shun): the discharge of blood and tissue from the uterus at the end of the menstrual cycle.

meristem tissue (mer´i-stem): undifferentiated plant tissue that is capable of dividing and producing new cells.

mesentery (mes´en-ter´e): a fold of peritoneal membrane that attaches an abdominal organ to the abdominal wall.

mesoderm (mes´o-derm): the middle one of the three primary germ layers.

mesophyll (mes´o-fil): the middle tissue layer of a leaf containing cells that are active in photosynthesis.

mesothelium (mes´o-the´le-um): a simple squamous epithelial tissue that lines body cavities and covers visceral organs; also called *serosa.*

metabolism (me-tab´o-lizm): the chemical changes that occur within a cell.

metacarpus (met´ah-kar´pus): the region of the hand between the wrist and the digits, including the five bones that support the palm of the hand.

metamorphosis (met´ah-mor´fo-sis): change in morphologic form, such as when an insect larva develops into the adult or as a tadpole develops into an adult frog.

metastasis (me-tas´tah-sis): the spread of a disease from one organ or body part to another.

metatarsus (met´ah-tar´sus): the region of the foot between the ankle and the digits containing five bones.

microbiology (mi-kro-bi-ol´o-je): the science dealing with microscopic organisms, including fungi, protozoa, and viruses.

microspore (mi´kro-spor): a spore in seed plants that develops into a pollen grain, the male gametophyte.

microvilli (mi´kro-vil´i): microscopic, hairlike projections of cell membranes on certain epithelial cells.

midbrain: the portion of the brain between the pons and the forebrain.

middle ear: the middle of the three ear chambers, containing the three auditory ossicles.

migration: movement of organisms from one geographical site to another.

mimicry (mim´ik-re): a protective resemblance of an organism to another.

mitosis (mi-to´sis): the process of cell division, in which the two daughter cells are identical and contain the same number of chromosomes.

mitral valve (mi´tral): the left atrioventricular heart valve; also called the bicuspid valve.

mixed nerve: a nerve containing both motor and sensory nerve fibers.

molecule (mol´e-kule): a minute mass of matter, composed of a combination of atoms that form a given chemical substance or compound.

molting (molt´ing): periodic shedding of an epidermal derived structure.

monocot (mon´o-kot): a type of angiosperm in which the seed has only a single cotyledon; also called *monocotyledon.*

motor neuron (nu´ron): a nerve cell that conducts action potential away from the central nervous system and innervates effector organs (muscles and glands); also called *efferent neuron.*

motor unit: a single motor neuron and the muscle fibers it innervates.

mucosa (mu-ko´sah): a mucous membrane that lines cavities and tracts opening to the exterior.

muscle (mus´el): an organ adapted to contract; three types of muscle tissue are cardiac, smooth, and skeletal.

mutation (mu-ta´shun): a variation in an inheritable characteristic, a permanent transmissible change in which the offspring differ from the parents.

mutualism (mu´tu-al-izm´): a beneficial relationship between two organisms of different species.

myelin (me´e-lin): a lipoprotein material that forms a sheathlike covering around nerve fibers.

myocardium (mi´o-kar´de-um): the cardiac muscle layer of the heart.

myofibril (mi´o-fi´bril): a bundle of contractile fibers within muscle cells.

myoneural junction (mi´o-nu´ral): the site of contact between an axon of a motor neuron and a muscle fiber.

myosin (mi´o-sin): a thick filament protein that together with actin causes muscle contraction.

nail: a hardenend, keratinized plate that develops from the epidermis and forms a protective covering on the dorsal surfaces of the digits.

nares (na´rez): the openings into the nasal cavity; also called *nostrils.*

nasal cavity (na´zal): a mucosa-lined space above the oral cavity, which is divided by a nasal septum and is the first chamber of the respiratory system.

nasal septum (sep´tum): a bony and cartilaginous partition that separates the nasal cavity into two portions.

natural selection: the evolutionary mechanism by which better adapted organisms are favored to reproduce and pass on their genes to the next generation.

nephron (nef´ron): the functional unit of the kidney, consisting of a glomerulus, glomerular capsule, convoluted tubules, and the loop of the nephron.

nerve: a bundle of nerve fibers outside the central nervous system.

neurofibril node (nu´ro-fi´bril): a gap in the myelin sheath of a nerve fiber; also called the *node of Ranvier.*

neuroglia (nu-rog´le-ah): specialized supportive cells of the central nervous system.

neurolemmocyte (nu´ri-lem-o´site): a specialized neuroglia cell that surrounds an axon fiber of a peripheral neuron and forms the neurilemmal sheath; also called the *Schwann cell.*

neuron (nu´ron): the structural and functional unit of the nervous system, composed of a cell body, dendrites, and an axon; also called a *nerve cell.*

neutron (nu´tron): a subatomic particle in the nucleus of an atom that has a weight of one atomic mass unit and carries no charge.

neutrophil (nu´tro-fil): a type of phagocytic white blood cell.

niche (nich): the position and functional role of an organism in its ecosystem.

nipple: a dark pigmented, rounded projection at the tip of the breast.

nitrogen fixation: a process carried out by certain organisms, such as by soil bacteria, whereby free atmospheric nitrogen is converted into ammonia compounds.

node: location on a stem where a leaf is attached.

node of Ranvier (rah-ve-a´): see *neurofibril node.*

notochord (no´to-kord): a flexible rod of tissue that extends the length of the back of an embryo.

nucleic acid (nu-kle´ik as´id): an organic molecule composed of joined nucleotides, such as RNA and DNA.

nucleus (nu´kle-us): a spheroid body within the eukaryotic cell that contains the chromosomes of the cell.

nurse cells: specialized cells within the testes that supply nutrients to developing spermatozoa; also called *sertoli cells* or *sustentacular cells.*

nut: a hardened and dry single-seeded fruit.

olfactory (ol-fak´to-re): pertaining to the sense of smell.

oocyte (o´o-site): a developing egg cell.

oogenesis (o´o-jen´e-sis): the process of female gamete formation.

oogonium: a unicellular female reproductive organ of various protists that contains a single or several eggs.

optic (op´tik): pertaining to the eye and the sense of vision.

optic chiasma (ki-as´mah): an X-shaped structure on the inferior aspect of the brain where there is a partial crossing over of fibers in the optic nerves.

optic disc: a small region of the retina where the fibers of the ganglion neurons exit from the eyeball to form the optic nerve; also called the *blind spot.*

oral: pertaining to the mouth; also called *buccal.*

organ: a structure consisting of two or more tissues, which performs a specific function.

organelle (or´gan-el´): a minute structure of the eukaryotic cell that performs a specific function.

organism (or´gah-nizm): an individual living creature.

orifice (or´i fis): an opening into a body cavity or tube.

origin (or´i-jin): the place of muscle attachment onto the more stationary point or proximal bone; opposite the insertion.

osmosis (os-mo´sis): the diffusion of water from a solution of lesser concentration to one of greater concentration through a semipermeable membrane.

ossicle (os´si-k´l): one of the three bones of the middle ear.

osteocyte (os´te-o-site): a mature bone cell.

osteon (os´te-on): a group of osteocytes and concentric lamellae surrounding a central canal within bone tissue; also called a *haversian system.*

oval window: See *vestibular window.*

ovarian follicle (o-va´re-an fol´li-k´l): a developing ovum and its surrounding epithelial cells.

ovary (o´vah-re): the female gonad in which ova and certain sexual hormones are produced.

oviduct (o´vi-dukt): the tube that transports ova from the ovary to the uterus; also called the *uterine tube* or *fallopian tube.*

ovipositor: a structure at the posterior end of the abdomen in many female insects for laying eggs.

ovulation (o´vu-la´shun): the rupture of an ovarian follicle with the release of an ovum.

ovule (o´vul): the female reproductive structure in a seed plant that contains the megasporangium where meiosis occurs and the female gametophyte is produced.

ovum (o´vum): a secondary oocyte after ovulation but before fertilization.

palate (pal´at): the roof of the oral cavity.

palisade layer (pal´i-sad): the upper layer of the mesophyll of a leaf that carries out photosynthesis.

palmar (pal´mar): pertaining to the palm of the hand.

pancreas (pan´kre-as): organ in the abdominal cavity that secretes gastric juices into the GI tract and insulin and glucagon into the blood.

pancreatic islets (pan´kre-at´ik): a cluster of cells within the pancreas that forms the endocrine portion of the pancreas; also called *islets of Langerhans.*

papillae (pah-pil´e): small nipplelike projections.

paranasal sinus (par´ah-na´zal si´nus): an air chamber lined with a mucous membrane that communicates with the nasal cavity.

parasite (par´ah-sit): an organism that resides in or on another from which it derives sustenance.

parasympathetic (par´ah-sim´pah-thet´ik): pertaining to the division of the autonomic nervous system concerned with activities that are antagonistic to sympathetic division of the autonomic nervous system.

parathyroids (par´ah-thi´roids): small endocrine glands that are embedded on the posterior surface of the thyroid glands and are concerned with calcium metabolism.

parenchyma (pah-reng´ki-mah): the principal structural cells of plants.

parietal (pah-ri´e-tal): pertaining to a wall of an organ or cavity.

parotid gland: (pah-rot´id): one of the paired salivary glands on the sides of the face over the masseter muscle.

parturition (par´tu-rish´un): the process of childbirth.

pathogen (path´o-jen): any disease-producing organism.

pectin: an organic compound in the inter-cellular layer and primary wall of plant cell walls; the basis of fruit jellies.

pedicel: the stalk of a flower in an inflorescence.

pectoral girdle (pek´to-ral): the portion of the skeleton that supports the upper extremities.

pelvic (pel´vik): pertaining to the pelvis.

pelvic girdle: the portion of the skeleton to which the lower extremities are attached.

penis (pe´nis): the external male genital organ, through which urine passes during urination and which transports semen to the female during coitus.

perennial: a plant that lives through several to many growing seasons.

pericardium (per´i-kar´de-um): a protective serous membrane that surrounds the heart.

pericarp: the fruit wall that forms from the wall of a mature ovary.

perineum (per´i-ne´um): the floor of the pelvis, which is the region between the anus and the scrotum in the male and between the anus and the vulva in the female.

periosteum (per´e-os´te-um): a fibrous connective tissue covering the outer surface of bone.

peripheral nervous system (pe-rif´er-al): the nerves and ganglia of the nervous system that lie outside of the brain and spinal cord.

peristalsis (per´i-stal´sis): rhythmic contractions of smooth muscle in the walls of various tubular organs, which move the contents along.

peritoneum (per´i-to-ne´um): the serous membrane that lines the abdominal cavity and covers the abdominal viscera.

petal (pet´al): the leaf of a flower, which is generally colored.

petiole (pet´e-ol): structure of a leaf that connects the blade to the stem.

phagocyte (fag´o-site): any cell that engulfs other cells, including bacteria, or small foreign particles.

phalanx (fa´lanks) (*pl.* **phalanges**): a bone of the finger or toe.

pharynx (far´inks): the organ of the GI tract and respiratory system located at the back of the oral and nasal cavities and extending to the larynx anteriorly and the esophagus posteriorly; also called the *throat.*

phenotype (fe´no-type): the appearance of an organism caused by the genotype and environmental influences.

pheromone (fer´o-mon): a chemical secreted by one organism that influences the behavior of another.

phloem (flo´em): vascular tissue in plants that transports nutrients.

photoperiodism (fo´to-pe´re-od-izm): the response of an organism to periods of light and dark.

photosynthesis (fo´to-sin´the-sis): the process of using the energy of the sun to make carbohydrates from carbon dioxide and water.

phototropism: plant growth or movement in response to a directional light source.

physiology (fiz´e-ol´o-je): the science that deals with the study of body functions.

phytoplankton: microscopic, free-floating, photosynthetic organisms that are the major producers in freshwater and marine ecosystems.

pia mater (pi´ah ma´ter): the innermost meninx that is in direct contact with the brain and spinal cord.

pineal gland (pin´e-al): a small cone-shaped gland located in the roof of the third ventricle.

pistil (pis´t´l): the reproductive structure of a flower that is composed of the stigma, style, and ovary.

pith: a centrally located tissue within a dicot stem.

pituitary gland (pi-tu´i-tar´e): a small, pea-shaped endocrine gland situated on the inferior surface of the brain that secretes a number of hormones; also called the *hypophysis* and commonly called the *master gland*.

placenta (plah-sen´tah): the organ of metabolic exchange between the mother and the fetus.

plankton (plank´ton): aquatic, free-floating microscopic organisms.

plasma (plaz´mah): the fluid, extracellular portion of circulating blood.

plastid (plas´tid): an organelle of a plant that assists chloroplasts in photosynthesis.

platelets (plate´lets): fragments of specific bone marrow cells that function in blood coagulation: also called *thrombocytes*.

pleural membranes (ploor´al): serous membranes that surround the lungs and line the thoracic cavity.

plexus (plek´sus): a network of interlaced nerves or vessels.

plica circulares (pli´kah ser-ku-lar´is): a deep fold within the wall of the small intestine that increases the absorptive surface area.

pollen grain (pol´en): male gametophyte generation of seed plants.

pollination (pol´i-na´shun): the delivery by wind, water, or animals of pollen to the stigma of a pistil in flowering plants, leading to fertilization.

polypeptide (pol´e-pep´tid): a molecule of many amino acids linked by peptide bonds.

pons (ponz): the portion of the brain stem just above the medulla oblongata and anterior to the cerebellum.

population (pop´u-la´shun): all the organisms of the same species in a particular location.

posterior (dorsal) (pos-ter´e-or): toward the back.

predation (pre-da´shun): the consumption of one organism by another.

pregnancy: a condition where a female has a developing offspring in the uterus.

prenatal (pre-na´tal): the period of offspring development during pregnancy; before birth.

prey: organisms that are food for a predator.

producers: organisms within an ecosystem that synthesize organic compounds from inorganic constituents.

prokaryote (pro´kar-e-ot´): organism, such as a bacterium, that lacks the specialized organelles characteristic of complex cells.

proprioceptor (pro´pre-o-sep´tor): a sensory nerve ending that responds to changes in tension in a muscle or tendon.

prostate (pros´tate): a walnut-shaped gland surrounding the male urethra just below the urinary bladder that secretes an additive to seminal fluid during ejaculation.

protein (pro´te-in): a macromolecule composed of one or several polypeptides.

prothallus (pro´thal-us): a heart-shaped structure that is the gametophyte generation of a fern.

proton (pro´ton): a subatomic particle of the atom nucleus that has a weight of one atomic mass unit and carries a positive charge; also a hydrogen ion.

proximal (prok´si-mal): closer to the midline of the body or origin of an appendage: opposite of *distal*.

puberty (pu´ber-te): the period of development in which the reproductive organs become functional.

pulmonary (pul´mo-ner´e): pertaining to the lungs.

pupil: the opening through the iris that permits light to enter the vitreous chamber of the eyeball and be refracted by the lens.

radial symmetry: symmetry around a central axis so that any half of an organism is identical to the other.

receptacle: the tip of the axis of a flower stalk that bears the floral organs.

receptor (re-sep´tor): a sense organ or a specialized end of a sensory neuron that receives stimuli from the environment.

rectum (rek´tum): the terminal portion of the GI tract, between the sigmoid colon and the anal canal.

reflex arc: the basic conduction pathway through the nervous system, consisting of a sensory neuron, interneuron, and a motor neuron.

regeneration: regrowth of tissue or the formation of a complete organism from a portion.

renal (re´nal): pertaining to the kidney.

renal corpuscle (kor´pus´l): the portion of the nephron consisting of the glomerulus and a glomerular capsule.

renal pelvis: the inner cavity of the kidney formed by the expanded ureter and into which the calyces open.

renewable resource: a commodity that is not used up because it is continually produced in the environment.

replication (re-pli-ka´shun): the process of producing a duplicate; a copying or duplication, such as DNA replication.

respiration: (res´pi-ra´shun): the exchange of gases between the external environment and the cells of an organism; the metabolic activity of cells resulting in the production of ATP.

rete testis (re´te tes´tis): a network of ducts in the center of the testis, site of spermatozoa production.

retina (ret´i-nah): the inner layer of the eye that contains the rods and cones.

retraction (re-trak´shun): the movement of a body part, such as the mandible, backward on a plane parallel with the ground; the opposite of *protraction*.

rhizome: an underground stem in some plants that stores photosynthetic products and gives rise to aboveground stems and leaves.

rod: a photoreceptor in the retina of the eye that is specialized for colorless, dim light vision.

root: the anchoring subterranean portion of a plant that permits absorption and conduction of water and minerals.

root cap: end mass of parenchyma cells which protects the apical meristem of a root.

root hair: unicellular epidermal projection from the root of a plant, which functions in absorption.

rotation (ro-ta´shun): the movement of a bone around its own longitudinal axis.

rugae (ru´je): the folds or ridges of the mucosa of an organ.

sagittal (saj´i-tal): a vertical plane through the body that divides it into right and left portions.

salinity: saltiness in water or soil; a measure of the concentraiton of dissolved salts.

salivary gland (sal´i-ver-e): an accessory digestive gland that secretes saliva into the oral cavity.

sarcolemma (sar´ko-lem´ah): the cell membrane of a muscle fiber.

sarcomere (sar´ko-mere): the portion of a skeletal muscle fiber between the two adjacent Z lines that is considered the functional unit of a myofibril.

savanna: open grassland with scattered trees.

Schwann cell (shwahn): see *neurolemmocyte*.

sclera (skle´rah): the outer white layer of connective tissue that forms the protective covering of the eye.

sclerenchyma (skle-reng´ki-mah): supporting tissue in plants composed of hollow cells with thickened walls.

scolex (sko´leks): head region of a tapeworm.

scrotum (skro´tum): a pouch of skin that contains the testes and their accessory organs.

sebaceous gland (se-ba´shus): an exocrine gland of the skin that secretes *sebum*, an oily protective product.

secondary growth: plant growth in girth from secondary or lateral meristems.

seed: a plant embryo with a food reserve that is enclosed in a protective seed coat; seeds develop from matured ovules.

semen (se´men): the secretion of the reproductive organs of the male, consisting of spermatozoa and additives.

semicircular ducts: tubular channels within the inner ear that contain the receptors for equilibrium; also called *semicircular canals*.

semilunar valve (sem´e-lu´nar): crescent-shaped heart valves, positioned at the entrances to the aorta and the pulmonary trunk.

seminal vesicles (sem´i-nal ves´i-k´lz): a pair of accessory male reproductive organs lying posterior and inferior to the urinary bladder, which secrete additives to spermatozoa into the ejaculatory ducts.

sensory neuron (nu´ron): a nerve cell that conducts an impulse from a receptor organ to the central nervous system; also called *afferent neuron*.

sepal (se´pal): outermost structure of a flower beneath the petal; collectively called the *calyx*.

serous membrane (se´rus): an epithelial and connective tissue membrane that lines body cavities and covers viscera; also called *serosa*.

sesamoid bone (ses´ah-moid): a membranous bone formed in a tendon in response to joint stress.

sessile (ses´il): organisms that lack locomotion and remain stationary, such as sponges and plants.

shoot: portion of a vascular plant that includes a stem with its branches and leaves.

sigmoid colon (sig´moid ko´lon): the S-shaped portion of the large intestine between the descending colon and the rectum.

sinoatrial node (sin´o-a´tre-al): a mass of cardiac tissue in the wall of the right atrium that initiates the cardiac cycle; the SA node; also called the *pacemaker*.

sinus (si´nus): a cavity or hollow space within a body organ such as a bone.

skeletal muscle: a type of muscle tissue that is multinucleated, occurs in bundles, has crossbands of proteins, and contracts either in a voluntary or involuntary fashion.

small intestine: the portion of the GI tract between the stomach and the cecum, functions in absorption of food nutrients.

smooth muscle: a type of muscle tissue that is nonstriated, composed of fusiform, single-nucleated fibers, and contracts in an involuntary, rhythmic fashion within the walls of visceral organs.

solute (sol´ut): a substance dissolved in a solvent to form a solution.

solvent (sol´vent): a fluid such as water that dissolves solutes.

somatic (so-mat´ik): pertaining to the nonvisceral parts of the body.

somatic cells: all the cells of the body of an organism except the germ cells.

sorus (so´rus): a cluster of sporangia on the underside of fern pinnae.

species (spe´shez): a group of morphologically similar (common gene pool) organisms that are capable of interbreeding and producing fertile offspring and are reproductively isolated.

spermatic cord (sper´mat´ik): the structure of the male reproductive system composed of the ductus deferens, spermatic vessels, nerve, cremasteric muscle, and connective tissue.

spermatogenesis (sper´mah-to-jen´e-sis): the production of male sex gametes, or spermatozoa.

spermatozoon (sper´mah-to-zo´on): a sperm cell, or gamete.

sphincter (sfingk´ter): a circular muscle that constricts a body opening or the lumen of a tubular structure.

spinal cord (spi´nal): the portion of the central nervous system that extends from the brain stem through the vertebral canal.

spinal nerve: one of the thirty-one pairs of nerves that arise from the spinal cord.

spiracle (spir´ah-k´l): a respiratory opening in certain animals such as arthropods and sharks.

spirillum (*pl.* **spirilla**): a spiral-shaped bacterium.

spleen: a large, blood-filled organ located in the upper left of the abdomen and attached by the mesenteries to the stomach.

spongy bone (spun´je): a type of bone that contains many porous spaces; also called *cancellous bone.*

spore: a reproductive cell capable of developing into an adult organism without fusion with another cell.

sporangium (spo´ran´je-um): an organ within which spores are produced.

sporophyll: a sporangium-bearing leaf.

stamen (sta´men): the structure of a flower which is composed of a filament and an anther, where pollen grains are produced.

starch: carbohydrate molecule synthesized from photosynthetic products; common food storage substance in plants.

stigma (stig´mah): the upper portion of the pistil of a flower.

stele: the vascular tissue and pith or ground tissue at the central core of a root or stem.

stoma (sto´mah): an opening in a plant leaf through which gas exchange takes place.

stomach: a pouchlike digestive organ between the esophagus and the duodenum.

style: the long slender portion of the pistil of a flower.

submucosa (sub´mu-ko´sah): a layer of supportive connective tissue that underlies a mucous membrane.

succession: the sequence of ecological stages by which a particular biotic community gradually changes until there is a community of climax vegetation.

sucrose: a disaccharide (double sugar) consisting of a linked glucose and fructose molecule; the principal transport sugar in plants.

superior vena cava (ve´nah ka´vah): a large systemic vein that collects blood from regions of the body superior to the heart and returns it to the right atrium.

surfactant (ser-fak´tant): a substance produced by the lungs that decreases the surface tension within the alveoli.

suture (su´chur): a type of fibrous joint articulating between bones of the skull.

symbiosis: a close association between two organisms where one or both species derive benefit.

sympathetic (sim´pah-thet´ik): pertaining to that part of the autonomic nervous system concerned with activities antagonistic to the parasympathetic.

synapse (sin´aps): a minute space between the axon terminal of a presynaptic neuron and a dendrite of a postsynaptic neuron.

syngamy: union of gametes in sexual reproduction; fertilization.

synovial cavity (si-no´ve-al): a space between the two bones of a synovial diarthrotic joint, filled with synovial fluid.

system: a group of body organs that function together.

systole (sis´to-le): the muscular contraction of the ventricles of the heart during the cardiac cycle.

systolic pressure (sis´tol´ik): arterial blood pressure during the ventricular systolic phase of the cardiac cycle.

taproot: a plant root system in which a single root is thick and straight.

target organ: the specific body organ that a particular hormone affects.

tarsus (tahr´sus): pertaining to the ankle; the proximal portion of the foot that contains the seven tarsal bones.

taxonomy (tak-son´o-me): the science of describing, classifying, and naming organisms.

tendo calcaneous (ten´do kal-ka´ne-us): the tendon that attaches the calf muscles to the calcaneous bone.

tendon (ten´dun): a band of dense regular connective tissue that attaches muscle to bone.

testis (tes´tis): the primary reproductive organ of a male, which produces spermatozoa and male sex hormones.

tetrapod: a four-appendaged vertebrate, such as amphibian, reptile, bird, or mammal.

thoracic (tho-ras´ik): pertaining to the chest region.

thoracic duct: the major lymphatic vessel of the body, which drains lymph from the entire body except the upper right quadrant and returns it to the left subclavian vein.

thorax (tho´raks): the chest.

thymus gland (thi´mus): a bilobed lymphoid organ positioned in the upper mediastinum, posterior to the sternum and between the lungs.

tissue: an aggregation of similar cells and their binding intercellular substance, joined to perform a specific function.

tongue: a protrusible muscular organ on the floor of the oral cavity.

toxin: a poisonous compound.

trachea (tra´ke-ah): a tubule in the respiratory system of some invertebrates; the airway leading from the larynx to the bronchi in the respiratory system of vertebrates; also called the *windpipe.*

tract: a bundle of nerve fibers within the central nervous system.

trait: a distinguishing feature studied in heredity.

transpiration (tran´spi-ra´shun): the evaporation of water from a leaf, which pulls water from the roots through the stem to the leaf.

transverse colon (ko´lon): a portion of the large intestine that extends from right to left across the abdomen between the hepatic and splenic flexures.

tricuspid valve (tri-kus´pid): the heart valve between the right atrium and the right ventricle.

turgor pressure (tur´gor): osmotic pressure that provides rigidity to a cell.

tympanic membrane (tim-pan´ik): the membranous eardrum positioned between the outer and middle ear; also called the *tympanum,* or the *ear drum.*

umbilical cord (um-bil´i-kal): a cordlike structure containing the umbilical arteries and vein, which connects the fetus with the placenta.

umbilicus (um-bil´i-kal): the site where the umbilical cord was attached to the fetus: also called the *navel.*

ureter (u-re´ter): a tube that transports urine from the kidney to the urinary bladder.

urethra (u-re´thrah): a tube that transports urine from the urinary bladder to the outside of the body.

urinary bladder (u´re-ner´e): a distensible sac in the pelvic cavity which stores urine.

uterine tube (u´ter-in): the tube through which the ovum is transported to the uterus and where fertilization takes place: also called the *oviduct* or *fallopian tube.*

uterus (u´ter-us): a hollow, muscular organ in which a fetus develops. It is located within the female pelvis between the urinary bladder and the rectum.

uvula (u´vu-lah): a fleshy, pendulous portion of the soft palate that blocks the nasopharynx during swallowing.

vacuole: a fluid-filled organelle.

vagina (vah-ji´nah): a tubular organ that leads from the uterus to the vestibule of the female reproductive tract and receives the male penis during coitus.

vascular tissue: plant tissue composed of xylem and phloem, functioning in transport of water, nutrients, and photosynthetic products throughout the plant.

vegetative: plant parts not specialized for reproduction; asexual reproduction.

vein: a blood vessel that conveys blood toward the heart.

ventral (ven´tral): toward the front surface of the body: also called *anterior.*

vertebrate (ver´te-brat): an animal that possesses a vertebral column.

vestibular folds: the supporting folds of tissue for the vocal folds within the larynx.

vestibular window: a membrane-covered opening in the bony wall between the middle and inner ear, into which the footplate of the stapes fits; also called *oval window.*

viscera (vis´er-ah): the organs within the abdominal or thoracic cavities.

vitreous humor (vit´re-us hu´mor): the transparent gell that occupies the space between the lens and retina of the eye.

vocal folds: folds of the mucous membrane in the larynx that produce sound as they are pulled taut and vibrated; also called *vocal cords.*

vulva (vul´vah): the external genitalia of the female that surround the opening of the vagina; also called the *pudendum.*

wood: interior tissue of a tree composed of secondary xylem.

xylem (zi´lem): vascular tissue in plants that transports water and minerals.

zoospore: a flagellated or ciliated spore produced asexually by some protists.

zygote (zi´gote): a fertilized egg cell formed by the union of a sperm and an ovum.

Index

A

Aardvark, 126
A band, 7
Abdomen, 106, 107, 108, 109, 114, 129, 182
Abdominal
  aponeuris, 135
  extensor, 109
  flexor, 109
  oblique, 135, 156, 157, 158, 159, 166, 169, 174, 176, 181
Abductor
  digiti minimi, 134
  longus, 136
  pollicis brevi, 134
  pollicis longus, 136
Absorption, 70, 128
Accessory digestive organs, 128
Acetabulum, 133, 155
Acetate, 18
Achnanthes flexalla, 24
Acontia, 91
Acorn, 81
Acoustic meatus, 131, 132
Acrasiomycota, 23
Acromiodeltoid, 169, 174
Acromion process, 163
Acromiotrapezius, 174, 175
Actin, 84
Actinomyces (Actinomycetes), 19
Adiantum, 60
Adductor, 100, 181
  brevis dorsalis, 156
  femoris, 176
  longus, 134, 158, 159, 176
  magnus, 134, 157, 158, 159
Adipose tissue, 7, 8, 129
Adipocyte, 7
Adrenal gland, 139, 146, 171, 172
Adventitia, 9
Aeciospore, 42
Aecium, 42
Aerial stem, 54, 55
Aerobe, 18
African sleeping sickness, 23
Agaricus rodmani, 41
Air bladder, 30, 31
Air pore, 47
Agnatha, 119, 121–122, 149
  Petromyzon marinus, 119, 121, 122, 149
Akinete, 20
Albumen, 15
Albumin gland, 99
Alga, 1, 18, 23–35, 36, 43–44
  blue-green, 19
  brown, 30–32
  golden-brown, 23, 24
  green, 27, 28, 30, 36
  hummock, 31
  red, 33
  symbiotic association with fungi, 43–44
  yellow-green, 23, 24
Allantois, 128
Allium, 12
Allomyces, 35
Alternate leaf, 75
Alternation of generations, 23, 33, 45
Alveolus, 128, 144
Amanita pantherina, 41
Ambulacral
  groove, 115, 116, 117
  ridge, 116, 117

Ambystoma gracile, 123
Amoeba proteus, 25
Ameloblast, 142
American pit viper, 123
Amnion, 128
Amoeboid
  locomotion, 23
  plasmodium, 23
Ammonia, 19
Amoebocyte, 85
Amphibia, 84, 119, 123, 155–161
  leopard frog, 155–161
  Litoria infrafrenata, 123
Amphioxus, 84, 119, 120
Ampulla, 116, 117, 119
Ampullary gland, 172
Anabaena, 19, 20
Anal pore, 26
Anaphase, 11, 12, 13, 14
Anapsida, 162–165
Anchorage, 70
Anconeus, 134, 156, 157
Angiosperm, 45, 68–83
Animal role in seed dispersal, 81
Angular bone, 162
Animalia (kingdom), 84–127
  Annelida, 84, 102–105
  Arthropoda, 84, 106–114
  Chordata, 84, 119–127
  Cnidaria, 84, 87–91
  Echinodermata, 84, 115–118
  Mollusca, 84, 98–101
  Nematoda, 84, 96–97
  Platyhelminthes, 84, 92–95
  Porifera, 84, 85–86
  Rotifera, 98
Ankle, 155, 180
Annulus, 42, 59, 60, 61
Annelida, 84, 102–105
  Hirudinea, 102, 105
  Oligochaeta, 102, 104
  Polychaeta, 102
Anoplura, 111
Ant, 106
Anteater, 126
Antebrachium, 129, 155, 173
Antenna, 98, 99, 106, 108, 110, 114
Antennule, 108, 110
Anterior chamber, 139, 140
Anther, 68, 77, 78, 79, 80
Antheridium, 24, 28, 30, 31, 32, 46, 48, 51, 53, 59, 61
  receptacle, 46, 47, 48
Antheroceros, 50
Anthocerotae, 45
Anthozoa, 87, 91
  Acropora, 91
  brain coral, 91
  mushroom coral, 91
  rose coral, 91
  sea anemone, 91
  sea fan, 87
Antibiotic, 36, 40
Antigen, 8
Antipodal cell, 83
Antrum, 148
Anura, 155–161
  leopard frog, 155–161
Anus, 16, 17, 96, 98, 100, 101, 103, 105, 106, 107, 108, 115, 116, 118, 119, 120, 148, 154, 164
Aorta, 100, 101, 120, 122, 141, 142, 146, 151, 152, 163, 169, 170, 171, 172, 182, 183

Aortic arch, 119, 141, 142, 160, 170, 178, 182, 183, 185
Apex, 142, 178
Apical meristem, 70, 71, 73, 83
Apicomplexa, 23, 25
Apocrine sweat gland, 129
Apothecium, 38, 42
Apople, 86
Appendage
  echinodermata, 115
  feeding, 108
  paired, 119
  thoracic, 108
Apple, 81, 82
Aquapharyngeal bulb, 118
Arachnida, 106, 107
  Argiope, 107
  Dugesiella, 107
  Latrodectus mactans, 107
  Pandinus, 107
Arachnoid membrane, 137
Araucaria, 64, 66
Arbor vitae, 137
Archaebacteria, 18, 19
Archegonium, 46, 48, 49, 51, 52, 59, 61, 62
Archenteron, 16, 17
Argiope, 107
Aristotle's lantern, 118
Arm, 101, 175
  sea star, 115, 116, 117
Armpit, 129
Arrector pilli muscle, 129
Arteriole, 128, 146
Artery, 128
  afferent branchial, 150, 152
  annular, 152
  aorta, 141, 152, 177
  aortic arch, 141, 161, 169, 177
  ascending aorta, 141, 177, 183
  axillary, 141, 167, 177, 183
  azygos, 182
  basilar, 137
  brachial, 141, 177
  brachiocephalic trunk, 141, 163, 177, 178
  carotid, 137, 141, 142, 152, 160, 161, 163, 167, 169, 170, 177, 182
  caudal, 169
  caudal-mesenteric, 169
  celiac trunk, 141, 152, 169, 177, 183
  celiacomesenteric trunk, 161
  coronary, 142, 169
  cranial-mesenteric, 169
  cutaneous, 161
  deep femoral, 141, 177
  efferent branchial, 152
  efferent hyoidian, 152
  epigastric, 152, 161, 177
  femoral, 136, 141, 152, 161, 177, 183
  gastric, 152, 169, 171, 177
  gastroduodenal, 171
  gastrohepatic, 152
  gastrosplenic, 152
  genital, 152
  gonadal, 169, 177
  heart, 141
  hepatic, 152, 171, 177
  iliac, 141, 152, 169, 171, 177, 182, 183

  iliolumbar, 169, 171, 177
  intestinal, 152
  intraintestinal, 152
  lumbar, 169
  lumbodorsal, 182
  mesenteric, 141, 152, 161, 170, 171, 177, 183
  occipital, 161
  olfactory, 152
  ophthalmic, 152
  pancreatico-mesenteric, 152
  peroneal, 141, 161
  phrenic, 169
  popliteal, 141
  principal human, 141
  pulmocutaneous, 161
  pulmonary, 142, 163, 177, 182, 183
  pyloric, 152
  radial, 141, 177
  renal, 141, 146, 169, 172, 177, 182, 183, 184
  sciatic, 161
  spermatic, 171
  splenic, 171, 177
  stapedial, 152
  subclavian, 141, 142, 152, 161, 163, 169, 177, 183
  suprarenal, 177
  systemic arch, 161
  testicular, 141
  thoracic, 177
  thyrocervical, 177
  thyroid, 177
  tibial, 141, 161
  ulnar, 141, 177
  umbilical, 182, 184
  urogenital, 161
  vertebral, 137
Arthropoda, 84, 106–114
  Arachnida, 106, 107
  Chilopoda, 106
  Crustacea, 106, 108–110
  Diplopoda, 106
  Insecta, 106, 111–114
  Merostomata, 106
Articular bone, 162, 163
Artifactual space, 140, 142
Ascaris, 96, 97
Ascocarp, 36, 38, 39
Ascomycota (ascomycetes), 36, 37–40
  Claviceps purpurea, 39
Ascon body type (*Leucosolenia*), 85
Ascospore, 36, 37, 38, 39
Ascus, 37, 38, 39
Asexual reproduction, 11, 12, 23, 36
Aspergillus, 37, 40
Asplenium, 59
  pinna, 59
  sori, 59
Aster, 13, 78
Asteraceae, 78
Asterias, 115–117
Asteroidea, 115–117
  Asterias, 115–117
Asterosclereid, 4
Astragalus, 155
Atlantic coast, 106
Atlas, 133, 168, 173
ATP, 2
Atriopore, 120
Atrioventricular valve, 142, 178
Atrium, 100, 120, 142, 151, 152, 163, 167, 170, 178, 180, 182, 185

Auditory meatus, 174, 184
Aurelia, 90
Auricle (ear), 9, 92
Auricle (heart), 10, 142, 161, 178
Autonomic nervous system (ANS), 128
Aves, 119, 124–125, 166–167
  Anseriformes, 124
  Apodiformes, 124
  Charadriformes, 124
  Ciconiformes, 124
  Columbiformes, 124, 165–166
  Coracliformes, 124
  Falconiformes, 124
  Galliformes, 124
  Gaviiformes, 124
  Passeriformes, 124
  Pelecaniformes, 124
  Phoenicopteriformes, 124
  Piciformes, 124
  Podicipediformes, 124
  Psittaciformes, 124
  Spheniciformes, 124
  Strigiformes, 124
  Struthioniformes, 124
Avian taxonomy, 125
Axilla, 129
Axillary bud, 70, 73
Axis, 133, 168, 173
Axon, 8, 139

B

Bacillum megaterium, 22
Bacillus, 18, 19, 22
Bacterium, 1, 11, 12, 18, 19, 20, 21, 22
  flagellated, 21
  gram-negative, 19
  gram-positive, 19, 22
  green, 19
  nitrogen-fixing, 19, 20
  photosynthetic, 19
  purple, 19
Bald cypress tree, 44
Barberry leaf, 42
Barb, 125
Barbule, 125
Barley smut spore, 3
Basal disc (foot), 87, 88
Basement membrane, 8
Basidia, 36, 42
Basidiocarp, 36, 41
  of mushrooms, 41
Basicioma, 40
Basidiomycota (basidiomycetes), 36, 40–42
  Clavariaceae, 41
  Ganoderma applanatum, 41
  Geastrum saccutum, 41
  Lycoperdon ericetorum, 41
Basidiopore, 42
Basidiospore, 40, 42
Basilar
  artery, 137
  membrane, 140, 144
Basophil, 8
Basswood (linden), 75
Bat, 126
Beak, 162
Bean
  garden, 80, 82, 83
  lima, 82
  seed, 82
  seed coat, 82
  string, 82
Bear, 126

Beer, 36
Beetle, 106, 111
Berry, 81
Biceps
  brachii, 134, 136, 166, 169, 181
  femoris, 134, 135, 156, 157, 174, 175, 181
Bicuspid valve, 185
Bilateral symmetry, 84
Bile duct, 171
Bipedal vertebrate, 129
Bipolar neuron, 139, 140
Biramous appendage, 106
Bird, 84. *See also* Aves
Bird skin, 125
Bivalvia, 98, 100
  freshwater clam, 100
  shell, 100
Blackberry, 81
Black widow spider. *See Latrodectus mactans*
Bladder, 94
  cloacal, 98
  swim, 119, 154
  urinary, 154, 160, 164, 172, 179, 183
Blade, 31, 32
Blastocoel, 16, 17
Blastopore, 16, 17
Blastostyle, 89
Blastula, 16, 17, 84, 86, 88
Blattariae, 111
Blood, 1, 128
  cell, 1, 128
    red, 22, 25
    type, 1
    white, 22
  sucking, 105
  See also Vessel; Vein
Blueberry, 81
Blue spruce. *See Picea pungens*
Boletus, 41
Body tube, 101
Bone, 8, 128
  marrow, 128
  matrix, 6, 8
  tissue, 129
Bordered pit, 4
Borella recurrentis, 22
Brachialis, 134, 136, 175, 181
Brachiocephalic trunk, 142, 170, 181
Brachioradialis muscle, 134, 136
Brachium, 155, 173
Brain, 98, 103, 110, 122, 128, 137, 138, 139, 153, 154
Branchial,
  basket, 149
  heart, 101
  tube, 122
  vein, 101
Branchium, 129
Bread, 36
Bread mold, 37
Breast, 144, 147
Brittle star, 115
Brome grass, 42
Bronchus, 143, 144
Brood chamber, 108
Brown salamander. *See Ambystoma Gracile*
Bryophyta, 45, 46–53
  Anthoceotae, 45
  Hepaticae, 45
  Musci, 45
Buccal
  bulb, 101

cavity, 92, 121, 149
funnel, 121, 122
papillae, 121
Bud, 51, 52, 72, 73, 87, 88, 95
axillary, 73
lateral, 72, 73
terminal, 72, 73
zone, 95
Budding, 11, 12, 36
Bulb, 72, 73
scale, 73
Bulbourethral gland, 147, 172
Bundle sheath, 76
Burdock, 81
Buttercup. *See Ranunculus*
Butterfly, 106, 111, 113
compound eye, 113
curled tongue, 113
Buttock, 129

C

Cabbage, 34, 72
root, 34
Caecum, 94, 101, 116, 117
gastric, 114
hepatic (liver), 120
intestinal, 94, 116
pyloric, 116, 117, 154
Calamus, 125
Calcaneum, 155
Calcaneus, 130
Calciferous gland, 103
Calf, 129
Calyptra, 47, 51
Calyx, 77, 115, 146
major, 146
minor, 146
Camas, 69
Cambarus, 108–110
Cambium, 65, 71, 74, 75
fascicular, 74
interfascicular, 74
Canada, 106
Canal
carotid, 132
cell, 48
central, 129
circular, 90, 117
excretory, 95
excurrent, 85
granule, 95
Haversian, 129
hypoglossal, 132
incurrent, 85
periradial, 90
purkinje, 139
radial, 85, 86, 90, 116, 117
ring, 116, 118, 119
semicircular, 153
stellate, 139
stone, 116, 117
supraorbital, 153
system, 90
transverse, 116
Canaliculi, 6, 129
Canine tooth, 131
Capillary, 6, 128, 139
Capitulum, 133
Capsella bursapastoris, 83
Capsule, 47, 49, 51, 52, 53
Carapace, 107, 108, 162, 163
Carbohydrate, 1, 2, 45, 84
Carbon dioxide, 18, 76
Cardiac muscle, 8
Cardiovascular system, 128
Carina, 143
Carotenoid, 45
Carotid
artery, 137, 141, 142, 152,
163, 167, 169, 170, 177,
182, 183
body, 161
canal, 132
Carpal bone, 130, 155, 163,
168, 173
Carpometacarpal bone, 166
Carpospore, 33
Carposporophyte, 33
Carrot, 70
Caryopsis, 81, 83
Cartilage
annular, 122
branchial basket, 121, 122
costal, 130, 133
cranial, 122
cricoid, 143
hyaline, 143
hypobranchial, 150
lingual, 122
Meckel's, 150
palatopterygoquadrate,
150
thyroid, 143
tracheal, 143

Cat, 173–179
arteries, 177
GI tract, 178
heart, 178
muscles, 174–176
planes of reference, 173
skeleton, 173
skull, 174
superficial structures, 173
urogenital system, 179
veins, 177
Caterpillar, 112
Caudal muscle, 175
Caudofemoralis, 166, 174,
175, 176
Cecum, 93, 171
Celery, 72
Celiac trunk, 171
Cell, 1–10
animal, 5–10, 11, 13
barley smut spore, 3
body, 7
chief, 145
chlorogogue, 104
cycle, 11
diploid, 11
division, 1, 11–14, 28, 72
epithelial, 127, 143, 144
eukaryotic, 1
fiber, 2
flame, 84
genetic material, 1
goblet, 144
guard, 76
homeostasis, 1
interstitial (of Leydig), 148
marker, 13
membrane, 1, 2, 3, 5, 6, 7,
8, 13, 16, 25, 26, 36
mesenchyme, 16
nurse, 148
parenchyma, 2, 4, 74
parietal, 145
plant, 1–4
prokaryotic, 1
sex, 11
sieve, 65
subsidiary, 76
sugar cane leaf, 3
sustentacular, 148
synergid, 83
tracheid, 2, 4
vegetative, 28
wall, 1, 2, 3, 4, 11, 12, 14,
23, 25, 29, 36, 45, 84
worn, 2
zymogenic, 145
Celiac trunk, 182
Cellulose, 1, 2, 23, 25, 45
Centipede, 106
Central nervous system
(CNS), 128
Centriole, 2, 5, 11, 13
Centromere, 12, 13, 14
Centrosome, 2
Cephalic vein, 135
Cephalochordata, 119, 120
Amphioxus, 119, 120
Cephalopoda, 98, 101
Loligo (squid), 101
Nautilus, 101
Cephalothorax, 106, 107, 108
Ceramium, 33
Ceratium hirundinella, 25
Cercaria, 93, 94
free-swimming, 93, 94
Cerebellum, 137, 139, 184
arbor vitae, 137
shark, 153
Cerebral
aqueduct, 138
arterial circle, 137
fissure, 137
occipital lobe, 137
temporal lobe, 137
Cerebrum, 137, 138, 184
frontal region, 184
occipital region, 184
parietal region, 184
temporal region, 184
Cervical region, 129, 170
Cervix, 148, 172
Cestoda, 92, 95
Chalaza, 83
Chelonia dissection, 162–165
Chamaeleo jacksoni, 123
Chara, 30
Cheese, 36
Chelate leg, 106
Chelicerae, 106, 107
Chicken egg, 15
Chilaria, 106
Chiliped, 108, 109, 110
Chilopoda, 106
Chinese liver fluke, 94

Chiton. See *Chiton stokesii*
Chiton stokesii, 99
Chipmunk, 126
Chlamydia, 19
Chlamydomonas, 27
Chlorobium, 19
Chlorogogue cell, 104
Chlorophyll, 18, 45, 76
Chlorophyta, 23, 27–30
Chloroplast, 1, 2, 3, 23, 24, 26,
27, 29, 45
envelope, 3
Choanocyte, 85, 86
Chondrichthyes, 119, 150–153
characteristics, 119
dogfish shark, 149–153
Chondrocranium, 150, 153
Chondrocyte, 9
Chondropython viridis, 165
Chordae tendineae, 142, 178
Chordata (Chordates), 84,
119–127
Cephalochordata, 119, 120
Urochordata, 119
Vertebrata, 119
Choroid, 139, 140
Chromatid, 11, 12, 13, 14
Chromatin, 1, 2
Chromophil, 139
Chromophobe, 139
Chromosome, 1, 11, 12, 13, 14
homologous, 14
Chrysophyta, 23, 24
Cicada, 111
Ciliary
body, 139, 140
processes, 140
Ciliophora, 23, 26
Cilium, 2, 6, 8, 18, 23, 26, 92,
119, 143, 144
Cingulate gyrus, 138
Circle of Willis, 137
Circular
furrow, 104
muscle, 92, 104
Circulatory system, 128
Circumesophageal
connective, 110
Circulatory system, 169
Cirrus, 120
oral, 120
Cisternae, 6
Clam, 84, 98, 100
freshwater, 100
heart region, 100
shell, 100
Clasper, 150, 151
Clavariaceae, 41
Claviceps purpurea, 39
Clavicle, 130, 135, 144, 155,
168
Clavobrachialis, 174
Clavotrapezius, 174
Claw, 107, 114, 173
Cleavage, 84
Cleistothecium, 38, 40
Clitellum, 103, 104
Clitoris, 148, 179
Cloaca, 118, 151, 164
Clonorchis sinesis, 94
Clonorchis sinensis, 94
Closterium, 29
Clostridium, 19
Clover, 74
stem, 74
Club moss, 45, 55–57
Lycopodium, 55, 57
Clubroot, 34
Clypeus, 114
Cnidaria, 84, 87–91
Anthozoa, 87, 91
body types, 87
Hydrozoa, 87, 88–89
medusa, 87
polyp, 87
Scyphozoa, 87, 90
Cnidocyst, 87, 88
Cnidocyte, 84
Crinoidea, 115
sea lilly, 115
Coagulating gland, 172
Coccygeoiliacus, 156, 157
Coccyx, 130, 133
Cochlea, 140
Cochlear
duct, 140
nerve, 137, 140
Cockroach, 111
Coconut, 81
Coccus, 18
Coelenteron, 87, 88, 91
Coelom, 84, 102, 104, 105,
116, 117
pouch, 16
sac, 16

Coenosarc, 89
Coleoptera, 111
Coleorhiza, 83
Coleoptile, 83
Coleus, 73
Collagenous
fiber, 9
filament, 6
Collar, 101
Collembola, 111
Collenchyma, 3
Colliculi, 138
Colon, 179
Columbia, 166–167
Columbine, 69
Columella, 37, 50, 53
Columnar epithelium, 8, 9
Comatricha typhoidea, 35
Common coracoarcual, 151
Compound eye, 106
Conduction
eye, 139
nutrients, 70
passageway, 1
water, 70
Cone, 45, 62, 64
axis, 67
megasporangiate, 62, 66,
67
microsporangiate, 62, 64,
67
ovulate, 64, 67
staminate, 67
Conidia, 39, 40
Conidiophore, 40
Coniferophyta (conifer), 45,
62, 64–67
Araucaria, 64
Picea pungens, 64
Pinus, 64, 65, 66, 67
Podocarpus, 64
Conjugation, 29
fungi, 36, 37
tube, 29
Conjunctiva, 139, 140
Connective tissue, 1, 8, 10, 145
Constrictor, 150, 151
Contractile sheath, 96
Conus ateriosus, 151, 152
Coprinus, 42
Coracobrachialis muscle, 134
Coracoid bone, 166
Coral, 84, 91
brain, 91
mushrom, 91
rose, 91
Cork, 75
linden, 75
Corm, 57, 72
Corn, 42, 69, 72, 81
root, 72
smut-infected, 42
stem, 72
Cornea, 139, 140
Corolla, 77, 78
Corona, 98
glandis, 147
radiata, 148
Coronal
plane, 128, 173
suture, 131
Coronoid fossa, 133
Corpora
cavernosa penis, 147
quadrigemina, 138
Corpus
albicans, 148
callosum, 138
genu, 138
splenium, 138
truncus, 138
luteum, 148
spongiosum penis, 147
Corpuscle, 19
degenerating thymic, 143
Corrugator muscle, 134
Cortex, 54, 55, 65, 70, 71, 72,
74, 75, 143, 146, 148
Cortinarius, 42
Cosmarium, 29
Costal cartilage, 130
Cotyledon, 65, 69, 82, 83
Cow liver fluke, 93
Coxae, 107, 114
Coxal endite, 107
Crab, horseshoe, 106
Cranial nerve, 128, 137
Cranium, 166
Crayfish. *See Cambarus*
Cricket, 111, 112
molting, 112
Crista, 2
Cristose lichen, 43, 44
Crop, 99, 103, 104, 114, 166,
167

Cross sectional plane, 128,
173
Crustacea, 84, 106, 108–110
Cambarus, 108–110
Daphnia, 108
Ctenocephalides, 111, 113
mouthparts, 113
Cubital fossa, 129
Cuboidal epithelium, 8
Cucumaria, 118, 119
dissection, 119
Cucurbita maxima, 4
Cumulus oophorus, 148
Cutaneous abdominis, 156,
157
Cuticle, 45, 84, 96, 97
Cyanobacterium, 18, 19, 20,
21
Cycadophyta (cycad), 45, 62
Cycas revoluta, 62
Zamia, 62
Cycas revoluta, 62
Cyclotella, 24
Cymbella, 24
Cymbium, 107
Cyrtomium falcatum, 59, 61
Cystocarp, 33
Cytokinesis, 11
Cytoplasm, 1, 2, 5, 11, 21, 23
extension, 2, 25
Cytoskeleton, 23

D

Dactylozooid, 89
Damsel fly, 111
Dandelion, 78, 81
Daphnia, 108
Dart sac, 99
Daughter cell, 11, 13, 27
Deep digital flexor, 181
Deltoid, 134, 135, 156, 157,
158, 181
tuberosity, 133
Denaturation, 18
Dendrite, 7, 8
Dendritic ending, 140
Dentary, 154, 155, 162, 163,
165, 166
Dentin, 142
Depressor anguli oris muscle,
134
Dermal
branchiae, 115, 116
tissue system, 1, 3
Dermaptera, 111
Dermis, 129
Desmid, 29
Desmosome, 84
Diaphragm, 119, 144, 145,
170, 171, 178, 180
Diatom, 23, 24
freshwater, 24
Diatoma, 24
Dicot (*Dicotyledonae*), 45, 69,
70, 71, 73, 74, 80
Diencephalon, 153
Diffusion, 23
Digastric muscle, 181
Digestive
accessory organs, 128
gland, 99, 100, 109, 110,
116, 117
system, 128
Digit, 155, 166, 168, 180
Digital extensor, 181
Dikaryotic cell, 40
Diplobacillus, 18
Diplococcus, 18
Diploid cell, 11, 15, 31, 40, 45,
59, 64, 68, 84
Diplopoda, 106
Diptera, 111
Disc, intervertebral, 9
DNA (deoxyribonucleic acid),
1, 2, 11
synthesis, 11
Dogfish shark. *See Squalus
acanthias*
Dolphin, 126
Dorsal cirrus, 99
Dorsalis scapulae, 156, 157
Dorsoventral muscle, 92
Dorsum of hand, 129
Downy mildew, 35
Drupe, 81
Duck, 124
Ductus arteriosus
fetal pig, 183
Ductus deferens, 95, 99, 103,
147, 172, 184
Dugesiella, 107
defensive posture, 107

Duodenum, 151, 160, 170,
171
Dura mater, 137

E

Ear, 9, 128, 140, 165
Earthstar, 41
Earthworm. *See Lumbricus*
Earwig, 111
Eccrine sweat gland, 129
Ecdysis, 112
Echinoderm, 11, 12, 84
Echinodermata, 84, 115–118
Asteroidea, 115–117
Crinoidea, 115
Echinodea, 115, 118
Holothuroidea, 115, 118
Ophiuroidea, 115
Echinodea, 115, 118
Arbacia (sea urchin), 118
sand dollar, 115
sea urchin, 115
Ectoderm, 17, 88
Ectoplasm, 25
Egg, 11, 15, 16, 17, 27, 28, 30,
31, 32, 35, 46, 48, 51, 52,
59, 64, 80, 83, 87, 88, 90,
93, 94, 97, 112, 119, 163
Ejaculatory duct, 93, 96, 97
Elastic
cartilage, 9
fiber, 9
Elater, 46, 47, 49, 50, 58
Elephant, 126
Elbow, 129, 180
Embryo, 45, 46, 49, 51, 62, 63,
64, 68, 82, 83
sac, 80, 83
Enamel, 142
Endocrine system, 128
Endoderm, 17, 88, 92
Endodermis, 54, 55, 66, 71, 72
Lycopodium, 55
Pinus, 66
Psilotum nudum, 54
Tmesipteris, 54
Endoplasmic reticulum, 2, 5,
6, 23
Endomysium, 10
Endoplasm, 25
Endoskeleton, 116
ossicle, 116
Endosperm, 68, 83
nucleus, 68
sac, 68
starchy, 83
Endospore, 18, 22
Endostyle, 120
Endothelial cell, 6
Enteron, 17
Enzyme, 2, 18
Epaxial muscle, 154
Ephyra, 90
Epibranchial groove, 120
Epicondyle, 133
Epicotyl, 83
Epidermal scale, 119
Epidermis, 3, 45, 47, 50, 53,
54, 55, 61, 66, 70, 71, 72,
74, 75, 76, 82, 84, 87, 88,
92, 100, 104, 116, 120, 129,
140
Epididymis, 147, 170, 171,
179, 184
Epiglottis, 143, 178
Epigynum, 107
Epiphysis, 153
Epiphyte, 45
Epithelial tissue, 1, 6, 8, 9,
127, 139, 140, 143, 144,
146
columnar, 8, 9, 143, 144,
146
cuboidal, 8
germinal, 148
glandular, 148
pigmented, 139, 140
pseudostratified ciliated
columnar, 143, 144
stratified squamous, 8,
140, 144, 148
surface, 145
Equator, 13, 14
Erysiphe graminis, 39
Erythrocyte, 6, 7
Escherichia, 19, 21
conjugation, 21
Esophageal sphincter, 178
Esophagus, 93, 94, 100, 101,
103, 110, 114, 120, 122,
163, 164, 166, 170, 171,
178, 182, 183
Ethmoidal sinus, 138

Ethmoid bone, 131, 132
  cribriform plate, 132
  crista galli, 132
  perpendicular plate, 131
Eubacteria, 18, 19
Euglena, 26
Euglenophyta, 23, 26
Eukaryotic
  cell, 1
  organism, 23, 45, 84
Eunotia, 24
Euspongia, 85
Exoccipital bone, 155, 162
Exoskeleton
  cricket, 112
Extensor
  brevis, 166
  carpi radialis, 136, 156, 181
    brevis, 136
    longus, 136, 175
  carpi ulnaris, 134, 136, 156, 181
  cruris, 158, 159
  digitorum, 134, 136
  digitorum communis, 156, 175, 181
  digitorum lateralis, 175
  digitorum minimi, 136
  dorsi communis, 175
  metacarpi radialis, 166
  pollicis muscle, 136
    brevis, 136
    longus, 136
  retinaculum, 134, 136
Extraembryonic membrane, 128
Extremity (human), 129
  lower, 129
  upper, 129
Eye, 17, 99, 101, 106, 107, 108, 113, 114, 128, 137, 139, 149, 150, 153, 155
  butterfly, 113
  compound, 106, 108, 109, 113, 114
  dogfish shark, 150
  fetal pig, 184
  flea, 113
  grasshopper, 114
  human, 139, 140
    structure, 139
  lamprey, 121, 122
  leopard frog, 155
  muscles, 137
  shark, 153
  stalk, 99, 108
  turtle, 162
Eyelash, 140
Eyelid, 140, 165, 173
Eyespot, 92, 98, 116, 120

F

Facial nerve, 137, 150
Falciform ligament, 145
Falcon, 124
False rib, 133
Fang, 107
Fasciculus
  cuneatus, 138
  gracilis, 138
Fasciola hepatica, 93
  life cycle, 93
Fascioa magnus, 93
Fat, 18, 84
  body, 160
  saturated, 18
Feather, 119, 125
  contour, 125
  structure, 125
Feather star, 115
Feces, 93
Feeding polyp, 88
Femoral artery, 136
Femur, 107, 114, 130, 133, 155, 163, 166, 168, 173
Fern, 45, 59–61
  Adiantum, 60
  Asplenium, 59
  Cyrtomium falcatum, 61
  life cycle, 59
  Polypodium, 60
  Polystichum, 60
Fertilization, 11, 15, 16, 31, 45, 46, 59, 62, 63, 64, 68, 80
  diagram of angiosperm, 80
  membrane, 16
  pore, 24
Fetal pig, 180–184
Fiber cell, 2, 4
Fibril, 2
Fibroblast, 148
Fibrocartilage, 9

Fibrous root system, 70
Fibula, 130, 163, 166, 168, 173
Fibulare, 155
Fiddle head, 59
Fig, 81
Filament, 24, 27, 28, 29, 69, 77, 78, 79
  coenocytic, 24
  conjugated, 29
  empty, 27
  Oedogonium, 28
  vegetative, 28
  Ulothrix, 27
  Zygnema, 29
Filbert, 81
Filter-feeder, 119
Fimbriae, 148
Fin, 101
  anal, 154
  caudal, 120, 121, 149, 150, 154
  dorsal, 120, 121, 149, 150, 154
  pectoral, 149, 150, 151, 154
    levator, 150
  pelvic, 149, 150, 154
  rays, 120
  ventral, 120
Fish, 84, 119
  bony, 119
  cartilaginous, 119
Fission, 11, 12
Flagellum, 2, 21, 22, 23, 25, 26, 87
Flame bulb, 98
Flamingo, 124
Flatworm, 11, 12, 84
Flax stem, 4
Flea. See *Ctenocephalides*
Flexor
  carpi radialis, 134, 136
  carpi ulnaris, 134, 136, 166
  digitorum brevis, 156
  digitorum superficialis, 166
Float, 31
Floating, rib, 133
Floral tube, 79, 82
Floret, 77
Flower, 68, 70, 73, 77–79
  apple, 82
  Asteraceae, 78
  diagram of angiosperm, 77
  Gladiolus, 79
  grass, 77
  pea (*Pisum*), 78
  pear, 79
  rose, 79
  strawberry, 78
  tomato, 78
  tulip, 77
Fluke, 92, 93, 94
Fly, 111
Foliage, 70
Flying lemur, 126
Foliose lichen, 43, 44
  Spanish moss, 44
Follicle, 148
Follicular
  cell, 148
  fluid, 148
Fons, 114
Foot, 47, 48, 49, 50, 98, 99, 100, 155, 162, 173
  pentadactyl, 162
Foramen
  incisive, 132
  infraorbital, 131
  lacerum, 132
  magnum, 132
  mental, 131
  nutrient, 133
  obturator, 133
  ovale, 132
Fossa, 129
  coronoid, 133
  cubital, 129
  glenoid, 155
  mandibular, 132
  olecranon, 133
  popliteal, 129
Fornix, 138
Fovea centralis, 139
Fragaria, 78
Fragmentation, 11, 12, 20
Frog, 17
Frontal
  bone, 130, 131, 132, 162, 174
  plane, 128, 173
  sinus, 132, 138
Frontalis muscle, 134
Frontoparietal bone, 155

Fruit, 45, 78, 79, 80–83
  aggregate, 78, 81
  angiosperm, 80–83
  multiple, 81
  pea, 78
  peanut, 82
  pear, 79
  seed dispersal, 81
  simple, 81
  strawberry, 78
  tomato, 78, 82
  wall, 82
Fruiting body, 23, 38
  Helvella, 38
  Monolina frusticola, 38
  Morchella, 38
  Peziza repanda, 38
Fucus, 30, 31, 32
  antheridium, 32
  conceptacle, 32
  female conceptacle, 32
  life cycle, 31
Fuligo, 34
Fungus, 1, 11, 23, 36–44
  conjugation, 36, 37
  coral, 41
  Penicillin, 40
  shelf, 41
  symbiotic association with alga, 43–44
Funiculus, 83

G

Gallbladder, 146, 160, 164, 178
Gametangia, 24, 30, 35, 37, 45
Gamete, 11, 14, 15, 23, 29, 35, 45, 84, 128
  diploid, 11, 15, 45
  haploid, 11, 14, 15, 45, 84
Gametophyte, 45, 46, 47, 49, 50, 51, 52, 53, 59, 61, 62, 63, 64
Ganglion, 84, 94, 99, 101
  cerebral, 94, 99
  neuron, 139, 140
  pedal, 99, 101
  sympathetic, 183
  visceral, 101
Ganoderma applanatum, 41
Garden bean. See *Phaseolus*
Garden spider. See *Argiope*
Gastric
  filament, 90
  gland, 98, 145
  muscle, 109
  pit, 145
  pouch, 90
Gastrocnemius, 134, 136, 156, 157, 158, 159, 166, 176, 181
Gastrocoel, 16, 17
Gastrodermis, 84, 87, 88, 92
Gastrolith, 109
Gastropoda, 98, 99
  slug, 99
  snail, 99
Gastrovascular cavity, 87, 88, 92
Gastrozooid, 89
Gastrula, 16, 17
Geastrum saccutum, 41
Gene, 11, 114
Genetic variation, 11
Genitalia, 147, 148
Geochelone sulcata, 123
Geranium, 76
Germinal center, 143
Germination, 45, 80–83
  angiosperm, 80–83
Germ layer, 84
Gill, 17, 40, 41, 42, 100, 101, 106, 109, 110, 115, 119, 150, 151, 154
  bar, 120
  book, 106
  cleft, 151
  mushroom, 42
  pouch, 122
  slit, 84, 119, 120, 121, 122, 149, 150, 153
  skin, 115
Ginkgo biloba, 63
Ginkgophyta, 45, 62, 63
  Ginkgo biloba, 63
Girdle, 98
  pectoral, 150, 151, 154
  pelvic, 133, 150, 154
GI tract, 128, 178
Gizzard
  earthworm, 103, 104
  grasshopper, 114
Gladiolus, 72, 79
Gland, 1

Glandular
  acini, 148
  epithelium, 148
Glans penis, 147
Glenodinium monensis, 25
Glenoid fossa, 155
Glossopharyngeal nerve, 137
Glume, 77
Gluteal region, 129, 156, 157
Gluteus maximus, 134, 135, 136, 174, 176
Gluteus medius, 175, 176, 181
  region, 134, 135
Gluteus minimis, 174
Glycogen, 84
Gnathobase, 106
Gnetophyta, 62
Goblet cell, 144, 146
Golgi apparatus, 2, 5, 23
Gonad, 84, 89, 90, 91, 100, 116, 117, 118, 119, 122, 128, 154
Gonangium, 88, 89
Gonium, 27
Gonotheca, 89
Gonozooid, 89
Gracilis, 134, 136, 176, 181
  major, 158, 159
  minor, 156, 157, 158, 159
Grana, 2, 3
  membrane, 2
Granule, 139
Grape, 72, 81
Grass, 70, 77, 81
Grasshopper, 114
Grebe, 124
Green gland, 109, 110
  duct, 109
Green tree python, See *Chondropython viridis*
Growth, 11, 71
  line, 100
  phases, 11
  ring (*Pinus*), 66
Ground tissue system, 1, 3, 74
Guberaculum, 147
Gull, 124
Gullet, 26
Gut, 102
Gymnosperm, 45, 62–67
Gyrus, 137

H

Hagfish, 119
Hair, 127
  bulb, 127, 129
  dendritic ending, 140
  eyelash, 140
  follicle, 127, 129, 140
  germinal cells, 127
  root, 127
  shaft, 127, 127
Hallobacterium, 19
Hand, 127
  dorsum, 129
Haploid cell, 11, 14, 15, 40, 45, 59, 64, 68, 84
Hatching King snake, 123
Haversian
  canal, 129
  system, 129
H band, 7
Head
  garden spider, 107
  human, 129
  lamprey, 121
  Planaria, 92
  slug, 99
  turtle, 162
Heart, 17, 99, 100, 101, 103, 128, 154, 160, 163, 164
  branchial, 101
  cat, 177, 178
  crayfish, 109
  earthworm, 103, 104
  fetal pig, 180, 183
  four-chambered, 119
  human, 142, 143, 144, 145
    internal structure, 142
  leopard frog, 160
  perch, 154
  pigeon, 167
  rat, 169, 170, 171
  sheep, 185
  systemic, 101
  three-chambered, 119
  ventral, 119
  water flea, 108
Heat pit, 165
Hedge privet. See *Ligustrum*
Helvella, 38
Hermaphroditic duct, 99
Hemiptera, 111
Hemitrichia, 34
Gland, 1

Hemocoel, 84
Hepaticae, 45, 46–49
Hepatic duct, 171
Hereditary instruction, 11
Heron, 124
Heterocercal tail, 149, 150
Heterocyst, 20
Hilum, 82
Hindgut, 17, 108
Hinge ligament, 100
Hirudinea, 102, 105
Histone, 1
Holdfast, 28, 30, 31
Holothuroidea, 115, 118
  Cucumaria, 118
Homeothermous, 119
Homeostasis requirements, 1
Homoptera, 111
Honeybee, 112
  developmental stages, 112
Hood, oral, 120
Hook, 95
Hooklet, 125
Hormone, 1, 128
  receptor, 1
Hornwort, 45, 50
  Anthoceros, 50
Horseshoe crab. See *Limulus*, 106
Horsetail, 45, 58
  Equisetum, 58
Human liver fluke, 94
Humerus, 133, 130, 133, 155, 163, 167, 168, 173
Hummingbird, 124
Hyaline cartilage, 9
Hydra, 11, 84, 87, 88
Hydranth, 89
Hydrictyon, 30
Hydrosinus, 122
Hydrostatic skeleton, 84
Hydrotheca, 88
Hydrozoa, 87, 88–89
  hydra, 87, 88
  obelia, 88, 89
  Physalia (Portuguese man-of-war), 89
Hymen, 148
Hymenoptera, 111
Hyoid bone, 173
Hypaxial muscle, 151, 154
Hypha, 36, 37, 40, 42, 44
  dikaryotic, 40
  fungal, 44
Hypobranchial
  cartilage, 150
  groove, 120
Hypocotyl, 82, 83
  axis, 83
Hypodermis, 129
Hypostome, 87, 88, 89
Hypothalamus, 138
Hyrax, 126

I

I band, 7
Ileum, 151, 171
Iliac crest, 133, 135
Iliacus internus, 157, 181
Iliolumbar, 156
Iliopsoas muscle, 134, 136
Iliotibialis, 166
Iliotibial tract, 136
Iliotrochantericus, 166
Ilium, 130, 133, 155, 166, 168, 173
  crest, 133, 135
Immune reaction, 1
Incisive foramen, 132
Incisor, 131, 132
Indonesian giant tree frog. See *Kitoria infrafrenata*
Indusium, 60, 61
  false, 60
Inflorescence, 77
Infraorbital foramen, 131
Infraspinatus, 134, 135, 175
Infundibular stalk, 138
Inguinal ligament, 135, 136
Ink sac, 101
Insect, 84, 111, 112, 113, 114
Integument, 62, 63, 82, 83
Intercostal, 181
Intermediate mass, 138
Internode, 70, 73
Interparietal bone, 174
Interosseous muscle, 134, 136
Interphase, 11, 12
Interstitial cell, 148
Interthalamic adhesion, 138
Intertrochanteric crest, 133
Intertubercular groove, 133
Interventricular septum, 142, 178

Intestinal gland, 146
Intestine, 92, 96, 97, 98, 99, 100, 102, 103, 104, 105, 154
  Amphioxus, 120
  cat, 179
  crayfish, 109
  fetal pig, 180
  grasshopper, 114
  human, 145, 146
    ileum, 146
    jejunum, 146
  lamprey, 122
  leopard frog, 160
  lumen, 104
  perch, 154
  pigeon, 167
  rat, 170, 171
  sea cucumber, 118, 119
  sea urchin, 118
  turtle, 164
Isoetes melanopoda, 57
Invertebrate, 1, 84
  host, 92, 94
Iris, 69, 72
Iris (eye), 139, 140
Ischium, 130, 133, 155, 163, 166, 168, 173
Isoptera, 111

J

Jackrabbit, 126
Jackson's Chameleon. See *Chamaeleo jacksoni*
jaw, dogfish, 150
Jellyfish, 84, 87, 90
  Aurelia, 90
Joint, 128
  sacroiliac, 133
Jugal bone, 162
Junction, 84
  gap, 84
  tight, 84

K

Kangaroo, 126
Karyogamy, 40
Karyokinesis, 11
Kelp, 30–32
Kidney (nephridium), 99, 100, 101, 103, 104, 105, 120, 122, 139, 146, 151, 154, 150, 160, 170, 171, 172, 179, 182, 183, 184
Kingfisher, 124
Knee, 180
Klebsiella pneumoniae, 22

L

Labial palp, 100, 114
Labium, 107, 114, 148
  majus, 148
  minus, 148
Labrum, 114
Lacewing, 111
Lacrimal bone, 131
Lactobacillus, 19
Lacteal, 146
Lacuna, 6, 9, 129
  osteocytes in, 129
Lambdoidal suture, 131
Lamellae
  interstitial, 129
  osteon, 129
Lamina, 31, 70, 76
Lamina propria, 9, 140, 143, 144, 145, 146, 148
  Laminara, 30
Lamprey. See *Petromyzon marinus*
Lancelet. See *Amphioxus*
Larch. See *Larix*
Larix, 66
Larva, 84, 97, 112, 119
  appearance, 112
  bipinnaria, 16
  brachiolaria, 16
  food requirements, 112
  free-swimming, 119
  honeybee, 112
Larynx, 143, 178, 180, 183
Lateral
  bud, 72, 73
  line, 96, 97, 149
Latissimus dorsi, 134, 135, 156, 157, 166, 169, 174, 175, 181
Latrodectus mactans (black widow spider), 107
Leaf
  angiosperm, 75–76
  arrangement, 75

complexity, 75
epidermis, 76
gap, 73
ginkgo, 63
lamina (blade), 70, 76
margin, 70, 75
midrib, 70
node, 70
primordium, 73
scar, 72
surface features, 76
uses, 75
venation, 69, 70, 75
Leech, 102, 105
Leg, 106
    walking, 107, 108, 109, 110
Legume, 81, 82
Lemma, 77
Lemur, 126
Lens, 122, 139, 140
Lenticel, 72, 73
Leopard frog. *See Rana pipiens*
Lepidoptera, 111
Lepidosauria, 165
Leucon body type
    (*Euspongia*), 85, 86
Leukocyte, 8
    types, 8
Leucosolenia, 85
Leydig cell, 148
Lichen, 36, 43–44
    crustose, 43, 44
    foliose, 43, 44
    friticose, 43
    thallus, 43, 44
Ligament, 148
    round, 148
Ligule, 56, 57, 101
Ligustrum, 76
    leaf, 76
Lilac, 80
    pollen grains, 80
Lilium, 83
Lily, 80, 83
    embryo sac, 83
    pollen, 14, 80
Lima bean. *See* bean
Limb bud, 17
Limulus, 106
Linea alba, 151, 158, 169, 176
Linden (basswood), 75
Lingual muscle, 122
Linum, 4
Lip, 96
Litoria infrafrenata, 123
Liver, 17, 94, 101
    Amphioxus, 120
    cat, 178, 179
    dogfish shark, 151
    fetal pig, 180, 183
    human, 94, 128, 144, 145,
        146
    lamprey, 122
    leopard frog, 160
    perch, 154
    pigeon, 167
    rat, 170, 171
    squid, 101
    turtle, 165
Liverwort, 45, 46–49, 50
    life cycle, 46
    Marchantia, 46, 47
    Pelia, 49
    Porella, 49
Lobster, 106
Locule, 83
Loligo (squid), 101
Longissimus dorsi, 156, 157
Longitudinal
    fissure, 184
    muscle, 92, 96, 97, 104,
        119, 145, 146
    nerve, 95
Loon, 124
Louse, 111
Lumbar region (human), 129
Lumbodorsal fascia, 174, 175
Lumbricus, 103–105
    cocoon, 105
    surface anatomy, 103
Lumen, 1, 6, 9, 92, 97, 143,
    144, 145, 146, 148
    bronchus, 144
    capillary, 6
    gut, 102
    intestine, 104, 146
    pharynx, 92
    seminiferous tubule, 148
    stomach, 145
    trachea, 9
    vagina, 148
Lung, 106, 107, 119, 143, 144,
    145, 160, 167, 170, 171,
    180, 183

Lxodidae, 107
Lycogala, 34
Lycoperdon ericetorum, 41
Lycopersicon esculentum, 78
Lycophyta, 45, 55–57
Lycopod, 56, 57
    Isoetes, 57
    Selaginella, 56
Lycopodium, 55, 57
    aerial stem, 55
    branch, 55
    rhizome, 55
Lymph node, 128, 143
    cortex, 143
    medulla, 143
Lymph nodule, 143, 146
Lymphocyte, 8, 143
Lysosome, 2, 5

M

Macronucleus, 26
Macrosclerid, 82
Madreporite, 115, 116, 117,
    118
Maidenhair fern, 60
Maidenhair tree. *See Ginkgo
    biloba*
Malar, 174
Malaria, 23, 25
Malpighian tubule, 114
Mammalia, 84, 119, 126–127,
    168–185
    Artiodactyla, 126
    Carnivora, 126
    cat, 173–179
    Cetacea, 126
    characteristics, 119
    Chiroptera, 126
    Dermoptera, 126
    Edentata, 126
    fetal pig, 180–184
    Hyracoidea, 126
    Insectivora, 126
    Lagomorpha, 126
    Marsupialia, 126
    Monotremata, 126
    Perissodactyla, 126
    Pholidota, 126
    Primates, 126
    Proboscidea, 126
    rat, 168–172
    Rodentia, 126
    sheep, 185
    Sirenia, 126
    skin, 127
    Tubulidentata, 126
Mammalian taxonomy, 127
Mammary gland, 119
Mammilary body, 138
Mandible, 101, 106, 109, 114
    cat, 173, 174, 178
    condyloid proces, 174
    coronoid process, 174
    grasshopper, 114
    human, 130, 131, 132
        angle, 131
        condyloid process, 131
        coronoid process, 131
    rat, 168
Mandibular
    adductor, 150, 151
    fossa, 132
    muscle, 109, 110
    notch, 131
Manatee, 126
Manganese, 23
Mantle, 99, 100, 101
Manubrium, 89, 130, 133
Manus, 163, 173
Maple, 81
Marchantia, 46, 47, 48, 49, 50
Masseter, 134, 181
Mastax, 98
Mastoid process, 174
Matrix, 6, 8, 9
Maxilla, 107, 109, 114
    grasshopper, 114
    human, 130, 131, 132
    leopard frog, 155
    turtle, 162
Maxillary pulp, 114
Maxilliped, 108, 109
Meckel's cartilage, 150
Median plane, 128, 173
Mediastinum, 144
Medulla
    oblongata, 137, 138, 153,
        184
    renal, 146
Medusa, 87, 88, 89, 90
    bud, 88, 89
Megagametophyte, 64, 68
Megasporangium, 56, 57, 64
Megaspore, 45, 56, 64, 68

Megasporophyll, 56, 62, 67
Meiosis, 14, 15, 23, 31, 40, 45,
    46, 51, 59, 64
Meiospore, 46, 49, 51, 53, 58
Membranous sac, 2
Meninges, 137
Mental foramen, 131
Merismopedia, 20
Meristem, 73
Merostomata, 106
    Limulus, 106
Merozite, 25
Mesencephalon, 17, 153
Mesenchyme, 16, 92
Mesentery, 161
Mesoderm, 17
Mesoglea, 84, 87, 88
Mesonephric duct, 151
Mesophyll, 76
    palisade, 76
    spongy, 76
Mesothorax, 114
Metabolism, 1, 11, 18
    byproduct, 18
Metacarpal bone, 130, 155,
    163, 168, 173
Metacercaria, 93, 94
Metamorphosis, 84, 112
Metaphase, 11, 12, 13, 14
Metapleural fold, 120
Metatarsal bone, 130, 155,
    163, 168, 173
Metatarsus
    garden spider, 107
Metathorax, 114
Metencephalon, 153
Methane, 18
Methanobacteria, 18
Methanogen, 18, 19
Methanol, 18
Metridium, 91
Micrococcus luteus, 22
Microfilament, 2, 23
Micronucleus, 26
Micropyle, 63, 83
Microsporangium, 56, 57, 62,
    63, 64, 67, 68
Microspore, 45, 56, 64
Microsporophyll, 56, 62
Microtubule, 2, 5, 6, 11, 23
    9 + 2 arrangement, 6
Microvilli, 146
Midbrain, 138
    tegmentum, 138
Midgut, 108, 120
Midrib, 76
Midsaggital plane, 128, 173
Millipede, 106
Miracidium, 93, 94
Mite, 106
Mitochondrion, 2, 3, 5, 7, 23
Mitosis, 2, 11, 12, 13, 14, 23,
    45, 71, 84
Mnium, 52, 53
Molar, 131, 132
Mold, 36, 37–40
    Aspergillus, 40
Molecule
    carbohydrate, 1
    regulatory, 1
    transport, 1
Mollusca, 84, 98–101
    Bibalbia, 100
    Cephalopoda, 101
    Gastropoda, 99
    Monoplacophora, 98
    Polyplacophora, 99
Molting, 112
Monera, 18–22
Monocot (*Monocotyledonae*),
    45, 68–70, 74
    comparison with dicot, 69
    examples, 69
Monocotyledonae. See monocot
Monocyte, 8
Monolina fructicola, 38
Monoplacophora, 98
Mons pubis, 148
Morchella, 36, 38
    esculenta, 36
Morel, 36, 37–40
Mosquito, 25
Moss, 45, 51–53
Mouth
    Amphioxus, 120
    Ascaris, 96
    clam, 100
    dogfish shark, 150, 151
    earthworm, 103
    horseshoe crab, 106
    hydrozoa, 87, 88, 89
    lamprey, 121, 122
    larva, 16
    leech, 105
    frog, 17

liver fluke, 93, 94
Planaria, 92
rotifer, 98
sandworm, 102
sea star, 116, 117
sea urchin, 118
snail, 99
water flea, 108
Mucosa, 146
Mucosal ridge, 145
Mucous gland, 99
Mucus, 99
    slug locomotion, 99
Musci, 45, 52, 52–53
    Mnium, 52
    sea anemone, 91
    Sphagnum, 52
Muscle
    cell, 96
    circular, 145
    eye, 137
    oblique, 145
    tissue, 1, 7, 8, 10
    skeletal, 7, 10
Muscular system, 128, 134–136
    antebrachium, 136
    forearm, 136
    gluteal region, 135
    leopard frog, 156–159
    medial brachium, 136
    thigh, 136
    trunk, 135
Muscularis
    externa, 145
    mucosa, 145, 146
Mushroom, 36, 40–42
    Coprinus, 42
    gills, 42
    life cycle, 40
Mussel, 98
Mutant, 11
Mycelium, 36, 39
Mycoplasma, 19
Microphyll, 55
Myelin later, 139
Mylohyoid, 158, 159, 181
Myofiber, 10
Myofibril, 7
Myomere, 120, 121, 122
Myoseptum, 120, 122
Myosin, 84
Myotome, 17, 150
    epaxial, 150
    hypaxial, 150
    lateral bundle, 150
    ventral bundle, 150
Myxomycota, 23, 34–35

N

Naris, 173, 180
Nasal
    bone, 131, 132, 155, 174
    concha, 131, 132
    pit, 17
Nasopharyngel pouch, 122
Nautilus, 101
Neanthes (sandworm), 102
Neck, 95, 129, 133
    anatomical, 133
    surgical, 133
Needle, 64, 65, 67
Negative charge, 1
Neisseria, 19
Nematoda, 84, 96–97
    Ascaris, 96, 97
    Trichinella spiralis, 97
Neopilina, 98
Nephridium. *See* kidney
Nereocystis, 31
Nerve
    abducens, 137, 153
    auditory, 153
    branchial, 153
    cochlear, 137
    cord, 92, 104, 119
        central, 103
        dorsal, 120
    cranial, 153
    facial, 137, 153
    fiber, 104
    glossopharyngel, 137, 153
    hypobranchial, 153
    hypoglossal, 137
    hyomandibular, 153
    lateral, 153
    median, 183
    myelinated, 139
    oculomotor, 137, 153
    olfactory, 153
    ophthalmic, 153
    optic, 137, 140, 153
    radial, 175
    sciatic, 176
    spinal, 153

terminal, 153
trigeminal, 137, 153
trochear, 137, 153
ulnar, 183
vagus, 137, 153, 170
vestibular, 137
visceral, 153
Neural
    fold, 17
    hollow, 119
    plate, 17
    tube stage, 17
Neurocoel, 17
Neurofibril node, 139
Neuroglium, 8
Neuron, 7, 8, 128, 139
    bipolar, 139, 140
    ganglion, 139, 140
    photoreceptor, 140
Neuropodium, 102
Neuroptera, 111
Nidamental gland, 101
Nipple, 147
Nitrate, 19
Nitrite, 19
Nitrobacter, 19
Nitrogen, 19, 20
    fixation, 19, 20
Node of Ranvier, 139
Nose, 173, 180
    bridge, 173
Nostoc, 19
Nostril, 121, 122, 150, 155,
    162, 173
Notochord, 17, 84, 119, 120,
    121, 122, 149
Notopodium, 102
Nuchal
    crest, 174
    line, 132
Nucellus, 62, 63
Nucleic acid, 1, 2
Nucleolus, 1, 2, 5, 16, 28
Nucleoplast, 2
Nucleus, 1, 2, 3, 5, 6, 7, 8, 9,
    10, 16, 23, 25, 26, 28, 32,
    144
    envelope, 1, 2
    membrane, 1, 2, 5, 16
    pore, 2, 3
Nuphar, 4
    petiole, 4
Nurse cell, 148
Nutrient, 1
Nutrient foramen, 133
Nymph, 112

O

Oak, 75
    stem, 75
Obelia, 88, 89
    colony, 89
    feeding position, 89
    life cycle, 88
    medusa, 89
Oblique muscle, 153
Obturator foramen, 133
Occipital
    bone, 130, 131, 132
    condyle, 132
Occipitalis muscle, 134
Ocelli, 114
Octopode, 98
Ocular bristle, 113
Oculomotor nerve, 137
Oedogonium, 28
Olecranon fossa, 133
Olfactory
    bulb, 153
    crest, 101
    pit, 152
    sac, 122
    tract, 137, 153
Oligochaeta, 102, 103–105
    Lumbricus, 103–105
Omentum, 178
Onion, 12, 72, 73
Oocyte, 15, 148
Oogenesis, 15
Oogonium, 15, 24, 28, 30, 31,
    32, 35
Oomycota, 23, 35
Oospore, 35
Opercular bone, 154
Operculum, 51, 52, 53, 106,
    119
Ophiuroidea, 115
    brittle star, 115
Opisthoma, 107
Optic
    chiasma, 138
    disc, 139

lobe, 153
nerve, 137, 140
Oral
    arm, 90
    disc, 91
    evagination, 17
    plate, 17
    spine, 117
    sucker, 93, 94
Orbicularis oculi muscle, 134
Orbicularis oris muscle, 134
Orbit, 130, 154, 174
Orbital fissure, 131
Organelle, 1, 2, 11, 23
Organ of Corti, 140
Orthoptera, 111, 114
Oscillatoria, 19, 20
Osculum, 85, 86
Ossicle, 116, 118, 119
Osteichthyes, 119, 154
    characteristics, 119
    perch, 154
Osteocyte, 6, 129
    in lacuna, 129
Osteon, 129
    lamellae, 129
Ostiole, 32
Ostium, 85, 86, 91
Ostrich, 124
Ovary, 68, 77, 78, 79, 80, 82,
    83, 87, 88, 93, 94, 95, 96,
    97, 101, 109
    earthworm, 103
    fetal pig, 184
    grasshopper, 114
    human, 128, 147, 148
        structure, 148
    leopard frog, 160
    rat, 172
    turtle, 164
    wall, 78, 82
Oviduct, 93, 96, 97, 99, 101,
    103, 109, 114, 160, 172
Oviparous, 123
Ovipositor, 114
Ovotestis, 99
Ovoviviparous, 123
Ovulation, 148
Ovule, 62, 63, 64, 67, 68, 77,
    78, 79, 80, 82, 83
Ovuliferous scale, 67
Ovum, 15, 84, 128
Owl, 124
Oxygen, 1, 75
    replenishment, 75
Oyster, 98

P

Palate, 178
Palatine bone, 132, 155, 162
Palatopterygoquadrate
    cartilage, 150
Palea, 77
Palm, 129
Palmaris longus, 134, 136, 158
Palmar region (human), 129
Palmately compound leaf, 75
Palp, 102, 113
Palpebra, 137, 173
Pampiniform plexus, 171
Pancreas, 128, 151, 164, 179
Pandinus, 106, 107
Panoramic vision, 113
Papilla
    connective tissue, 148
    renal, 146
    secondary, 140
    vallate, 140
Papillary muscle, 142, 178
Pappus scale (plume), 78
Paramecium caudatum, 26
Paramylon granule, 26
Paraphyses, 32, 52, 53
Parapodium, 102
Parasphenoid bone, 155
Parasitism, 102
    blood-sucking, 102
Parenchyma, 2, 3, 4, 65, 71,
    74, 92
    leaf, 76
    ray, 2, 65
    storage, 65
Parietal bone, 130, 131, 132,
    162, 174
Parrot, 124
Patella, 107, 129, 130, 134,
    136, 168, 173
Patellar
    ligament, 134
    region (human), 129
Pathogen, 38
Pautron, 107
Pea, 78
Peach, 81

Peanut, 82
Pear. *See Pyrus*
Pectineus, 134, 136, 176, 181
Pectoral girdle, 150, 151
Pectoralis, 158, 159, 166, 181
    major, 134, 135, 169, 175
    minor, 169
Pedal
    disc, 91
    gland, 98
Pedicel, 77, 78, 79, 80, 82
Pedicellaria, 115, 116, 117, 118
Pedipalp, 106, 107
Peduncle, 77
Pelia, 49
Pelican, 124
Pellicle, 26
Pelvic girdle, 133, 150, 165
Pelvis, renal, 146
Penguin, 124
Penicillin, 36, 37, 40
Penis, 93, 99, 101, 147, 170, 171, 172, 179, 184
Pentaradial symmetry, 84, 115
Peptidoglycan, 18
Perch, 154
Perianth, 79
Pericapsular connective tissue, 143
Pericardial
    cavity, 151
    sac, 128
Pericardium, 100, 145
    sinus, 100
Pericarp, 33, 83
Perichondrium, 9
Pericycle, 55, 71
Periderm, 65
Peridinium wisconsiense, 25
Perineum, 148
Peripheral nervous system (PNS), 128
Perinysium, 10
Periopod, 108
Periostracum, 100
Perisarc, 89
Peristome, 52, 115, 117, 118
Perithecium, 38, 39
Peritoneum, 104
Peroneal nerve, 136
Peroneus, 156, 157
    brevis, 134
    longus, 134, 166
    tertius, 181
Perpetuation of life, 11–17
Pes, 163
Petal, 77, 78, 79
Petiole, 4, 70, 75, 76
Petromyson marinus, 121–122, 149
Peyer's patch, 146
Peziza repanda, 38
Phaeophyta, 23, 30–32
Phagocytosis, 23
Phalanx, 130, 155, 163, 166, 168, 173
Pharyngeal
    cavity, 92
    land, 122
    muscle, 103
Pharynx, 17, 91, 92, 93, 94, 96, 102, 103, 104, 105, 120, 122
    everted, 102
Phaseolus, 82, 83
    closeup, 82
    germination, 83
    seed coat, 82
Pheasant, 124
Philodina, 98
Phloem, 3, 4, 54, 55, 56, 65, 66, 70, 71, 72, 74, 75
Phospholipid, 2
Photoreceptor, 26
    neuron, 140
Photosynthesis, 1, 2, 18, 26, 36, 45, 76
    bacteria, 19, 26
    mesophyll, 66
    pigment, 1
    tissue, 47, 50, 52, 53
Physalia, 87, 89
Physarum, 35
Phytoplankton, 23
Picea pungens, 64
Pig (*see* Fetal Pig)
Pigment spot, 120
Pigeon, 124, 166–167
Pileus, 40, 42
Pinacocyte, 85, 86
Pine, 4, 64, 65, 66
    life cycle, 64
    pinion, 66
    seed, 64

sugar, 66
xylem, 4
Pineal
    body, 121
    organ, 122
Pineapple, 81
Pinna, 173
Pinnularia, 24
Pinnule, 59, 60, 61
Pinion pine. *See Pinus edulis*
Pinocytosis, 23
Pineal gland, 138
Pinus, 4, 64, 65, 66
    cone, 67
    edulis (pinion), 66
    growth ring, 66
    lambertiana (sugar), 66
    leaf (needle), 66
    life cycle, 64
    phloem, 65
    seed, 64, 65
    stem, 65, 66
    xylem, 4, 66
Piriformis, 156, 157
Pistil, 68, 77, 78, 79
Pisum (pea), 78
Pith, 45, 65, 70, 74, 75
Pituitary gland, 139, 138
Placenta, 128
Placoid scale, 119
Planes of reference (human), 128
Plankton, 23, 24
Plantae (kingdom), 45–83
    Anthophyta, 45, 68–83
Plantar surface, 129
Plant tissue, 1–4
    examples, 3
Planula, 88
Plasma membrane, 1, 2, 18
Plasmodial slime mold, 34, 35
Plasmodiophora brassicae, 34
Plasmodium vivax, 25
Plasmogamy, 40
Plastid, 2, 23
Plastron, 162, 163
Platyhelminthes, 84, 92–95
    Cestoda, 92, 95
    Trematoda, 92, 93–94
    Turbellaria, 92
Platypus, 126
Platysma muscle, 135
Plecoptera, 111
Pleopod, 108, 109
Pleural nerve, 101
Pleuron, 108
Plica circulares, 146
Plumule, 82
Pneumatophore, 89
Podocarpus, 64
Polar
    body, 15
    nucleus, 80, 83
Pole, 11, 13, 14
Pollen, 45, 79
    chamber, 80
    grain, 64, 68, 69, 80
    tube, 64, 68, 80, 83
Pollination, 45
    angiosperm, 80
Polychaeta, 102
    Neanthes, 102
    Sabellastarte indico, 102
Polymorphoclear leukocyte, 8
Polyp, 87, 88, 89, 90
    feeding, 89
    reproductive, 89
    stinging, 89
    three types, 89
Polyplacophora, 99
    Chiton Stokesii, 99
Polypodium, 60
Polysiphonia, 33
    cystocarp, 33
    gametophyte, 33
    tetrasporophyte, 33
Polystichum, 60
Pondlily, 4
    petiole, 4
Pons, 137, 138
Popliteal fossa, 129, 134
Popliteus muscle, 134
Poppy, 81
Pore, 52, 69, 84, 94, 99, 105
    excretory, 94, 99
    genital, 94, 95, 96, 97, 99, 109
Porella, 50
Porifera, 84, 85–86
    body types, 85
    Calcarea, 85, 86
    Desmospongiae, 86
Porphyra, 33
Portuguese man-of-war, 87, 89
Postabdomen, 107

Postelsia palmaeformis, 30
Posterior chamber, 139, 140
Postorbital bone, 162
Potato, 4, 72, 73
Powdery mildew. *See Erysiphe graminis*
Preabdomen, 107
Predentin, 142
Prefrontal bone, 162
Premaxilla, 154, 155, 162, 166, 174
Premolar, 131, 132
Prepuce, 147, 148, 171, 179
Preputial gland, 171, 172
Procambium, 73
Procoracoid, 163
Proglottid, 95
Prokaryotic cell, 1, 18
Pronator superficialis, 166
Prophase, 11, 12, 13, 14
Prosencephalon, 17
Prosoma, 107
Prosopyle, 85
Prostate, 147, 148, 171, 172, 179
Prostomium, 102, 103
Protein, 1, 2, 18
Proteus, 19
Prothallus, 61
Prothorax, 114
Protista, 23–35
    Acrasiomycota, 23
    Apicomplexa, 23, 25
    Chlorophyta, 23, 27–30
    Chrysophyta, 23, 24
    Ciliophora, 23, 26
    Dinoflagellata, 23, 25
    Euglenophyta, 23, 26
    Myxomycota, 23, 34–35
    Oomycota, 23, 35
    Phaeophyta, 23, 30
    Rhizopoda, 23, 25
    Rhodophyta, 23, 33
Protonephridia, 84
Protonema, 51, 52
Protozoa, 1, 11, 12, 23, 25, 26
Proventriculus, 167
Pseudomonas, 19, 21
Pseudocoel, 96, 97
Pseudopodium, 23, 25
Psilotophyta, 45, 53–54
    Psilotum nudum, 53–54
    Tmesipteris, 54
Psilotum nudum, 53–54
Psoas major, 181
Pterophyta, 45, 59–61
Pterygoid
    bone, 155
    muscle, 156
Pubic region (human), 129
Pubis, 130, 133, 155, 163, 166, 168, 173
Puccinia graminis, 42
Puffball, 41
Pulmocutaneous arch, 160
Pulmonary trunk, 178
Pulp, 142
Pupa, 112
Purkinje cell, 139
Pycnidium, 42
pygostyle, 166
Pyloric duct, 116
Pyramidalis muscle, 135
Pyrenoid, 29
Pyrus, 71, 79

Q

Quadrapedal vertebrate, 173
Quadrate bone, 162, 165
Quadratojugal bone, 162
Quadratus lumborum muscle, 146
Quadriceps femoris muscle, 134
Quercus, 75
Quill, 101
Quillwort, 45, 55–57

R

Rachilla, 77
Rachis, 125
Radial symmetry, 84
Radicle, 82, 83
Radio-ulna, 155
Radius, 130, 136, 163, 166, 168, 173
Radula, 101
Rana pipiens, 155–161
    musculature, 156–159
    legs, 157–169
Ranunculus, 71
    root, 71
Rat, 168–172

Ray flower, 78
Receptacle, 31, 32, 78, 79, 80
Rectal gland, 151
Rectum, 96, 101, 105, 114, 171, 184
Rectus, 137, 153
    abdominis, 134, 135, 158, 159, 169, 176, 181
        tendinous inscription, 135
    femoris, 134, 136, 176, 181
Rectus sheath, 134
Redia, 93, 94
Regeneration, 1
Renal
    artery, 146
    capsule, 146
    column, 146
    cortex, 146
    medulla, 146
    papilla, 146
    pelvis, 146
    pyramid, 146
Reproductive system, 128, 147
Reptilia, 84, 119, 123, 162–165
    Chelonia, 162–165
    Chamaeleo jacksoni, 123
    characteristics, 119
    Geochelone sulcata, 123
    Hatching King snake, 123
    lizard, 165
    snake, 165
    Squamata, 165
    turtle, 162–165
Resin duct, 65, 66
Respiration, 1
Respiratory
    conducting division, 128
    gases, 128
    system, 128, 144
    tree, 118, 119
Retina, 122, 139, 140
Retractor muscle, 91, 100, 101
Rheumatic fever, 21, 22
Rhizoblum, 19
Rhizoid, 37, 45, 46, 47, 61
Rhizome, 45, 55, 59, 72
Rhizopoda, 23, 25
Rhizopus, 37
Rhodophyta, 23, 33
Rhodospirillum, 19
Rhombencephalon, 17
Rhomboideus, 175
Rhomboideus
    capitis, 175, 181
    cervicis, 181
Rhopalium, 90
Rib, 130, 133, 135, 144, 154, 163, 165, 168, 173, 182
    cage, 130, 133
    false, 133
    floating, 133
    true, 133
Ribosome, 1, 2, 5, 6
Rickettsia, 19
RNA (ribonucleic acid), 1, 11, 18
    transfer, 18
    polymerase, 18
Rockweed, 30, 32
Rod, 139
Root ganglion, 138
Rotifera, 98
    Philodina, 98
Root, 70, 72, 83
    aerial, 70
    angiosperm, 70–72
    cap, 70, 71, 72
    diagrams, 71
    elongation region, 71
    fibrous system, 70
    hair, 70, 72
    maturation region, 71
    meristematic region, 71
    prop, 70
    taproot system, 70
Rose, 69, 79
Rostellum, 95
Rostrum, 108, 120, 150, 153
Rotifera, 98
Roundworm, 84
Runner, 12
Rust, 36, 40–42

S

Sabellastarte indico, 102
Saccharomyces cerevusuae, 37
Sacculus, 153
Scaroiliac joint, 133
Sacrospinalis, 175
Sacrum, 130, 133, 168
Sagittal suture, 174
Salivary gland, 98, 99, 169
Sand dollar, 84, 115

Sandworm. *See Neanthes*
Saprolegnia, 35
Saprophyte, 18
    heterotrophic, 18
Sarcomere, 7
Sarcoplasmic reticulum, 7
Sargassum, 30, 31
Sartorius, 134, 136, 158, 159, 166, 174, 176, 181
Savanah monitor. *See Varanus exanthematicus*
Scala
    tympani, 140
    vestibuli, 140
Scale, 73
    abdominal, 162
    anal, 162
    costal, 162
    femoral, 162
    gular, 162
    humeral, 162
    marginal, 162
    nuchal, 162
    pectoral, 162
    vertebral, 162
Scaphopoda, 98
Scapula, 130, 166, 168, 173, 175
Schistosome, 92
Sciatic notch, 133
Sclera, 139
Sclerenchyma, 4
Sclerotic bone, 166
Scolex, 92, 95
Scorpion. *See Pandinus*
Scrotum, 147, 170, 171, 179, 180
Scypha, 85
Scyphistoma, 90
Scyphozoa, 87, 90
Scytonema, 20
Sea anemone, 91
Sea
    cucumber, 84, 115, 118
    fan, 87
    feathers, 102
    lettuce, 29
    lily, 115
    palm, 30
    star, 16, 84, 115–117
    urchin, 84, 115
Sebaceous gland, 127, 129
    ductule, 127
Seed, 45, 63, 65, 78, 80–83
    angiosperm, 80–83
    coat, 63, 64, 65, 68, 83
    dispersal, 81
    germination, 80–83
    pine nut, 45
Segment (earthworm), 103
Segmented body, 106
Selaginella, 56
Sella turcica, 132
Semimembranosus, 134, 136, 156, 157, 159, 166, 176, 181
Seminal
    receptacle, 93, 94, 103, 104, 109
    vesicle, 96, 97, 103, 104, 147, 170, 171, 172
Seminiferous tubule, 148
    lumen, 148
Semitendinosis, 134, 135, 136, 159, 166, 176, 181
Sensory
    organ, 90, 119, 128, 153
    receptor, 129
Sepal, 77, 78, 79
Septum, 91, 103, 104
    pellucidum, 138
Serosa, 145, 146
Serous acini, 140
Serrate margin, 75, 76
Serratus, 175
    anterior, 134, 135
Seta, 47, 48, 49, 102, 103, 105, 108
    abdominal, 108
Sexual reproduction, 11, 23, 36, 84
Shaft (feather), 125
Shark, 149–153
    arterial system, 152
    head region, 153
    Squalus acanthias, 150–151
    Triakis semifasciata, 149
    venous system, 152
Sheep, 93, 126, 185
    heart, 185
Sheep liver fluke, 93
Shell
    clam, 100
    gland, 93, 95
    snail, 99

Shepherd's purse. *See Capsella bursapastoris*
Shoulder, 129, 175
Shrub, 70
Silica, 23, 45
Silverfish, 111
Simple leaf, 75
Sinus, 122, 132
    ethmoidal, 138
    genital, 152
    sphenoidal, 132, 138
    vascular, 142
    venous, 139, 151
Sinus venosus, 122, 152, 161
Siphon, 100, 101
Siphonaptera, 111
Skeletal
    muscle, 7, 8, 140
    myofibril, 7
    system, 128
Skeletomuscular system, 128
Skeleton
    branchial, 154
    cartilaginous, 119
    green tree python, 165
    human, 130
    leopard frog, 155
    perch, 154
    rat, 168
    turtle, 163
Skin (human), 129
Skull
    cat, 173, 174
    human, 131–132
    lamprey, 121
    rat, 168
Slug, 99
    locomotion, 99
Smut, 36, 40–42
Snail, 84, 93, 98, 99
Solanum tuberosum, 4
Soleus muscle, 134
Somite, 103
Soredium, 44
Sorus, 59, 60, 61
Southern hemisphere pine. *See Araucaria*
Spanish moss, 44
Sperm (spermatozoa), 11, 15, 28, 31, 32, 46, 51, 59, 61, 64, 80, 84, 87, 88, 128, 148
    duct, 109
    flagellated, 84
    nucleus, 64
Spermatia, 33
Spermatic
    cord, 147, 184
    fascia, 147
Spermatid, 15, 148
Spermatangia, 33
Spermathecal duct, 99
Spermatocyte, 15, 148
Spermatogenesis, 15
Spermatogenous tissue, 48, 53
Spermatogonium, 15, 148
Spermatophoric gland, 101
Spermiogenesis, 15
Sphenoidal sinus, 132, 138
Sphenoid bone, 131, 132
Spikelet, 77
Sphagnum, 52
Sphenophyta, 45, 58
Spicule, 85, 86, 96
Spinal
    cord, 122, 128, 137, 153, 184
    nerves, 128
Spinalis dorsi, 175
Spine
    horseshoe crab, 106
    perch, 154
        haemal, 154
        neural, 154
Spindle, 2, 11, 13, 14
Spines
    Arabacia, 118
    Asterias, 116
    movable, 106, 115
Spinneret, 107
Spinodeltoid, 174
Spinotrapezius, 174
Spiracle, 107, 114, 119, 149
Spiracular muscle, 150
Spiral organ. *See* organ of Corti
Spiral valve, 119
Spirillum, 18
Spirillum volutans, 22
Spirochaeta, 19
Spirochaete, 18, 19, 22
Spirogyra, 28, 29
    conjugation, 29
Spirulina, 19
Spleen, 142, 151, 164, 170, 171, 179, 180, 183

Splenius, 175, 181
   capitis, 134
Sponge, 84, 85, 86
   bath, 86
   freshwater, 86
   in natural habitat, 86
Spongocoel, 85, 86
Sporangiophore, 37, 58
Sporangium, 34, 35, 37, 40,
   45, 46, 49, 50, 53, 55, 57,
   58, 59, 60, 61
Spore, 37, 39, 40, 46, 47, 49,
   50, 53, 58, 59, 60, 61
Sporocyst, 93, 94
Sporogenous tissue, 48, 53
Sporophyll, 45, 55, 57, 63, 67
Sporophyte, 45, 46, 47, 48, 49,
   50, 51, 56, 59, 64, 65, 68
   developmental stages, 49
Sporulation, 11
Springtail, 111
Spur, 98
Spurred tortoise. *See*
   Geochelone sulcata
Squalus acanthias, 150–151
   axial musculature, 150
   external anatomy, 150
   gil, 150, 151
   heart, 151
   hypobranchial
      musculature, 151
   internal anatomy, 151
   jaw, 150
   skeleton, 150
   pectoral fin, 150
Squamata, 165
Squamosal suture, 131, 174
Squash, 4
   xylem, 4
Squid (*Loligo*), 84, 101
Stalk, 51, 53
Stamen, 68, 77, 78, 79
Staphylococcus, 18
Staphylococcus, 19, 22
Starch grain, 4, 71
Star fish. *See Asterias*
Stele, 54, 55, 56, 71, 72
   Lycopodium, 55
   Psilotum nudum, 54
   Selaginella, 56
   Tmesipteris, 54
Stellate
   cell, 139
   reticulum, 142
Stem, 58, 65, 70, 72, 73, 74,
   75, 76
   angiosperm, 72–75
   bulb, 72, 73
   corm, 72
   dicot, 74
   Equisetum, 58
   examples, 72
   monocot, 74
   leaf arrangement, 75
   Pinus, 65
   rhizome, 72
   tendril, 72
   tuber, 72
   underground, 73
   uses, 72
   woody, 72, 73, 75
Sterile jacket, 53
Sternal notch, 133
Sternohyoid, 169, 181
Sternomastoid, 169, 174, 181
Sternonucleomastoid, 134, 135
Sternoradialis, 158
Sternum
   cat, 173
   garden spider, 107
   human, 130, 133, 135
   leopard frog, 155
   pigeon, 166
   rat, 168
Stigonema, 21
Stigma, 68, 77, 78, 79, 80, 82
Stinging apparatus
   (scorpion), 107
Stipe, 31, 32, 40, 42
Stolon, 37
Stoma, 45, 50, 53, 54, 62, 66,
   76
Stomach
   Arabacia, 118
   cardiac, 110, 116
   cardiac region, 145
   cat, 178
   curvature, 145
   dogfish shark, 151
   fetal pig, 183
   fundic area, 145
   grasshopper, 114
   human, 144, 145
   larva, 16
   leopard frog, 160

perch, 154
   pyloric, 110, 116
   rat, 170, 171
   rotifer, 98
   snail, 99
   squid, 101
   turtle, 164
Stonefly, 111
Storage tissue, 47
Stratified squamous
   epithelium, 8, 140, 144,
   148
Stratum
   basale, 129
   corneum, 127, 129
Strawberry, 12
Strepsiptera, 111
Strep throat, 21
Streptobacillus, 18
Streptococcus, 18
Streptococcus, 19
   pyogenes, 21, 22
Strobilus, 55, 56, 57, 58, 63
   axis, 58
Stroma, 3, 39
Style, 68, 77, 78, 79, 80, 82
Stylops, 111
Subdural space, 137
Subepidermal sclerid, 82
Subgenital pit, 90
Submucosa, 145, 146
Subscapularis, 175
Sucker, 95, 98, 105, 115
Suction cup, 123
Sugar cane leaf, 3
Sulcus, 137, 138
Sulfobulus, 19
Sulfur granule, 21
Sunflower, 69
Superficial digital flexor, 136
Supraangular bone, 152
Suprabranchial chamber, 100
Supracondylar ridge, 133
Supraoccipital bone, 162
Supraorbital margin, 131
Supraspinatus, 175, 181
Supratemporal bone, 165
Suprarenal gland. *See* adrenal
   gland
Surirella, 24
Suspensory ligament, 139
Sustentacular cell, 148
Sweat duct, 129
Swimmeret, 108, 109
Sycon body type (*Scypha*), 85
Sympathetic chain ganglia, 182
Symphysis pubis, 130
Synangium, 53
Synapsis, 14
Syngamy, 31, 46, 51
Synergid cell, 83
Systemic
   arch, 160
   heart, 101

T

Tadpole, 17
Taenia pisiformis, 95
   proglottid, 95
   scolex, 95
Tail
   cat, 173
   fetal pig, 180
   tadpole, 17
   turtle, 162, 164
Tapeworm, 92, 95
Taproot system, 70
Tarantula. *See Dugesiella*
Tarsal bone, 130, 155, 163,
   168, 173
Tarsalis, 158
Tarsometatarsal bone, 166
Tarsus, 107, 114
Taste bud, 140
Teat, 180
Telophase, 11, 12, 13, 14
Telson, 106, 108
Temporal
   bone, 131, 132, 174
process, 132
Temporalis, 156
Tendo calcaneus, 134, 156,
   158
Tendon, 10
Tendril, 72
Tensor fasciae, 174
Tensor fasciae latae, 134, 136,
   175, 176, 181
Tentacle, 84, 87, 88, 89, 90,
   91, 98, 99, 101, 102, 115,
   118, 119
Teres major, 135, 175, 181
Teres minor, 135, 175

Tergum, 108
Terminal
   bud, 70, 72, 73
   budscale scar, 72, 73
Testis, 87, 88, 93, 94, 95, 96,
   97, 101, 103, 110, 120, 128,
   147, 148, 151, 160, 170,
   171, 172. 179, 184
Thalamus, 138
Thallus, 43, 46
Thermoacidophile, 18, 19
Thermoplasma, 19
Thigh, 129, 173, 176
Thoracic
   cavity, 128
   region, 170
Thorax, 106, 144, 129
Thrush, 124
Thylakoid membrane, 3
Thymus, 143, 170, 180, 183
   fetal, 143
Thyroid, 143, 163, 180, 183
Thysanura, 111
Tibia, 107, 114, 130, 155, 163,
   168, 173
Tibiale, 155
Tibialis
   anterior, 134, 158, 159
   anticus longus, 156, 166
   posterior, 158, 159
Tibial nerve, 136
Tibiofibula, 155
Tibiotarsal bone, 166
Tick, 106, 107
   Ixodidae, 107
Tissue, 1–10
   adipose, 7, 8
   animal, 5–10
   connective, 1, 8, 10
   epithelial, 1, 6, 8
   maintenance, 11
   muscle, 1, 7, 8, 10
   nervous, 1, 8
   plant, 1–4
   principal types, 1
   repair, 11
Toadstool, 36, 40–42
Toe, 98
Tomato. *See Lycopersicon*
   esculentum
Tongue, 113, 140, 149, 154,
   178, 180
   support, 121, 149
Tooth
   calcareous, 118
   canine, 131, 132
   developing, 142
   horny, 121, 122
   incisor, 131, 132
   molar, 131, 132
   premolar, 131, 132
Tooth shell, 98
Toucan, 124
Touch-me-not, 81
Trabeculae carneae, 142, 143
Trachea, 9, 106, 143, 170, 178
   epithelium, 9
      lining, 144
   hyaline cartilage, 9
Tracheid, 2, 4, 65
Tradescantia, 76
   leaf, 76
Transfusion tissue, 66
Transverse
   abdominis, 176, 181
   colon, 145
   plane, 128, 173
   process, 155
   septum, 150
Trapezius, 134, 135, 175, 181
Trematoda, 92, 93–94
   Clonorsis sinensis, 94
   Clonorchis sinensis, 94
   Fasciola magna, 93
   Fasciola hepatica, 93
Treponema, 19
Triakis semifasciata, 149
Triangle of ausculation, 135
Triceps
   brachii, 134, 136, 166, 169,
      174, 175, 181
   femoris, 156, 157, 159
Trichocyst, 26
Trichome, 73
Trichinella spiralis, 97
Tricuspid valve, 185
Trigeminal nerve, 137
Trillium, 69
Triticum, 72, 74, 83
   grain (caryopsis), 83
   root, 72
   root hairs, 72
   stem, 72
Prochanter, 107, 114, 133
Trochea, 133

Trochlear nerve, 137
Trophozoite, 25
Truffle, 36, 37–40
   Tuber, 36
Truncus arteriosus, 160
Trunk (human), 145
T-tubule, 7
Tube foot, 115, 116, 117, 118,
   119
   extended, 115
   retracted, 115
   with suckers, 115
   without suckers, 115
Tuber (angiosperm stem), 72
Tuber, 36
Tubercle, 133
Tubeworm. *See Sabellastarte*
   indico
Tulip, 77
Tunicate, 119
Tunica vaginalis, 171
Turbellaria, 92
   Planaria, 92
Turnip, 70
Turtle, 162–165
Tympanic bulla, 174
Tympanic membrane. 114,
   155
Typhlosole, 103, 104

U

Ulna, 130
Ulothrix, 27
Ulva, 29
Ulna, 163, 166, 168, 173
Umbilical
   artery, 180, 183
   cord, 128, 180
   vein, 183
Umbilicus muscle, 134, 135
Umbo, 100
   clam, 100
Undulate leaf, 75
Ureter, 128, 146, 154, 164,
   171, 172, 179, 182, 183, 184
Urethra, 128, 147, 148, 172,
   179, 184
   opening, 148
Urinary system, 128
Urine, 128
Urochordata, 119
Urogenital
   orifice, 172, 184
   pore, 151
   sinus, 184
   system, 128, 172, 179, 184
Uropod, 108
Urostyle, 155
Ustilago maydis, 42
Uterine
   cervix, 148
   ostium, 148
   tube, 147, 148
Uterus, 93, 94, 95, 96, 97, 128,
   147, 148, 160, 164, 172,
   179, 184

V

Vacuole, 1, 2, 3, 7, 8, 23, 25,
   26
   contractile, 23, 25, 26
   fat, 8
   food, 25, 26
   lipid-filled, 7
Vagina, 95, 96, 97, 147, 148,
   172, 179, 184
   human, 148
   orifice, 148, 172
   rugae, 148
Vane (feather), 125
Vas deferens, 93, 94, 95, 96,
   97, 101, 103, 110, 172, 184
Vascular
   bundle, 4, 69, 74
   cambium, 3, 70
   ray, 67
   sinus, 142
   tissue, 1, 3, 70, 73
Vastus
   internus, 158
   lateralis, 134, 135, 136, 176
   medialis, 134, 136, 176, 181
Vaucheria, 24
Vein, 4, 70, 76, 128
   abdominal, 151, 160, 161,
      163
   afferent renal, 152
   afferent branchial, 152
   axillary, 141, 177
   basilic, 141
   brachial, 141, 152, 161, 177

brachiocephalic, 141, 177,
   178, 183
   cardiac, 161
   cardinal, 152
   caudal, 152, 177
   cephalic, 141, 177
   cloacal, 152
   cubital, 141, 177
   cutaneous, 161
   deep femoral, 141
   diaphragm, 177
   dorsolumbar, 161
   efferent renal, 152
   facial, 177
   femoral, 141, 152, 161, 177
   femoral circumflex, 141
   gastric, 161
   gonadal, 169
   hepatic, 141, 152, 161,
      169, 177
   hepatic portal, 141, 151
   iliac, 141, 152, 171, 177
   iliolumbar, 169, 171, 177
   innominate, 161
   intestinal, 151, 161
   jugular, 141, 152, 161, 169,
      170, 182, 183
   lingual, 161
   maxillary, 161
   mesenteric, 141
   ovarian, 177
   pelvic, 161
   phrenic, 177
   principal human, 141
   pulmonary, 161
   radial, 141
   renal, 146, 152, 171, 177,
      179, 182, 184
   saphenous, 141, 183
   spermatic, 141, 171, 177
   subclavian, 141, 152, 177
   subscapular, 161, 169, 177
   thyrocervical, 183
   tibial, 141
   transverse, 177
   transverse-subscapular, 177
   ulnar, 141
   urinogenital, 161
   vena cava, 141, 183
Velum, 120, 122
Vena cava, 100, 101, 141, 142,
   146, 161, 169, 170, 171,
   177, 178, 182, 183, 185
Venation
   palmate, 75
   parallel, 75
   pinnate, 75
Venous sinus, 139
Venter, 52
Ventral
   nerve, 96, 97
   scale, 47
   sucker, 93, 94
Ventricle, 100, 138, 152
   cat, 178
   clam, 100
   dogfish shark, 151, 152
   fetal pig, 180, 182
   pigeon, 167
   human, 138, 142
   lamprey, 122
   rat, 170
   sheep, 185
   turtle, 163
Venule, 128, 146
Veranus exanthematicus, 165
Vertebra, 119
   caudal, 163, 165, 173
   cervical, 119, 130, 133,
      163, 166, 168, 173
   lumbar, 130, 133, 168, 173
   perch, 154
   thoracic, 130, 133, 144,
      168, 170, 173
   trunk, 150, 165
Vertebral
   artery, 137
   column, 133, 135, 154
Vertebrata (Vertebrate), 1, 84,
   119–185
   Agnatha, 119, 121–122, 149
   Amphibia, 119, 123,
      155–161
   Aves, 119, 124–125,
      166–167
   bipedal, 129
   characteristics, 119
   Chondrichthyes, 119,
      149–153
   dissections, 149–185
   host, 92, 94
   Mammalia, 126–127,
      168–185
   Osteichthyes, 119, 154
   Reptilia, 119

Vertex, 114, 162–165
Vesicle
   plant, 2
   polian, 117, 118
Vessel, 102, 103, 104, 128
   blood, 128, 129, 144
      cutaneous, 129
   dorsal, 102
   great, 142
   lymphatic, 128
   ovarian, 148
   pulmonary, 143
   subneural, 104
   ventral, 103, 104
Vessel element, 2, 4, 75
Vestibular gland, 172
Vibrio, 18
Villus, 146
Violet, 69
Visceral arch, 150
Vitellarium, 98
Vitelline membrane, 15
Vitreous chamber, 139
Vocal sac, 160
Volvox, 27
Vomer, 131, 132, 155
von Ebner's gland, 140

W

Walrus, 126
Wasp, 111
Waste
   elimination, 1
   fecal, 18
Water
   flea. *See Daphnia*
   lily, 69
   mold, 35
Wheat, 39, 69, 72, 83
   grain (caryopsis), 83
   root, 72
   rust, 42
   stem, 74
Wheel organ, 120
Whisk fern, 45, 53, 54
   Psilotum nudum, 53, 54
   Tmesipteris, 54
White rust, 35
Whorl, 75
Wine, 36
Wood
   growth ring, 67
   spring, 66, 75
   summer, 67, 75
Woody plant, 70, 72, 73, 75
Worm, 84
Wrist, 180

X

Xiphisternum, 168
Xiphoid process, 133, 135
Xylem, 3, 4, 54, 55, 56, 65, 66,
   70, 71, 72, 74, 75

Y

Yeast, 11, 36, 37–40
   Baker's, 37
   Candida, 37
Yolk, 15, 17, 93, 94
   duct, 93, 94
   gland, 93, 94, 95
   plug, 17
   reservoir, 93
   sac, 128
Yucca, 4
   leaf, 4
Yucca brevifolia, 4

Z

Zamia, 62
   cones, 62
   ovule, 62
Zea mays, 72, 74
Z line, 7
Zonular fiber, 139
Zoosporangium, 28
Zoospore, 23, 27, 28, 29
Zygnema, 29
Zygomatic
   arch, 132, 174
   bone, 130, 131, 132, 174
   muscle, 174
Zygomycota, 36, 37
Zygosporangium, 37
Zygote, 11, 15, 27, 29, 31, 46,
   49, 51, 59, 64, 68, 84
   colony, 27
Zymogenic cell, 145